Computational Mathematics: Concepts and Applied Principles

Computational Mathematics: Concepts and Applied Principles

Edited by
Lawrence Grattan

Willford Press
www.willfordpress.com

Published by Willford Press,
118-35 Queens Blvd., Suite 400,
Forest Hills, NY 11375, USA

ISBN: 978-1-68285-688-8

Cataloging-in-Publication Data

Computational mathematics : concepts and applied principles / edited by Lawrence Grattan.
p. cm.
Includes bibliographical references and index.
ISBN 978-1-68285-688-8
1. Mathematics--Data processing. 2. Computer science--Mathematics.
3. Computational complexity. I. Grattan, Lawrence.
QA76.95 .C66 2019
510.285 53--dc23

For information on all Willford Press publications
visit our website at www.willfordpress.com

Willford Press

Contents

Preface

Computational mathematics is the science that integrates computing and mathematics, for research. This is particularly significant in the areas of science that can be aided by computation. It involves the use of numerical methods, algorithms and symbolic computations. Computational mathematics is used in solving mathematical and scientific problems and to aid research in logic, number theory, cryptography, discrete mathematics and computational algebraic topology. Besides these, it is crucial in finance, economics and accounting. Some of the focus areas of computational mathematics are numerical linear algebra, Monte Carlo methods, stochastic finite elements, numerical analysis, etc. There has been rapid progress in computational mathematics and its applications are finding their way across multiple fields. This book elucidates new techniques and their applications in a multidisciplinary manner. Coherent flow of topics, student-friendly language and extensive use of examples make this book an invaluable source of knowledge.

Various studies have approached the subject by analyzing it with a single perspective, but the present book provides diverse methodologies and techniques to address this field. This book contains theories and applications needed for understanding the subject from different perspectives. The aim is to keep the readers informed about the progresses in the field; therefore, the contributions were carefully examined to compile novel researches by specialists from across the globe.

Indeed, the job of the editor is the most crucial and challenging in compiling all chapters into a single book. In the end, I would extend my sincere thanks to the chapter authors for their profound work. I am also thankful for the support provided by my family and colleagues during the compilation of this book.

Editor

1

Relaxed resolvent operator for solving a variational inclusion problem

Iqbal Ahmad, Mijanur Rahaman, Rais Ahmad*

Department of Mathematics, Aligarh Muslim University, India

Abstract In this paper, we introduce a new resolvent operator and we call it relaxed resolvent operator. We prove that relaxed resolvent operator is single-valued and Lipschitz continuous and finally we approximate the solution of a variational inclusion problem in Hilbert spaces by defining an iterative algorithm based on relaxed resolvent operator. A few concepts like Lipschitz continuity and strong monotonicity are used to prove the main result of this paper. Thus, no strong conditions are used. Some examples are constructed.

Keywords Relaxed; Inclusion; Space; Algorithm; Lipschitz

1. Introduction and Preliminaries

Variational inequality is an inequality involving a functional, which has to be solved for all possible values of a given variable, belonging usually to a convex set. The theory of variational inequalities was initially developed to deal with equilibrium problems, precisely the Signorini problem. After that it has been extended and generalized to study a wide class of problems arising in mechanics, physics, optimization and control, nonlinear programming, economics, finance, regional structural,transformation, elasticity, and applied sciences, etc., see e.g., [1, 2, 4, 12, 15, 16, 17, 18, 19, 20] and references therein.

A useful and important generalization of variational inequalities is called a variational inclusion which was introduced by *Hassouni* and *Moudafi* [10] and includes mixed variational inequalities as special cases. Many problems concerning variational inclusions are solved by using the concept of maximal monotonicity and its generalized concepts such as H-monotonicity [6], H-accretivity [5] etc., see e.g., [7, 8, 9, 13, 21] and references therein. Most of the splitting methods are based on the resolvent operator of the form $[I + \lambda M]^{-1}$, where M is a set-valued monotone mapping, λ is a positive constant and I is the identity mapping.

In this paper, we introduce a new resolvent operator of the form $[(I - H) + \lambda M]^{-1}$, where H is a relaxed Lipschitz continuous mapping, M is a set-valued monotone mapping, λ is a positive constant and I is an identity mapping. We call this new resolvent operator as relaxed resolvent operator and prove that it is single-valued and Lipschitz continuous. We define an iterative algorithm based on relaxed resolvent operator to solve a variational inclusion problem. Convergence of the iterative sequences generated by the iterative algorithm is also discussed. Some examples are constructed.

*Correspondence to: Rais Ahmad (Email: raisain_123@rediffmail.com). Department of Mathematics, Aligarh Muslim University, Aligarh 202002, India

Throughout this paper, we suppose that X is a real Hilbert space endowed with a norm $\|\cdot\|$ and an inner product $\langle\cdot,\cdot\rangle$, d is the metric induced by the norm $\|\cdot\|$, 2^X (respectively,$CB(X)$) is the family of all nonempty (respectively, closed and bounded) subsets of X, and $D(\cdot,\cdot)$ is the Hausdorff metric on $CB(X)$ defined by

$$D(P,Q) = \max\left\{\sup_{x\in P} d(x,Q), \sup_{y\in Q} d(P,y)\right\},$$

where $d(x,Q) = \inf_{y\in Q} d(x,y)$ and $d(P,y) = \inf_{x\in P} d(x,y)$.

The following definitions are needed in the sequel.

Definition 1.1
A mapping $g : X \to X$ is said to be

(i) Lipschitz continuous if, there exist a constant $\lambda_g > 0$ such that

$$\|g(x) - g(y)\| \leq \lambda_g \|x - y\|, \forall x, y \in X;$$

(ii) monotone,if

$$\langle g(x) - g(y), x - y\rangle \geq 0, \forall x, y \in X;$$

(iii) strongly monotone if, there exists a constant $\xi > 0$ such that

$$\langle g(x) - g(y), x - y\rangle \geq \xi\|x - y\|^2, \forall x, y \in X;$$

(iv) relaxed Lipschitz continuous if, there exists a constant $r > 0$ such that

$$\langle g(x) - g(y), x - y\rangle \leq -r\|x - y\|^2, \forall x, y \in X.$$

.

Definition 1.2
A mapping $N : X \times X \times X \to X$ is said to be Lipschitz continuous with respect to first argument if, there exists a constant λ_{N_1} such that

$$\|N(x_1, x_2, x_3) - N(y_1, x_2, x_3)\| \leq \lambda_{N_1}\|x_1 - y_1\|, \forall x_1, y_1, x_2, x_3 \in X.$$

Similarly, we can define the Lipschitz continuity of N in rest of the arguments.

Definition 1.3
A set-valued mapping $A : X \to CB(X)$ is said to be D-Lipschitz continuous if, there exists a constant δ_A such that

$$D(A(x), A(y)) \leq \delta_A\|x - y\|, \forall x, y \in X.$$

2. Relaxed Resolvent operator

We begin this section with the introduction of relaxed resolvent operator and demonstrate some of its properties.

Definition 2.1
Let $H : X \to X$ be a mapping and $I : X \to X$ be an identity mapping. Then, a set-valued mapping $M : X \to 2^X$ is a said to be $(I - H)$-monotone if, M is monotone, H is relaxed Lipschitz continuous and

$$[(I - H) + \lambda M](X) = X,$$

where $\lambda > 0$ is a constant.

Definition 2.2
Let $H : X \to X$ be relaxed Lipschitz continuous mapping and $I : X \to X$ be an identity mapping. Suppose that $M : X \to 2^X$ is a set-valued, $(I - H)$-monotone mapping. The relaxed resolvent operator $R_{\lambda,M}^{(I-H)} : X \to X$ associated with I,H and M is defined by

$$R_{\lambda,M}^{I-H}(x) = [(I - H) + \lambda M]^{-1}(x), \forall x \in X, \tag{1}$$

where $\lambda > 0$ is a constant.

Example 2.1
Let $X = S^2$, the space of all 2×2 real symmetric matrices equipped with inner product $\langle A, B\rangle = tr(A \cdot B)$, $\forall A, B \in X$, and let α, β be two positive real numbers such that $\beta \le \alpha$. Let $H : X \to X$ be a mapping defined by

$$H\left(\begin{bmatrix} x_1 & a \\ a & x_2 \end{bmatrix}\right) = \begin{bmatrix} -\alpha x_1 & a \\ a & -\alpha x_2 \end{bmatrix}, \forall x_1, x_2, a \in \mathbb{R}.$$

For $x = \begin{bmatrix} x_1 & a \\ a & x_2 \end{bmatrix} \in X$, we calculate,

$$\begin{aligned}
&\langle H(x) - H(y), x - y\rangle \\
&= \left\langle \begin{bmatrix} -\alpha(x_1 - y_1) & 0 \\ 0 & -\alpha(x_2 - y_2) \end{bmatrix}, \begin{bmatrix} (x_1 - y_1) & 0 \\ 0 & (x_2 - y_2) \end{bmatrix} \right\rangle \\
&= tr\left(\begin{bmatrix} -\alpha(x_1 - y_1) & 0 \\ 0 & -\alpha(x_2 - y_2) \end{bmatrix} \cdot \begin{bmatrix} (x_1 - y_1) & 0 \\ 0 & (x_2 - y_2) \end{bmatrix} \right) \\
&= -\alpha(x_1 - y_1)^2 - \alpha(x_2 - y_2)^2 \\
&= -\alpha[(x_1 - y_1)^2 + (x_2 - y_2)^2] \\
&\le -\beta[(x_1 - y_1)^2 + (x_2 - y_2)^2].
\end{aligned}$$

Also,

$$\begin{aligned}
\|x - y\|^2 &= \langle x - y, x - y\rangle \\
&= \left\langle \begin{bmatrix} x_1 - y_1 & 0 \\ 0 & x_2 - y_2 \end{bmatrix}, \begin{bmatrix} x_1 - y_1 & 0 \\ 0 & x_2 - y_2 \end{bmatrix} \right\rangle \\
&= tr\left(\begin{bmatrix} x_1 - y_1 & 0 \\ 0 & x_2 - y_2 \end{bmatrix} \cdot \begin{bmatrix} x_1 - y_1 & 0 \\ 0 & x_2 - y_2 \end{bmatrix} \right) \\
&= (x_1 - y_1)^2 + (x_2 - y_2)^2.
\end{aligned}$$

From above it follows that

$$\langle H(x) - H(y), x - y\rangle \le -\beta\|x - y\|^2, \forall x, y \in X.$$

i.e., H is β-relaxed Lipschitz continuous.
Suppose that $M : X \to 2^X$ is defined by

$$M\left(\begin{bmatrix} x_1 & a \\ a & x_2 \end{bmatrix}\right) = \begin{bmatrix} \alpha x_1 & a \\ a & \alpha x_2 \end{bmatrix}, \forall x_1, x_2, a \in \mathbb{R}.$$

We calculate

$$\begin{aligned}
&\langle M(x)-M(y),x-y\rangle\\
=&\left\langle M\left(\begin{bmatrix} x_1 & a\\ a & x_2\end{bmatrix}\right)-M\left(\begin{bmatrix} y_1 & a\\ a & y_2\end{bmatrix}\right),\begin{bmatrix} x_1-y_1 & 0\\ 0 & x_2-y_2\end{bmatrix}\right\rangle\\
=&\left\langle\begin{bmatrix} \alpha(x_1-y_1) & 0\\ 0 & \alpha(x_2-y_2)\end{bmatrix},\begin{bmatrix} x_1-y_1 & 0\\ 0 & x_2-y_2\end{bmatrix}\right\rangle\\
=&\ tr\left(\begin{bmatrix} \alpha(x_1-y_1) & 0\\ 0 & \alpha(x_2-y_2)\end{bmatrix}\cdot\begin{bmatrix} x_1-y_1 & 0\\ 0 & x_2-y_2\end{bmatrix}\right)\\
=&\ \alpha(x_1-y_1)^2+\alpha(x_2-y_2)^2,
\end{aligned}$$

which implies that

$$\langle M(x)-M(y),x-y\rangle\geq 0,\forall x,y\in X.$$

i.e., M is monotone.

Now, we show that for $\lambda=1$, every element $x=\begin{bmatrix} \alpha x_1 & a\\ a & \alpha x_2\end{bmatrix}\in 2^X$ has a pre-image $y=\begin{bmatrix} \frac{\alpha x_1}{1+2\alpha} & a\\ a & \frac{\alpha x_2}{1+2\alpha}\end{bmatrix}\in X$.

$$\begin{aligned}
[(I-H)+M](y) &= (I-H)(y)+M(y)\\
&= y-H(y)+M(y)\\
&= \begin{bmatrix} \frac{\alpha x_1}{1+2\alpha} & a\\ a & \frac{\alpha x_2}{1+2\alpha}\end{bmatrix}-\begin{bmatrix} \frac{-\alpha^2 x_1}{1+2\alpha} & a\\ a & \frac{-\alpha^2 x_2}{1+2\alpha}\end{bmatrix}+\begin{bmatrix} \frac{\alpha^2 x_1}{1+2\alpha} & a\\ a & \frac{\alpha^2 x_2}{1+2\alpha}\end{bmatrix}\\
&= \begin{bmatrix} \frac{\alpha(1+2\alpha)x_1}{1+2\alpha} & a\\ a & \frac{\alpha(1+2\alpha)x_2}{1+2\alpha}\end{bmatrix}\\
&= \begin{bmatrix} \alpha x_1 & a\\ a & \alpha x_2\end{bmatrix}\in 2^X.
\end{aligned}$$

It follows that

$$[(I-H)+M](X)=X,$$

i.e., M is $(I-H)$-monotone mapping.

Now, we prove some of the properties of relaxed resolvent operator defined by (1).

Theorem 2.2

Let $H:X\to X$ be a r-relaxed Lipschitz continuous mapping, $I:X\to X$ be an identity mapping and $M:X\to 2^X$ be a set-valued $(I-H)$-monotone mapping. Then the operator $[(I-H)+\lambda M]^{-1}$ is single-valued, where $\lambda>0$ is a constant.

Proof

For any $z\in X$ and a constant $\lambda>0$, let $x,y\in[(I-H)+\lambda M]^{-1}(z)$. Then,

$$\lambda^{-1}[z-(I-H)(x)]\in M(x);$$

$$\lambda^{-1}[z-(I-H)(y)]\in M(y).$$

Since M is monotone, we have

$$\begin{aligned}\langle -(I-H)(x)+z+(I-H)(y)-z, x-y\rangle &\geq 0;\\ -\langle (I-H)(x)-(I-H)(y), x-y\rangle &\geq 0;\\ -\langle x-H(x)-y+H(y), x-y\rangle &\geq 0;\\ \langle x-H(x)-y+H(y), x-y\rangle &\leq 0;\\ \langle x-H(x)-y+H(y), x-y\rangle &\leq 0;\\ \langle x-y, x-y\rangle - \langle H(x)-H(y), x-y\rangle &\leq 0.\end{aligned}$$

Since H is r-relaxed Lipschitz continuous, we have

$$0 \geq \langle x-y, x-y\rangle - \langle H(x)-H(y), x-y\rangle \geq \|x-y\|^2 + r\|x-y\|^2 \geq 0,$$

it follows that $(1+r)\|x-y\|^2 = 0$, which implies that $x = y$. Thus $[(I-H)+\lambda M]^{-1}$ is single-valued. □

Theorem 2.3
Let $H : X \to X$ be a r-relaxed Lipschitz continuous mapping, $I : X \to X$ be an identity mapping and $M : X \to 2^X$ be a set-valued, $(I-H)$-monotone mapping. Then the resolvent operator $R_{\lambda,M}^{I-H} : X \to X$ is $\frac{1}{[1+r]}$-Lipschitz continuous, i.e.,

$$\|R_{\lambda,M}^{I-H}(x) - R_{\lambda,M}^{I-H}(y)\| \leq \frac{1}{[1+r]}\|x-y\|, \forall x, y \in X.$$

Proof
Let x and y be any given point in X. If follow from (1) that

$$\begin{aligned}R_{\lambda,M}^{I-H}(x) &= [(I-H)+\lambda M]^{-1}(x),\\ R_{\lambda,M}^{I-H}(y) &= [(I-H)+\lambda M]^{-1}(y).\end{aligned} \tag{2}$$

It follows that

$$\begin{aligned}\frac{1}{\lambda}\left[x-(I-H)(R_{\lambda,M}^{I-H}(x))\right] &\in M\left(R_{\lambda,M}^{I-H}(x)\right),\\ \frac{1}{\lambda}\left[y-(I-H)(R_{\lambda,M}^{I-H}(y))\right] &\in M\left(R_{\lambda,M}^{I-H}(y)\right).\end{aligned} \tag{3}$$

Since M is $(I-H)$-monotone i.e., M is monotone, we have

$$\begin{aligned}&\frac{1}{\lambda}\left\langle x-(I-H)(R_{\lambda,M}^{I-H}(x)) - (y-(I-H)(R_{\lambda,M}^{I-H}(y))), R_{\lambda,M}^{I-H}(x) - R_{\lambda,M}^{I-H}(y)\right\rangle \geq 0,\\ &\frac{1}{\lambda}\left\langle x-y-\{(I-H)(R_{\lambda,M}^{I-H}(x)) - (I-H)(R_{\lambda,M}^{I-H}(y))\}, R_{\lambda,M}^{I-H}(x) - R_{\lambda,M}^{I-H}(y)\right\rangle \geq 0.\end{aligned} \tag{4}$$

It follows that

$$\begin{aligned}&\left\langle x-y, R_{\lambda,M}^{I-H}(x) - R_{\lambda,M}^{I-H}(y)\right\rangle\\ \geq\ &\left\langle (I-H)(R_{\lambda,M}^{I-H}(x)) - (I-H)(R_{\lambda,M}^{I-H}(y)), R_{\lambda,M}^{I-H}(x) - R_{\lambda,M}^{I-H}(y)\right\rangle.\end{aligned} \tag{5}$$

By Cauchy-Schwartz inequality, (5) and r-relaxed Lipschitz continuity of H, we have

$$\begin{aligned}
& \left\|x-y\right\|\left\|R^{I-H}_{\lambda,M}(x)-R^{I-H}_{\lambda,M}(y)\right\| \\
\geq\ & \left\langle x-y, R^{I-H}_{\lambda,M}(x)-R^{I-H}_{\lambda,M}(y)\right\rangle \\
\geq\ & \left\langle R^{I-H}_{\lambda,M}(x)-H(R^{I-H}_{\lambda,M}(x))-R^{I-H}_{\lambda,M}(y)+H(R^{I-H}_{\lambda,M}(y)), R^{I-H}_{\lambda,M}(x)-R^{I-H}_{\lambda,M}(y)\right\rangle \\
=\ & \left\langle R^{I-H}_{\lambda,M}(x)-R^{I-H}_{\lambda,M}(y), R^{I-H}_{\lambda,M}(x)-R^{I-H}_{\lambda,M}(y)\right\rangle - \\
& \left\langle H(R^{I-H}_{\lambda,M}(x))-H(R^{I-H}_{\lambda,M}(y)), R^{I-H}_{\lambda,M}(x)-R^{I-H}_{\lambda,M}(y)\right\rangle \\
\geq\ & \left\|R^{I-H}_{\lambda,M}(x)-R^{I-H}_{\lambda,M}(y)\right\|^2 + r\left\|R^{I-H}_{\lambda,M}(x)-R^{I-H}_{\lambda,M}(y)\right\|^2 \\
=\ & (1+r)\left\|R^{I-H}_{\lambda,M}(x)-R^{I-H}_{\lambda,M}(y)\right\|^2.
\end{aligned} \tag{6}$$

Thus, we have

$$\left\|R^{I-H}_{\lambda,M}(x)-R^{I-H}_{\lambda,M}(y)\right\| \leq \frac{1}{[1+r]}\|x-y\|,$$

i.e., the relaxed resolvent operator $R^{I-H}_{\lambda,M}$ is $\frac{1}{[1+r]}$-Lipschitz continuous. □

In support of Theorem 2.2, we have the following example.

Example 2.4
Let $X=\mathbb{R}^2$ with usual inner product. Let $H: X \to X$ be a mapping defined by

$$H(x) = (-2x_1, -4x_2), \forall x = (x_1, x_2) \in X,$$

and the mapping $M: X \to 2^X$ be defined by

$$M(x) = (4x_1, 2x_2), \forall x = (x_1, x_2) \in X.$$

Then, it easy to check that H is 2-relaxed Lipschitz continuous and M is monotone. In addition, it is easy to verify that for $\lambda = 1$, $[(I-H)+\lambda M](X) = X$, which shows that M is $(I-H)$-monotone mapping. Hence, the relaxed resolvent operator $R^{(I-H)}_{\lambda,M}: X \to X$ associated with I, H and M is of the form:

$$R^{I-H}_{\lambda,M}(x) = \left(\frac{x_1}{7}, \frac{x_2}{7}\right), \forall x = (x_1, x_2) \in X. \tag{7}$$

It is easy to see that the relaxed resolvent operator defined by (2.7) is single-valued.
Now, we prove that $R^{I-H}_{\lambda,M}$ is Lipschitz continuous.

$$\begin{aligned}
\left\|R^{I-H}_{\lambda,M}(x)-R^{I-H}_{\lambda,M}(y)\right\| &= \left\|\left(\frac{x_1}{7}, \frac{x_2}{7}\right)-\left(\frac{y_1}{7}, \frac{y_2}{7}\right)\right\| \\
&= \left\|\left(\frac{x_1-y_1}{7}, \frac{x_2-y_2}{7}\right)\right\| \\
&= \left[\frac{(x_1-y_1)^2}{49}+\frac{(x_2-y_2)^2}{49}\right]^{\frac{1}{2}} \\
&= \frac{1}{7}\left[(x_1-y_1)^2+(x_2-y_2)^2\right]^{\frac{1}{2}} \\
&\leq \frac{1}{3}\|x-y\|.
\end{aligned}$$

Hence, the resolvent operator $R^{I-H}_{\lambda,M}$ is $\frac{1}{3}$-Lipschitz continuous.

3. Formulation of the problem and iterative algorithm

In this section, we formulate a variational inclusion problem and an iterative algorithm based on relaxed resolvent operator to approximate the solution of our problem.

Let X be a real Hilbert space and $H, g : X \to X$, $N : X \times X \times X \to X$ be the single-valued mappings, $I : X \to X$ be an identity mapping. Suppose that $A, B, C : X \to CB(X)$ and $M : X \times X \to 2^X$ are the set-valued mappings such that M is $(I - H)$-monotone. We consider the problem of finding $x \in X$, $u \in A(x)$, $v \in B(x)$, $w \in C(x)$ and $g(x) \bigcap domM(.,x) \neq \emptyset$ such that

$$0 \in N(u, v, w) + M(g(x), x). \tag{8}$$

When $C \equiv 0$, $N(u, v, .) = N(u, v)$, then problem (8) reduces to find $x \in X$, $u \in A(x)$, $v \in B(x)$ such that

$$0 \in N(u, v) + M(g(x), x). \tag{9}$$

Problem (9) was introduced and studied by *Kazmi* and *Khan* [11].

In addition if $M(g(x), x) = M(g(x))$, then a similar analogue of problem (9) was introduced and studied by *Chang*, *Cho*, *Lee* and *Jung* [3], *Chang* [2], *Chang*, *Jim* and *Kim* [4].

It is clear that for suitable choices of mappings involved in the formulation of problem (8), one can obtain many variational inclusion problems studied in recent past.

By applying the relaxed resolvent operator, we establish an equivalence result for variational inclusion problem (8) and a nonlinear equation.

Lemma 3.1
Let $x \in X$, $u \in A(x)$, $v \in B(x)$ and $w \in C(x)$ is a solution of variational inclusion problem (3.1) if and only if (x, u, v, w) satisfies the equation:

$$g(x) = R^{I-H}_{\lambda, M(.,x)}[(I - H)g(x) - \lambda N(u, v, w)], \tag{10}$$

where

$$R^{I-H}_{\lambda, M(.,x)} = [(I - H) + \lambda M(.,x)]^{-1},$$

and $\lambda > 0$ is a constant.

Proof
The proof is a direct consequence of Definition 2.2. □

Theorem 3.2 (Nadler's Theorem [14])
Let (X, d) be a complete metric space. If $F : X \to CB(X)$ is a set-valued contraction mapping, then F has a fixed point.

Based on Lemma 3.1 and Theorem 3.2, we construct an iterative algorithm for finding approximate solutions of problem (8).

Iterative Algorithm 3.1
Let $g, H : X \to X$, $N : X \times X \times X \to X$ are the single-valued mappings, $I : X \to X$ be an identity mapping, and $A, B, C : X \to CB(X)$ are the set-valued mappings be such that for each $x \in X$, $Q(x_0) \subseteq g(x)$, where $Q : X \to 2^X$ be a set-valued mapping defined by

$$Q(x) = \bigcup_{u \in A(x)} \bigcup_{v \in B(x)} \bigcup_{w \in C(x)} [R^{I-H}_{\lambda, M(.,x)}[(I - H)g(x) - \lambda N(u, v, w)]], \tag{11}$$

where $M : X \times X \to 2^X$ be a set-valued mapping such that a fixed $x \in X$, $M(.,x)$ is $(I - H)$-monotone.

For any given $x_0 \in X$, $u_0 \in A(x_0)$, $v_0 \in B(x_0)$, and $w_0 \in C(x_0)$, let

$$z_0 = R^{I-H}_{\lambda,M(.,x_0)}[(I-H)g(x_0) - \lambda N(u_0,v_0,w_0)] \subseteq Q(x_0) \subseteq g(X).$$

Hence, there exist $x_1 \in X$ such that $z_0 = g(x_1)$. Since $u_0 \in A(x_0) \in CB(X)$, $v_0 \in B(x_0) \in CB(X)$ and $w_0 \in C(x_0) \in CB(X)$, by Theorem 3.2 there exist $u_1 \in A(x_1)$, $v_1 \in B(x_1)$ and $w_1 \in C(x_1)$ such that

$$\begin{aligned} \|u_1 - u_0\| &\leq D(A(x_1), A(x_0)), \\ \|v_1 - v_0\| &\leq D(B(x_1), B(x_0)), \\ \|w_1 - w_0\| &\leq D(C(x_1), C(x_0)). \end{aligned}$$

Let

$$z_1 = R^{I-H}_{\lambda,M(.,x_1)}[(I-H)g(x_1) - \lambda N(u_1,v_1,w_1)] \subseteq Q(x_1) \subseteq g(X).$$

Hence, there exist $x_2 \in X$ such that $z_1 = g(x_2)$. Continuing the above process inductively, we can define the iterative sequences $\{x_n\}$, $\{u_n\}$, $\{v_n\}$ and $\{w_n\}$ by the following scheme:

$$g(x_{n+1}) = R^{I-H}_{\lambda,M(.,x_n)}[(I-H)g(x_n) - \lambda N(u_n,v_n,w_n)], \tag{12}$$

$$u_{n+1} \in A(x_{n+1}), \|u_{n+1} - u_n\| \leq D(A(x_{n+1}), A(x_n)), \tag{13}$$

$$v_{n+1} \in B(x_{n+1}), \|v_{n+1} - v_n\| \leq D(B(x_{n+1}), B(x_n)), \tag{14}$$

$$w_{n+1} \in C(x_{n+1}), \|w_{n+1} - w_n\| \leq D(C(x_{n+1}), C(x_n)), \tag{15}$$

where $\lambda > 0$ is a constant and $n = 0, 1, 2, \cdots$.

4. Existence and convergence result

In this section, we prove an existence and convergence result for variational inclusion problem (8).

Theorem 4.1
Let X be a real Hilbert space and $g, H : X \to X$ be the single-valued mappings such that g is strongly monotone with constant ξ, Lipschitz continuous with constant λ_g and H is relaxed Lipschitz continuous with constant r and Lipschitz continuous with constant λ_H. Suppose that $N : X \times X \times X \to X$ is a single-valued mapping such that N is Lipschitz continuous in all the three arguments with constants λ_{N_1}, λ_{N_2} and λ_{N_3}, respectively and $A, B, C : X \to CB(X)$ be the set-valued mappings such that A is D-Lipschitz continuous with constant λ_A, B is D-Lipschitz continuous with constant δ_B and C is D-Lipschitz continuous with constant δ_C. Suppose that set-valued mapping $M : X \times X \to 2^X$ is such that for a fixed $x \in X$, $M(.,x)$ is $(I-H)$-monotone with respect to the first arguments, where $I : X \to X$ is the identity mapping and for each $x \in X$, $Q(x) \subseteq g(x)$, where Q is defined by (11). Suppose that there exists constants $\lambda > 0$ and $h > 0$ such that the following conditions hold:

$$\left\| R^{I-H}_{\lambda,M(.,x)}(z) - R^{I-H}_{\lambda,M(.,y)}(z) \right\| \leq h\|x - y\|, \quad \forall x, y, z \in X, \tag{16}$$

and

$$\lambda_g + \lambda_H \lambda_g + \lambda\lambda_{N_1}\delta_A + \lambda\lambda_{N_2}\delta_B + \lambda\lambda_{N_3}\delta_C < [\xi(1+r) - h]. \tag{17}$$

Then, there exist $x \in X$, $u \in A(x)$, $v \in B(x)$ and $w \in C(x)$ such that variational inclusion problem (8) is solvable. Moreover, $x_n \to x$, $u_n \to u$, $v_n \to v$ and $w_n \to w$, as $n \to \infty$, where $\{x_n\}$, $\{u_n\}$, $\{v_n\}$ and $\{w_n\}$ are the sequences defined in iterative Algorithm 3.1.

Proof
Using the strong monotonicity of g with constant ξ, we have

$$\begin{aligned}\|g(x_{n+1})-g(x_n)\|\|x_{n+1}-x_n\| &\geq \langle g(x_{n+1})-g(x_n), x_{n+1}-x_n\rangle \\ &\geq \xi\|x_{n+1}-x_n\|^2,\end{aligned}$$

which implies that

$$\|x_{n+1}-x_n\| \leq \frac{1}{\xi}\|g(x_{n+1})-g(x_n)\|. \tag{18}$$

By iterative Algorithm 3.1, Theorem 2.2 and condition (16), we have

$$\begin{aligned}
&\|g(x_{n+1})-g(x_n)\| \\
= &\Big\|R^{I-H}_{\lambda,M(\cdot,x_n)}[(I-H)g(x_n)-\lambda N(u_n,v_n,w_n)]-R^{I-H}_{\lambda,M(\cdot,x_{n-1})}[(I-H)g(x_{n-1}) \\
&-\lambda N(u_{n-1},v_{n-1},w_{n-1})]\Big\| \\
= &\Big\|R^{I-H}_{\lambda,M(\cdot,x_n)}[(I-H)g(x_n)-\lambda N(u_n,v_n,w_n)]-R^{I-H}_{\lambda,M(\cdot,x_n)}[(I-H)g(x_{n-1}) \\
&-\lambda N(u_{n-1},v_{n-1},w_{n-1})]+R^{I-H}_{\lambda,M(\cdot,x_n)}[(I-H)g(x_{n-1}) \\
&-\lambda N(u_{n-1},v_{n-1},w_{n-1})]-R^{I-H}_{\lambda,M(\cdot,x_{n-1})}[(I-H)g(x_{n-1})-\lambda N(u_{n-1},v_{n-1},w_{n-1})]\Big\| \\
\leq &\Big\|R^{I-H}_{\lambda,M(\cdot,x_n)}[(I-H)g(x_n)-\lambda N(u_n,v_n,w_n)] \\
&-R^{I-H}_{\lambda,M(\cdot,x_n)}[(I-H)g(x_{n-1})-\lambda N(u_{n-1},v_{n-1},w_{n-1})]\Big\|+\Big\|R^{I-H}_{\lambda,M(\cdot,x_n)}[(I-H)g(x_{n-1}) \\
&-\lambda N(u_{n-1},v_{n-1},w_{n-1})]-R^{I-H}_{\lambda,M(\cdot,x_{n-1})}[(I-H)g(x_{n-1})-\lambda N(u_{n-1},v_{n-1},w_{n-1})]\Big\| \\
\leq &\frac{1}{[1+r]}\|(I-H)g(x_n)-\lambda N(u_n,v_n,w_n)-((I-H)g(x_{n-1})-\lambda N(u_{n-1},v_{n-1},w_{n-1}))\| \\
&+h\|x_n-x_{n-1}\|.
\end{aligned} \tag{19}$$

Since g is Lipschitz continuous with constant λ_g, H is a Lipschitz continuous with constant λ_H, N is Lipschitz continuous in all three arguments with constants λ_{N_1}, λ_{N_2}, λ_{N_3}, respectively and A, B, C are D-Lipschitz continuous with constants δ_A, δ_B and δ_C, respectively, we have

$$\begin{aligned}
&\|(I-H)g(x_n)-(I-H)g(x_{n-1})-\lambda(N(u_n,v_n,w_n)-N(u_{n-1},v_{n-1},w_{n-1}))\| \\
\leq &\|(I-H)g(x_n)-(I-H)g(x_{n-1})\|+\lambda\|N(u_n,v_n,w_n)-N(u_{n-1},v_{n-1},w_{n-1})\| \\
\leq &\|g(x_n)-g(x_{n-1})\|+\|H(g(x_n))-H(g(x_{n-1}))\|+\lambda\|N(u_n,v_n,w_n)- \\
&N(u_{n-1},v_n,w_n)+N(u_{n-1},v_n,w_n)+N(u_{n-1},v_{n-1},w_n)-N(u_{n-1},v_{n-1},w_n) \\
&-N(u_{n-1},v_{n-1},w_{n-1})\|
\end{aligned}$$

$$\begin{aligned}
&\leq \|g(x_n)-g(x_{n-1})\| + \|H(g(x_n)) - H(g(x_{n-1}))\| + \lambda\|N(u_n,v_n,w_n) + \\
&\quad N(u_{n-1},v_n,w_n)\| + \lambda\|N(u_{n-1},v_n,w_n) - N(u_{n-1},v_{n-1},w_n)\| + \\
&\quad \lambda\|N(u_{n-1},v_{n-1},w_n) - N(u_{n-1},v_{n-1},w_{n-1})\| \\
&\leq \lambda_g\|x_n-x_{n-1}\| + \lambda_H\lambda_g\|x_n-x_{n-1}\| + \lambda\lambda_{N_1}\|u_n-u_{n-1}\| + \\
&\quad \lambda\lambda_{N_2}\|v_n-v_{n-1}\| + \lambda\lambda_{N_3}\|w_n-w_{n-1}\| \\
&\leq \lambda_g\|x_n-x_{n-1}\| + \lambda_H\lambda_g\|x_n-x_{n-1}\| + \lambda\lambda_{N_1}D(A(x_n),A(x_{n-1})) + \\
&\quad \lambda\lambda_{N_2}D(B(x_n),B(x_{n-1})) + \lambda\lambda_{N_3}D(C(x_n),C(x_{n-1})) \\
&\leq \lambda_g\|x_n-x_{n-1}\| + \lambda_H\lambda_g\|x_n-x_{n-1}\| + \lambda\lambda_{N_1}\delta_A\|x_n-x_{n-1}\| + \\
&\quad \lambda\lambda_{N_2}\delta_B\|x_n-x_{n-1}\| + \lambda\lambda_{N_3}\delta_C\|x_n-x_{n-1}\| \\
&\leq [\lambda_g + \lambda_H\lambda_g + \lambda\lambda_{N_1}\delta_A + \lambda\lambda_{N_2}\delta_B + \lambda\lambda_{N_3}\delta_C]\|x_n-x_{n-1}\|. \qquad (20)
\end{aligned}$$

Using (20), (19) becomes

$$\|g(x_{n+1})-g(x_n)\| \leq \left[\frac{[\lambda_g + \lambda_H\lambda_g + \lambda\lambda_{N_1}\delta_A + \lambda\lambda_{N_2}\delta_B + \lambda\lambda_{N_3}\delta_C]}{[1+r]} + h\right]\|x_n-x_{n-1}\|. \qquad (21)$$

Using (21), (18) becomes

$$\|x_{n+1}-x_n\| \leq \theta\|x_n-x_{n-1}\|,$$

where

$$\theta = \frac{1}{\xi}\left[\frac{\lambda_g + \lambda_H\lambda_g + \lambda\lambda_{N_1}\delta_A + \lambda\lambda_{N_2}\delta_B + \lambda\lambda_{N_3}\delta_C]}{[1+r]} + h\right]. \qquad (22)$$

By condition (17), we have $0 \leq \theta < 1$, thus $\{x_n\}$ is a Cauchy sequence in X and as X is complete, there exists $x \in X$ such that $x_n \to x$ as $n \to \infty$. From (13), (14) and (15) of Algorithm 3.1 and D-Lipschitz continuity of A, B and C with constants δ_A, δ_B and δ_C, respectively, we have

$$\|u_{n+1}-u_n\| \leq D(A(x_{n+1}),A(x_n)) \leq \delta_A\|x_{n+1}-x_n\|; \qquad (23)$$

$$\|v_{n+1}-v_n\| \leq D(B(x_{n+1}),B(x_n)) \leq \delta_B\|x_{n+1}-x_n\|; \qquad (24)$$

$$\|w_{n+1}-w_n\| \leq D(C(x_{n+1}),C(x_n)) \leq \delta_C\|x_{n+1}-x_n\|. \qquad (25)$$

It is clear from (23), (24) and (25) that $\{u_n\}$, $\{v_n\}$ and $\{w_n\}$ are also Cauchy sequences in X, so there exist u, v and w in X such that $u_n \to u$, $v_n \to v$ and $w_n \to w$. By using the continuity of the operators I, N, M, g, A, B, C, H, $R^{I-H}_{\lambda,M(.,x)}$ and iterative Algorithm 3.1, we have

$$g(x) = R^{I-H}_{\lambda,M(.,x)}[(I-H)g(x) - \lambda N(u,v,w)].$$

By Lemma 3.1, we conclude that (x,u,v) is a solution of problem (3.1). It remains to show that $u \in A(x)$, $v \in B(x)$ and $w \in C(x)$. In fact

$$\begin{aligned}
d(u,A(u)) &\leq \|u-u_n\| + d(u_n,A(x)) \\
&\leq \|u-u_n\| + D(A(x_n),A(x)) \\
&\leq \|u-u_n\| + \delta_A\|x_n-x\| \to 0, \quad as\ n\to\infty.
\end{aligned}$$

Hence $u \in A(x)$. Similarly, we can show that $v \in B(x)$ and $w \in C(x)$. This completes the proof. □

5. Conclusion

The resolvent operator techniques are applicable to solve several problem related to variational inequalities (inclusions), optimization problems, complementary problems etc..The aim of this work is to introduce a new type of resolvent operator based on relaxed Lipschitz continuity and monotonicity, and we call it as relaxed resolvent operator. We prove that relaxed resolvent operator is single valued and Lipschitz continuous. We define an iterative algorithm to approximate the solution of a variational inclusion problem. In our opinion, many other problems occurring in applied sciences may be solved by using relaxed resolvent operator in a different frame work.

REFERENCES

1. C. Baiocchi, A. Capelo: *Variational and quasi-variational inequalities*, John Wiley and Sons, New York (1984).
2. S.S. Chang: *Set-valued variational inclusions in Banach spaces*, Applied Mathematics and Computation, Vol. **248**, pp. 438-454 (2000).
3. S.S. Chang, Y.-J. Cho, B.S. Lee and I.H. Jung: *Generalized set-valued variational inclusions in Banach spaces*, Applied Mathematics and Computation, Vol. **246**, pp. 409-422 (2000).
4. S.S. Chang, J.K. Kim and K.H. Kim: *On the existence and iterative approximation problems of solutions for set-valued variational inclusions in Banach spaces*, Applied Mathematics and Computation, Vol. **268**, pp. 89-108 (2002).
5. Y.P. Fang and N.-J. Huang: *H-accretive operator and resolvent operator technique for variational inclusions in Banach spaces*, Applied Mathematics Latters, Vol. **17**, No. 6, pp. 647-653 (2004).
6. Y.P. Fang and N.-J. Huang: *H-monotone operator and resolvent operator technique for variational inclusions*, Applied Mathematics and Computation, Vol. **145**, No. 2-3, pp. 795-803 (2003).
7. Y.P. Fang, Y.-J. Chu, and J.K. Kim: *(H, η)-accretive operator and approximating solutions for systems of variational inclusions in Banach spaces*, to appear in Applied Mathematics Letters.
8. Y.P. Fang and N.-J. Huang: *Approximate solutions for non-linear variational inclusions with (H, η)-monotone operator*, Research Report, Sichuan University (2003).
9. Y.P. Fang, N.-J. Huang, and H.B. Thompson: *A new system of variational inclusions with (H, η)-monotone operators in Banach spaces*, Computers & Mathematics with Applications, Vol. **49**, No. 2-3, pp. 365-374 (2005).
10. A. Hassouni and A. Moudafi: *A perturbed algorithm for variational inclusions*, Journal of Mathematical Analysis and Application, Vol. **183**(3), pp. 706-712 (1994).
11. K.R. Kazmi, F.A. Khan: *Iterative approximation of a solution of multi-valued variational-like inclusion in Banach spaces: A P-η-proximal-point mapping approach*, Journal of Mathematical Analysis and Application, Vol. **325**, pp. 665-674 (2007).
12. D. Kinderlehrer, G. Stampacchia: *An introduction to variational inequalities and their applications*, Academic Press, New York (1980).
13. H.-Y. Lan, Y.J. Chu and R.U. Verma: *Non-linear relaxed cocoercive variational inclusions involving (A, η)-accertive mappings in Banach spaces*, Computers & Mathematics with Applications, Vol. **51**, No. 9-10, pp. 1529-1538 (2006).
14. Jr.S.B. Nadler: *Multivalued contraction mappings*, Pacific Journal of Mathematics, Vol **30**, pp. 475-488 (1992).
15. A. Nagurney: Network economics: *A variational inequality approach*, Kluwer Academic Publishers, Dordrecht (2002).
16. M.A. Noor, K.I. Noor and R. Kamal: *General variational inclusions involving defference of operators*, Journal of Inequalities and Applications, 2014. 2014:98.
17. M. Patriksson: *Non-linear programming and variational inequality problems: A unified approach*, Kluwer Academic Publishers, Dordrecht (1999).
18. A.H. Siddiqi and Q.H. Ansari: *An algorithm for a class of quasi-variational inequalities*, Journal of Mathematical Analysis and Application, Vol. **145**(2), pp. 413-418 (1990).
19. Y.K. Tang, S.S. Chang and S. Salahuddin: *A system of nonlinear set-valued variational inclusions*, Springer Plus, (2014) doi: 10.1186/2193-1801-3-318.
20. S. Wang: *On fixed point and variational inclusion problems*, Filomat, Vol. **29**(6), pp. 1409-1417 (2015).
21. D.L. Zhu and P. Marcotte: *Co-coercivity and its role in the convergence of iterative schemes for solving variational inequalities*, SIAM Journal of Optimization, Vol **6**, No. 3, pp. 714-726 (1996).

2

Generalized second-order parametric optimality conditions in semiinfinite discrete minmax fractional programming and second order $(\mathcal{F}, \beta, \phi, \rho, \theta, m)$-univexity

Ram U. Verma [1, *] and G. J. Zalmai [2]

[1]Department of Mathematics, University of North Texas, USA

[2]Department of Mathematics and Computer Science, Northern Michigan University, USA

Abstract This paper deals with mainly establishing numerous sets of generalized second-order parametric sufficient optimality conditions for a semiinfinite discrete minmax fractional programming problem, while the results on semiinfinite discrete minmax fractional programming problem are achieved based on some partitioning schemes under various types of generalized second-order ($\mathcal{F}$,β, ϕ, ρ, θ, m)-univexity assumptions.

Key Words Semiinfinite discrete minmax fractional programming, Generalized second-order ($\mathcal{F}$, β, ϕ, ρ, θ, m)-univex functions, Generalized sufficient optimality conditions.

1. Introduction and Preliminaries

Here in this paper, we plan to establish several sets of generalized parametric sufficient optimality conditions under various generalized $(\mathcal{F}, \beta, \phi, \rho, \theta, m)$-univexity assumptions for the following semiinfinite discrete minmax fractional programming problem of the form:

$$(P) \qquad \text{Minimize} \max_{1\leq i\leq p} \frac{f_i(x)}{g_i(x)}$$

subject to

$$G_j(x,t) \leq 0 \ \text{ for all } t \in T_j, \ \ j \in \underline{q},$$
$$H_k(x,s) = 0 \ \text{ for all } s \in S_k, \ \ k \in \underline{r},$$
$$x \in X,$$

where X is an open convex subset of $\mathbb{R}^n$, the n-dimensional Euclidean space, for each $j \in \underline{q} \equiv \{1, 2, \ldots, q\}$ and $k \in \underline{r}$, T_j and S_k are compact subsets of complete metric spaces, for each $i \in \underline{p}$, f_i and g_i are real-valued functions defined on $\mathbb{R}^n$, for each $j \in \underline{q}$, $x \to G_j(x,t)$ is a real-valued function defined on $\mathbb{R}^n$ for all $t \in T_j$, for each $k \in \underline{r}$, $x \to H_k(x,s)$ is a real-valued function defined on $\mathbb{R}^n$ for all $s \in S_k$, for each $j \in \underline{q}$ and $k \in \underline{r}$, $t \to G_j(x,t)$ and $s \to H_k(x,s)$ are continuous real-valued functions defined, respectively, on T_j and S_k for all $x \in \mathbb{R}^n$, and for each $i \in \underline{p}$, $g_i(x) > 0$ for all x satisfying the constraints of (P).

*Correspondence to: Ram Verma (Email: verma99@msn.com)

This communication deals with investigating primarily results on various second-order necessary and sufficient optimality conditions for several types of optimization problems for the semiinfinite discrete minmax fractional programming based on generalized $(\mathcal{F}, \beta, \phi, \rho, \theta, m)$-univexity assumptions, while problems for the discrete minmax fractional programming based on generalized $(\mathcal{F}, \beta, \phi, \rho, \theta, m)$-univexity assumptions were initiated in [4]. Here, we shall apply two types of partitioning schemes due to Mond and Weir [3] to the context of the new classes of generalized second-order uninvex functions to formulate and discuss numerous sets of generalized second-order sufficient optimality conditions for (P). To the best of our knowledge, all the second-order sufficient optimality results established in this paper are new in the area of semiinfinite discrete minmax fractional programming. In fact, it seems that results of this type have not yet appeared in any shape or form for any type of mathematical programming problems. The generalized optimality conditions established here are suitable to utilize in constructing some generalized second-order parametric duality models for (P) and proving numerous weak, strong, and strict converse duality theorems.

Section 1 of this paper is devoted first to some introductory materials on the semiinfinite discrete fractional programming and related developments, and then recalling some basic definitions and auxiliary results which will be needed in the sequel. In Section 2, we state and prove various second-order parametric sufficient optimality results for (P) using a variety of generalized $(\mathcal{F}, \beta, \phi, \rho, \theta, m)$-sounivexity assumptions along with some partitioning schemes. Finally, in Section 3 we summarize our main results and also point out some further research opportunities arising from certain modifications of the principal problem investigated in the present paper leading to the concluding remarks.

We also observe that all the optimality results obtained for (P) are also applicable, when appropriately specialized, to the following three classes of problems with discrete max, fractional, and conventional objective functions, which are particular cases of (P):

$$(P1) \qquad \underset{x \in \mathbb{F}}{\text{Minimize}} \max_{1 \leq i \leq p} f_i(x);$$

$$(P2) \qquad \underset{x \in \mathbb{F}}{\text{Minimize}} \frac{f_1(x)}{g_1(x)};$$

$$(P3) \qquad \underset{x \in \mathbb{F}}{\text{Minimize}}\ f_1(x),$$

where $\mathbb{F}$ (assumed to be nonempty) is the feasible set of (P), that is,

$$\mathbb{F} = \{x \in X : G_j(x) \leq 0,\ j \in \underline{q},\quad H_k(x) = 0,\ \ k \in \underline{r}\}.$$

We next introduce some new classes of generalized second-order univex functions, referred to as (strictly) $(\mathcal{F}, \beta, \phi, \rho, \theta, m)$-sounivex, (strictly) $(\mathcal{F}, \beta, \phi, \rho, \theta, m)$-pseudosounivex, and (prestrictly) $(\mathcal{F}, \beta, \phi, \rho, \theta, m)$-quasisounivex functions. They are further extensions of the classes of second-order (strictly) $(\phi, \eta, \rho, \theta, m)$-sonvex, (strictly) $(\phi, \eta, \rho, \theta, m)$-pseudosonvex, and (prestrictly) $(\phi, \eta, \rho, \theta, m)$-quasisonvex functions which were introduced recently in [4]. For brief accounts of the evolution of the generalized $\mathcal{F}$-convex and other related functions, the reader is referred to [4, 5]. We abbreviate "second-order **univex**" as **sounivex**. Let $f : X \to \mathbb{R}$ be a twice differentiable function.

Definition 1.1. *The function f is said to be* (strictly) $(\mathcal{F}, \beta, \phi, \rho, \theta, m)$-sounivex *at x^* if there exist functions $\beta : X \times X \to \mathbb{R}_+ \equiv (0, \infty)$, $\phi : \mathbb{R} \to \mathbb{R}$, $\rho : X \times X \to \mathbb{R}$, $\theta : X \times X \to \mathbb{R}^n$, a sublinear function $\mathcal{F}(x, x^*; \cdot) : \mathbb{R}^n \to \mathbb{R}$, and a positive integer m ($m \geq 1$ such that for each $x \in X$ ($x \neq x^*$) and $z \in \mathbb{R}^n$,*

$$\phi\big(f(x) - f(x^*)\big)(>) \geq \mathcal{F}\big(x, x^*; \beta(x, x^*)\nabla f(x^*)\big) + \frac{1}{2}\langle z, \nabla^2 f(x^*)z\rangle + \rho(x, x^*)\|\theta(x, x^*)\|^m,$$

where $\|\cdot\|$ is a norm on $\mathbb{R}^n$ and $\langle a, b\rangle$ is the inner product of the vectors a and b.

The function f is said to be (strictly) $(\mathcal{F}, \beta, \phi, \rho, \theta, m)$-sounivex *on X if it is* (strictly) $(\mathcal{F}, \beta, \phi, \rho, \theta, m)$-sounivex *at each $x^* \in X$.*

Definition 1.2. *The function f is said to be* (strictly) $(\mathcal{F},\beta,\phi,\rho,\theta,m)$-pseudosounivex *at x^* if there exist functions $\beta : X\times X\to\mathbb{R}_+$, $\phi:\mathbb{R}\to\mathbb{R}$, $\rho: X\times X\to\mathbb{R}$, $\theta: X\times X\to\mathbb{R}^n$, a sublinear function $\mathcal{F}(x,x^*;\cdot):\mathbb{R}^n\to\mathbb{R}$, and a positive integer m ($m\geq 1$ such that for each $x\in X$ ($x\neq x^*$) and $z\in\mathbb{R}^n$,*

$$\mathcal{F}\big(x,x^*;\beta(x,x^*)\nabla f(x^*)\big)+\frac{1}{2}\langle z,\nabla^2 f(x^*)z\rangle\geq -\rho(x,x^*)\|\theta(x,x^*)\|^m$$

$$\Rightarrow\quad \phi\big(f(x)-f(x^*)\big)(>)\geq 0.$$

The function f is said to be (strictly) $(\mathcal{F},\beta,\phi,\rho,\theta,m)$-pseudosounivex *on X if it is* (strictly) $(\mathcal{F},\beta,\phi,\rho,\theta,m)$-pseudosounivex *at each $x^*\in X$.*

Definition 1.3. *The function f is said to be* (prestrictly) $(\mathcal{F},\beta,\phi,\rho,\theta,m)$-quasisounivex *at x^* if there exist functions $\beta : X\times X\to\mathbb{R}_+$, $\phi:\mathbb{R}\to\mathbb{R}$, $\rho: X\times X\to\mathbb{R}$, $\theta: X\times X\to\mathbb{R}^n$, a sublinear function $\mathcal{F}(x,x^*;\cdot):\mathbb{R}^n\to\mathbb{R}$, and a positive integer m ($m\geq 1$ such that for each $x\in X$ and $z\in\mathbb{R}^n$,*

$$\phi\big(f(x)-f(x^*)\big)(<)\leq 0$$

$$\Rightarrow\quad \mathcal{F}\big(x,x^*;\beta(x,x^*)\nabla f(x^*)\big)+\frac{1}{2}\langle z,\nabla^2 f(x^*)z\rangle\leq -\rho(x,x^*)\|\theta(x,x^*)\|^m.$$

The function f is said to be (prestrictly) $(\mathcal{F},\beta,\phi,\rho,\theta,m)$-quasisounivex *on X if it is* (prestrictly) $(\mathcal{F},\beta,\phi,\rho,\theta,m)$-quasisounivex *at each $x^*\in X$.*

From the above definitions it is clear that if f is $(\mathcal{F},\beta,\phi,\rho,\theta,m)$-sounivex at x^*, then it is both $(\mathcal{F},\beta,\phi,\rho,\theta,m)$-pseudosounivex and $(\mathcal{F},\beta,\phi,\rho,\theta,m)$-quasisounivex at x^*, if f is $(\mathcal{F},\beta,\phi,\rho,\theta,m)$-quasisounivex at x^*, then it is prestrictly $(\mathcal{F},\beta,\phi,\rho,\theta,m)$-quasisounivex at x^*, and if f is strictly $(\mathcal{F},\beta,\phi,\rho,\theta,m)$-pseudosounivex at x^*, then it is $(\mathcal{F},\beta,\phi,\rho,\theta,m)$-quasisounivex at x^*.

In the proofs of the duality theorems, sometimes it may be more convenient to use certain alternative but equivalent forms of the above definitions. These are obtained by considering the contrapositive statements. For example, $(\mathcal{F},\beta,\phi,\rho,\theta,m)$-quasisounivexity can be defined in the following equivalent way:

The function f is said to be $(\mathcal{F},\beta,\phi,\rho,\theta,m)$-quasisounivex at x^* if there exist functions $\beta : X\times X\to\mathbb{R}_+$, $\phi:\mathbb{R}\to\mathbb{R}$, $\rho: X\times X\to\mathbb{R}$, $\theta: X\times X\to\mathbb{R}^n$, a sublinear function $\mathcal{F}(x,x^*;\cdot):\mathbb{R}^n\to\mathbb{R}$, and a positive integer m ($m\geq 1$ such that for each $x\in X$ and $z\in\mathbb{R}^n$,

$$\mathcal{F}\big(x,x^*;\beta(x,x^*)\nabla f(x^*)\big)+\frac{1}{2}\langle z,\nabla^2 f(x^*)z\rangle> -\rho(x,x^*)\|\theta(x,x^*)\|^m$$

$$\Rightarrow\quad \phi\big(f(x)-f(x^*)\big)>0.$$

Note that the new classes of generalized convex functions specified in Definitions 1.1 - 1.3 contain a variety of special cases that can easily be identified by appropriate choices of $\mathcal{F}$, β, ϕ, ρ, θ, and m. For example, if we let $\mathcal{F}\big(x,x^*;\nabla f(x^*)\big)=\langle\nabla f(x^*),\eta(x,x^*)\rangle$ and $\beta(x,x^*)\equiv 1$, then we obtain the definitions of (strictly) $(\phi,\eta,\rho,\theta,m)$-sonvex, (strictly) $(\phi,\eta,\rho,\theta,m)$-pseudosonvex, and (prestrictly) $(\phi,\eta,\rho,\theta,m)$-quasisonvex functions introduced recently in [4].

We conclude this section by recalling a set of second-order parametric necessary optimality conditions for (P). The form and features of this result will provide clear guidelines for formulating various sets of second-order parametric sufficient optimality conditions for (P).

Theorem 1.1. [4] Let $x^*\in\mathbb{F}$ and $\lambda^*=\max_{1\leq i\leq p} f_i(x^*)/g_i(x^*)$, for each $i\in \underline{p}$, let f_i and g_i be twice continuously differentiable at x^*, for each $j\in\underline{q}$, let the function $z\to G_j(z,t)$ be twice continuously differentiable at x^* for all $t\in T_j$, and for each $k\in\underline{r}$, let the function $z\to H_k(z,s)$ be twice continuously differentiable at x^* for all $s\in S_k$. If x^* is an optimal solution of (P), if the second order generalized Abadie constraint qualification holds at x^*, and if for any critical direction y, the set cone

$$\{\Big(\nabla G_j(x^*,t), \langle y, \nabla^2 G_j(x^*,t)y\rangle\Big) : t \in \hat{T}_j(x^*), j \in \underline{q}\}$$
$$+ \quad span\{\Big(\nabla H_k(x^*,s), \langle y, \nabla^2 H_k(x^*,s)y\rangle\Big) : s \in S_k, k \in \underline{r}\},$$
$$where\ \hat{T}_j(x^*) \equiv \{t \in T_j \ : \ G_j(x^*,t) = 0\},$$

is closed, then there exist $u^* \in U \equiv \{u \in \mathbb{R}^p : u \geq 0, \ \sum_{i=1}^p u_i = 1\}$ and integers ν_0^* and ν^*, with $0 \leq \nu_0^* \leq \nu^* \leq n+1$, such that there exist ν_0^* indices j_m, with $1 \leq j_m \leq q$, together with ν_0^* points $t^m \in \hat{T}_{j_m}(x^*)$, $m \in \underline{\nu_0^*}$, $\nu^* - \nu_0^*$ indices k_m, with $1 \leq k_m \leq r$, together with $\nu^* - \nu_0^*$ points $s^m \in S_{k_m}$ for $m \in \underline{\nu^*}\backslash\underline{\nu_0^*}$, and ν^* real numbers v_m^*, with $v_m^* > 0$ for $m \in \underline{\nu_0^*}$, with the property that

$$\sum_{i=1}^{p} u_i^*[\nabla f_i(x^*) - \lambda^*(\nabla g_i(x^*)] + \sum_{m=1}^{\nu_0^*} v_m^*[\nabla G_{j_m}(x^*,t^m)$$
$$+ \quad \sum_{m=\nu_0^*+1}^{\nu^*} v_m^* \nabla H_{k_m}(x^*,s^m) = 0, \tag{1.1}$$

$$\langle y, \Big[\sum_{i=1}^{p} u_i^*[\nabla^2 f_i(x^*) - \lambda^* \nabla^2 g_i(x^*)] + \sum_{m=1}^{\nu_0^*} v_m^* \nabla^2 G_{j_m}(x^*,t^m)$$
$$+ \quad \sum_{m=\nu_0^*+1}^{\nu^*} v_m^* \nabla^2 H_{k_m}(x^*,s^m)\Big]y\rangle \geq 0, \tag{1.2}$$

$$u_i^*[f_i(x^*) - \lambda^* g_i(x^*)] = 0, \ \ i \in \underline{p}, \tag{1.3}$$

where $\underline{\nu} \setminus \underline{\nu_0}$ is the complement of the set $\underline{\nu_0}$ relative to the set $\underline{\nu}$.

2. Generalized Second-Order Sufficient Optimality Conditions

In this section, we discuss several families of sufficient optimality results under various generalized $(\mathcal{F}, \beta, \phi, \rho, \theta, m)$-sounivexity hypotheses imposed on certain combinations of the problem functions. This is accomplished by employing a certain partitioning scheme which was originally proposed in [4] for the purpose of constructing generalized dual problems for nonlinear programming problems. For this we need some additional notations.

Let ν_0 and ν be integers, with $1 \leq \nu_0 \leq \nu \leq n+1$, and let $\{J_0, J_1, \ldots, J_M\}$ and $\{K_0, K_1, \ldots, K_M\}$ be partitions of the sets $\underline{\nu_0}$ and $\underline{\nu}\backslash\underline{\nu_0}$, respectively; thus, $J_i \subset \underline{\nu_0}$ for each $i \in \underline{M} \cup \{0\}$, $J_i \cap J_j = \emptyset$ for each $i, j \in \underline{M} \cup \{0\}$ with $i \neq j$, and $\cup_{i=0}^M J_i = \underline{\nu_0}$. Obviously, similar properties hold for $\{K_0, K_1, \ldots, K_M\}$. Moreover, if m_1 and m_2 are the numbers of the partitioning sets of $\underline{\nu_0}$ and $\underline{\nu}\backslash\underline{\nu_0}$, respectively, then $M = \max\{m_1, m_2\}$ and $J_i = \emptyset$ or $K_i = \emptyset$ for $i > \min\{m_1, m_2\}$.

In addition, we use the real-valued functions $z \to \Phi_i(z, \lambda, v, \bar{t}, \bar{s})$, $i \in \underline{p}$, $z \to \Phi(z, \lambda, u, v, \bar{t}, \bar{s})$, and $z \to \Lambda_\tau(z, v, \bar{t}, \bar{s})$ defined, for fixed $\lambda, u, v, \bar{t} \equiv (t^1, t^2, \ldots, t^{\nu_0})$, and $\bar{s} \equiv (s^{\nu_0+1}, s^{\nu_0+2}, \ldots, s^{\nu})$, on $\mathbb{R}^n$ as follows:

$$\Phi_i(z, \lambda, v, \bar{t}, \bar{s}) = f_i(z) - \lambda g_i(z) + \sum_{m \in J_0} v_m G_{j_m}(z, t^m) + \sum_{m \in K_0} v_m H_{k_m}(z, s^m), \ \ i \in \underline{p},$$

$$\Phi(z, \lambda, u, v, \bar{t}, \bar{s}) = \sum_{i=1}^{p} u_i[f_i(z) - \lambda g_i(z)] + \sum_{m \in J_0} v_m G_{j_m}(z, t^m) + \sum_{m \in K_0} v_m H_{k_m}(z, s^m),$$

$$\Lambda_\tau(z,v,\bar{t},\bar{s}) = \sum_{m\in J_\tau} v_m G_{j_m}(z,t^m) + \sum_{m\in K_\tau} v_m H_{k_m}(z,s^m), \quad \tau \in \underline{M}.$$

In the proofs of our sufficiency theorems, we shall make frequent use of the following auxiliary result which provides an alternative expression for the objective function of (P).

Lemma 1. *[7] For each $x \in X$,*

$$\varphi(x) \equiv \max_{1\le i\le p} \frac{f_i(x)}{g_i(x)} = \max_{u\in U} \frac{\sum_{i=1}^p u_i f_i(x)}{\sum_{i=1}^p u_i g_i(x)}.$$

Making use of the sets and functions defined above, we can now formulate our first collection of generalized sufficiency results for (P) as follows.

Theorem 2.1. *Let $x^* \in \mathbb{F}$, let $\lambda^* = \varphi(x^*)$, let the functions f_i, g_i, $i \in \underline{p}$, $z \to G_j(z,t)$, and $z \to H_k(z,s)$ be twice continuously differentiable at x^* for all $t \in T_j$ and $s \in S_k$, $j \in \underline{q}$, $k \in \underline{r}$, and assume that there exist $u^* \in U$ and integers ν_0 and ν, with $0 \le \nu_0 \le \nu \le n+1$, such that there exist ν_0 indices j_m, with $1 \le j_m \le q$, together with ν_0 points $t^m \in \hat{T}_{j_m}(x^*)$, $m \in \underline{\nu_0}$, $\nu - \nu_0$ indices k_m, with $1 \le k_m \le r$, together with $\nu - \nu_0$ points $s^m \in S_{k_m}$, $m \in \underline{\nu}\backslash\underline{\nu_0}$, and ν real numbers v_m^* with $v_m^* > 0$ for $m \in \underline{\nu_0}$, such that (1.1), (1.2) and (1.3) hold. Assume, further that any one of the following four sets of hypotheses is satisfied:*

(a) (i) *$z \to \Phi(z,\lambda^*,u^*,v^*,\bar{t},\bar{s})$ is prestrictly $(\mathcal{F},\beta,\bar{\phi},\bar{\rho},\theta,m)$-quasisounivex at x^*, and $\bar{\phi}(a) \ge 0 \Rightarrow a \ge 0$;*
(ii) *for each $\tau \in \underline{M}$, $z \to \Lambda_\tau(z,v^*,\bar{t},\bar{s})$ is strictly $(\mathcal{F},\beta,\tilde{\phi}_\tau,\tilde{\rho}_\tau,\theta,m)$-pseudosounivex at x^*, $\tilde{\phi}_\tau$ is increasing, and $\tilde{\phi}_\tau(0) = 0$;*
(iii) *$\bar{\rho} + \sum_{\tau=1}^M \tilde{\rho}_\tau \ge 0$;*

(b) (i) *$z \to \Phi(z,\lambda^*,u^*,v^*,\bar{t},\bar{s})$ is $(\mathcal{F},\beta,\bar{\phi},\bar{\rho},\theta,m)$-pseudosounivex at x^*, and $\bar{\phi}(a) \ge 0 \Rightarrow a \ge 0$;*
(ii) *for each $\tau \in \underline{M}$, $z \to \Lambda_\tau(z,v^*,\bar{t},\bar{s})$ is $(\mathcal{F},\beta,\tilde{\phi}_\tau,\tilde{\rho}_\tau,\theta,m)$-quasisounivex at x^*, $\tilde{\phi}_\tau$ is increasing, and $\tilde{\phi}_\tau(0) = 0$;*
(iii) *$\bar{\rho} + \sum_{\tau=1}^M \tilde{\rho}_\tau \ge 0$;*

(c) (i) *$z \to \Phi(z,\lambda^*,u^*,v^*,\bar{t},\bar{s})$ is prestrictly $(\mathcal{F},\beta,\bar{\phi},\bar{\rho},\theta,m)$- quasisounivex at x^*, and $\bar{\phi}(a) \ge 0 \Rightarrow a \ge 0$;*
(ii) *for each $\tau \in \underline{M}$, $z \to \Lambda_\tau(z,v^*,\bar{t},\bar{s})$ is $(\mathcal{F},\beta,\tilde{\phi}_\tau,\tilde{\rho}_\tau,\theta,m)$- quasisounivex at x^*, $\tilde{\phi}_\tau$ is increasing, and $\tilde{\phi}_\tau(0) = 0$;*
(iii) *$\bar{\rho} + \sum_{\tau=1}^M \tilde{\rho}_\tau > 0$;*

(d) (i) *$z \to \Phi(z,\lambda^*,u^*,v^*,\bar{t},\bar{s})$ is prestrictly $(\mathcal{F},\beta,\bar{\phi},\bar{\rho},\theta,m)$- quasisounivex at x^*, and $\bar{\phi}(a) \ge 0 \Rightarrow a \ge 0$;*
(ii) *for each $\tau \in \underline{M_1}$, $z \to \Lambda_\tau(z,v^*,\bar{t},\bar{s})$ is $(\mathcal{F},\beta,\tilde{\phi}_\tau,\tilde{\rho}_\tau,\theta,m)$-quasisounivex at x^*, for each $\tau \in \underline{M_2} \neq \emptyset$, $z \to \Lambda_\tau(z,v^*,\bar{t},\bar{s})$ is strictly $(\mathcal{F},\beta,\tilde{\phi}_\tau,\tilde{\rho}_\tau,\theta,m)$-pseudosounivex at x^*, and for each $\tau \in \underline{M}$, $\tilde{\phi}_\tau$ is increasing and $\tilde{\phi}_\tau(0) = 0$, where $\{\underline{M_1},\underline{M_2}\}$ is a partition of $\underline{M}$;*
(iii) *$\bar{\rho} + \sum_{\tau=1}^M \tilde{\rho}_\tau \ge 0$.*

Then x^* is an optimal solution of (P).

Proof
Let x be an arbitrary feasible solution of (P).

(a) : In view of the positivity of $\beta(x,x^*)$ and sublinearity of $\mathcal{F}(x,x^*;\cdot)$, it follows from (1.1) and (1.2) that

$$\mathcal{F}\Big(x,x^*;\beta(x,x^*)\Big\{\sum_{i=1}^{p}u_i^*[\nabla f_i(x^*)-\lambda^*\nabla g_i(x^*)]+\sum_{m\in J_0}v_m^*\nabla G_{j_m}(x^*,t^m)+\sum_{m\in K_0}v_m^*\nabla H_{k_m}(x^*,s^m)\Big\}\Big)$$
$$+\frac{1}{2}\Big\langle z^*,\Big[\sum_{i=1}^{p}u_i^*[\nabla^2 f_i(x^*)-\lambda^*\nabla^2 g_i(x^*)]+\sum_{m\in J_0}v_m^*\nabla^2 G_{j_m}(x^*,t^m)+\sum_{m\in K_0}v_m^*\nabla^2 H_{k_m}(x^*,s^m)\Big]z^*\Big\rangle$$
$$+\mathcal{F}\Big(x,x^*;\beta(x,x^*)\sum_{\tau=1}^{M}\Big[\sum_{m\in J_\tau}v_m^*\nabla G_{j_m}(x^*,t^m)+\sum_{m\in K_\tau}v_m^*\nabla H_{k_m}(x^*,s^m)\Big]\Big)$$
$$+\frac{1}{2}\Big\langle z^*,\sum_{\tau=1}^{M}\Big[\sum_{m\in J_\tau}v_j^*\nabla^2 G_{j_m}(x^*,t^m)+\sum_{m\in K_\tau}v_m^*\nabla^2 H_{k_m}(x^*,s^m)\Big]z^*\Big\rangle\geq 0. \quad (2.1)$$

Since $x,x^*\in\mathbb{F}$, for each $\tau\in\underline{M}$,

$$\begin{aligned}\Lambda_\tau(x,v^*,\bar{t},\bar{s})&=\sum_{m\in J_\tau}v_m^*G_{j_m}(x,t^m)+\sum_{m\in K_\tau}v_m^*H_{k_m}(x,s^m)\\&\leq 0\\&=\sum_{m\in J_\tau}v_m^*G_{j_m}(x^*,t^m)+\sum_{m\in K_\tau}v_m^*H_{k_m}(x^*,s^m)\\&=\Lambda_\tau(x^*,v^*,\bar{t},\bar{s}),\end{aligned}$$

and so using the properties of the function $\tilde{\phi}_\tau$, we get

$$\tilde{\phi}_\tau\big(\Lambda_\tau(x,v^*,\bar{t},\bar{s})-\Lambda_\tau(x^*,v^*,\bar{t},\bar{s})\big)\leq 0,$$

which because of (ii) implies that

$$\mathcal{F}\Big(x,x^*;\beta(x,x^*)\Big[\sum_{m\in J_\tau}v_m^*\nabla G_{j_m}(x^*,t^m)+\sum_{m\in K_\tau}v_m^*\nabla H_{k_m}(x^*,s^m)\Big]\Big)$$
$$+\frac{1}{2}\Big\langle z^*,\Big[\sum_{m\in J_\tau}v_m^*\nabla^2 G_{j_m}(x^*,t^m)+\sum_{m\in K_\tau}v_m^*\nabla^2 H_{k_m}(x^*,s^m)\Big]z^*\Big\rangle$$
$$<-\tilde{\rho}_\tau(x,x^*)\|\theta(x,x^*)\|^m.$$

Summing over τ and using the sublinearity of $\mathcal{F}(x,x^*;\cdot)$, we obtain

$$\mathcal{F}\Big(x,x^*;\beta(x,x^*)\sum_{\tau=1}^{M}\Big[\sum_{m\in J_\tau}v_m^*\nabla G_{j_m}(x^*,t^m)+\sum_{m\in K_\tau}v_m^*\nabla H_{k_m}(x^*,s^m)\Big]\Big)$$
$$+\frac{1}{2}\Big\langle z^*,\sum_{\tau=1}^{M}\Big[\sum_{m\in J_\tau}v_m^*\nabla^2 G_{j_m}(x^*,t^m)+\sum_{m\in K_\tau}v_m^*\nabla^2 H_{k_m}(x^*,s^m)\Big]z^*\Big\rangle$$
$$<-\sum_{\tau=1}^{M}\tilde{\rho}_\tau(x,x^*)\|\theta(x,x^*)\|^m. \quad (2.2)$$

Combining (2.1), (2.2) and using (iii) we get

$$\mathcal{F}\Big(x,x^*;\beta(x,x^*)\Big\{\sum_{i=1}^{p}u_i^*[\nabla f_i(x^*)-\lambda^*\nabla g_i(x^*)]+\sum_{m\in J_0}v_m^*\nabla G_{j_m}(x^*,t^m)$$
$$+\sum_{m\in K_0}v_m^*\nabla H_{k_m}(x^*,s^m)\Big\}\Big)+\frac{1}{2}\Big\langle z^*,\Big[\sum_{i=1}^{p}u_i^*[\nabla^2 f_i(x^*)-\lambda^*\nabla^2 g_i(x^*)]+\sum_{m\in J_0}v_m^*\nabla^2 G_{j_m}(x^*,t^m)$$
$$+\sum_{m\in K_0}v_m^*\nabla^2 H_{k_m}(x^*,s^m)\Big]z^*\Big\rangle>\sum_{\tau=1}^{M}\tilde{\rho}_\tau(x,x^*)\|\theta(x,x^*)\|^m>-\bar{\rho}(x,x^*)\|(\theta(x,x^*)\|^m, \quad (2.3)$$

which by virtue of (i) implies that $\bar{\phi}\big(\Phi(x,\lambda^*,u^*,v^*,\bar{t},\bar{s})-\Phi(x^*,\lambda^*,u^*,v^*,\bar{t},\bar{s})\big)\geq 0$, which in light of further properties of $\bar{\phi}$ reduces to

$$\Phi(x,\lambda^*,u^*,v^*,\bar{t},\bar{s})\geq\Phi(x^*,\lambda^*,u^*,v^*,\bar{t},\bar{s})=0,$$

where the equality follows from (1.3) and the feasibility of x^*. Since $x,x^*\in\mathbb{F}$ and $v_m^*>0,\ m\in\underline{\nu_0}$, we have

$$0\leq\sum_{i=1}^{p}u_i^*[f_i(x)-\lambda^* g_i(x)].$$

Now using this inequality and Lemma 2.1, we obtain $\varphi(x^*)\leq\varphi(x)$. Since x is arbitrary, we conclude that x^* is an optimal solution to (P).

(b): Based on the proof of part (a), we see that (ii) leads to the following inequality:

$$\mathcal{F}\Big(x,x^*;\beta(x,x^*)\sum_{\tau=1}^{M}\Big[\sum_{m\in J_\tau}v_m^*\nabla G_{j_m}(x^*,t^m)+\sum_{m\in K_\tau}v_m^*\nabla H_{k_m}(x^*,s^m)\Big]\Big)$$
$$+\frac{1}{2}\Big\langle z^*,\sum_{\tau=1}^{M}\Big[\sum_{m\in J_\tau}v_j^*\nabla^2 G_{j_m}(x^*,t^m)+\sum_{m\in K_\tau}v_m^*\nabla^2 H_{k_m}(x^*,s^m)\Big]z^*\Big\rangle$$
$$\leq-\sum_{\tau=1}^{M}\tilde{\rho}_\tau(x,x^*)\|\theta(x,x^*)\|^m. \quad (2.4)$$

Now combining this inequality with (2.1) and using (iii), we obtain

$$\mathcal{F}\Big(x,x^*;\beta(x,x^*)\Big\{\sum_{i=1}^{p}u_i^*[\nabla f_i(x^*)-\lambda^*\nabla g_i(x^*)]+\sum_{m\in J_0}v_m^*\nabla G_{j_m}(x^*,t^m)$$
$$+\sum_{m\in K_0}v_m^*\nabla H_{k_m}(x^*,s^m)\Big\}\Big)$$
$$+\frac{1}{2}\Big\langle z^*,\Big[\sum_{i=1}^{p}u_i^*[\nabla^2 f_i(x^*)-\lambda^*\nabla^2 g_i(x^*)]+\sum_{m\in J_0}v_m^*\nabla^2 G_{j_m}(x^*,t^m)$$
$$+\sum_{m\in K_0}v_m^*\nabla^2 H_{k_m}(x^*,s^m)\Big]z^*\Big\rangle\geq\sum_{\tau=1}^{M}\tilde{\rho}_\tau(x,x^*)\|\theta(x,x^*)\|^m\geq-\bar{\rho}(x,x^*)\|\theta(x,x^*)\|^M,$$

which applying (i) implies that $\bar{\phi}\big(\Phi(x,\lambda^*,u^*,v^*,\bar{t},\bar{s})-\Phi(x^*,\lambda^*,u^*,v^*,\bar{t},\bar{s})\big)\geq 0$. Thus, we conclude that x^* is an optimal solution to (P), while the proofs for (c) and (d) are similar to (a) and (b). □

Making use of the sets and functions defined above, we can now formulate our first collection of generalized second-order parametric sufficient optimality results for (P) as follows.

Theorem 2.2. *Let $x^* \in \mathbb{F}$, let $\lambda^* = \varphi(x^*)$, let the functions f_i, g_i, $i \in \underline{p}$, $z \to G_j(z,t)$, and $z \to H_k(z,s)$ be twice continuously differentiable at x^* for all $t \in T_j$ and $s \in S_k$, $j \in \underline{q}$, $k \in \underline{r}$, and assume that there exist $u^* \in U$ and integers ν_0 and ν, with $0 \le \nu_0 \le \nu \le n+1$, such that there exist ν_0 indices j_m, with $1 \le j_m \le q$, together with ν_0 points $t^m \in \hat{T}_{j_m}(x^*)$, $m \in \underline{\nu_0}$, $\nu - \nu_0$ indices k_m, with $1 \le k_m \le r$, together with $\nu - \nu_0$ points $s^m \in S_{k_m}$, $m \in \underline{\nu}\backslash\underline{\nu_0}$, and ν real numbers v_m^* with $v_m^* > 0$ for $m \in \underline{\nu_0}$, such that (1.1), (1.2) and (1.3) hold. Assume, further that any one of the following seven sets of hypotheses is satisfied:*

(a) *(i) for each $i \in I_+ \equiv \{i \in \underline{p} : u_i^* > 0\}$, $\xi \to \Phi_i(\xi, v^*, \lambda^*, \bar{t}, \bar{s})$ is $(\mathcal{F}, \beta, \bar{\phi}_i, \bar{\rho}_i, \theta, m)$-pseudosounivex at x^*, $\bar{\phi}_i$ is strictly increasing, and $\bar{\phi}_i(0) = 0$;*

(ii) for each $t \in \underline{M}$, $\xi \to \Lambda_t(\xi, v^*, \bar{t}, \bar{s})$ is $(\mathcal{F}, \beta, \tilde{\phi}_t, \tilde{\rho}_t, \theta, m)$-quasisounivex at x^*, $\tilde{\phi}_t$ is increasing, and $\tilde{\phi}_t(0) = 0$;

(iii) $\sum_{i\in I_+} u_i^* \bar{\rho}_i(x, x^*) + \sum_{t=1}^{M} \tilde{\rho}_t(x, x^*) \ge 0$ for all $x \in \mathbb{F}$;

(b) *(i) for each $i \in I_+$, $\xi \to \Phi_i(\xi, v^*, \lambda^*, \bar{t}, \bar{s})$ is prestrictly $(\mathcal{F}, \beta, \bar{\phi}_i, \bar{\rho}_i, \theta, m)$-quasisounivex at x^*, $\bar{\phi}_i$ is strictly increasing, and $\bar{\phi}_i(0) = 0$;*

(ii) for each $t \in \underline{M}$, $\xi \to \Lambda_t(\xi, v^*, \bar{t}, \bar{s})$ is strictly $(\mathcal{F}, \beta, \tilde{\phi}_t, \tilde{\rho}_t, \theta)$-pseudosounivex at x^*, $\tilde{\phi}_t$ is increasing, and $\tilde{\phi}_t(0) = 0$;

(iii) $\sum_{i\in I_+} u_i^* \bar{\rho}_i(x, x^*) + \sum_{t=1}^{M} \tilde{\rho}_t(x, x^*) \ge 0$ for all $x \in \mathbb{F}$;

(c) *(i) for each $i \in I_+$, $\xi \to \Phi_i(\xi, v^*, \lambda^*, \bar{t}, \bar{s})$ is prestrictly $(\mathcal{F}, \beta, \bar{\phi}_i, \bar{\rho}_i, \theta, m)$-quasisounivex at x^*, $\bar{\phi}_i$ is strictly increasing, and $\bar{\phi}_i(0) = 0$;*

(ii) for each $t \in \underline{M}$, $\xi \to \Lambda_t(\xi, v^*, \bar{t}, \bar{s})$ is $(\mathcal{F}, \beta, \tilde{\phi}_t, \tilde{\rho}_t, \theta, m)$-quasisounivex at x^*, $\tilde{\phi}_t$ is increasing, and $\tilde{\phi}_t(0) = 0$;

(iii) $\sum_{i\in I_+} u_i^* \bar{\rho}_i(x, x^*) + \sum_{t=1}^{M} \tilde{\rho}_t(x, x^*) > 0$ for all $x \in \mathbb{F}$;

(d) *(i) for each $i \in I_{1+}$, $\xi \to \Phi_i(\xi, v^*, \lambda^*, \bar{t}, \bar{s})$ is $(\mathcal{F}, \beta, \bar{\phi}_i, \bar{\rho}_i, \theta, m)$-pseudosounivex at x^*, for each $i \in I_{2+}$, $\xi \to \Phi_i(\xi, v^*, v^*, \lambda^*, \bar{t}, \bar{s})$ is prestrictly $(\mathcal{F}, \beta, \bar{\phi}_i, \bar{\rho}_i, \theta, m)$-quasisounivex at x^*, and for each $i \in I_+$, $\bar{\phi}_i$ is strictly increasing and $\bar{\phi}_i(0) = 0$, where $\{I_{1+}, I_{2+}\}$ is a partition of I_+;*

(ii) for each $t \in \underline{M}$, $\xi \to \Lambda_t(\xi, v^*, \bar{t}, \bar{s})$ is strictly $(\mathcal{F}, \beta, \tilde{\phi}_t, \tilde{\rho}_t, \theta, m)$-pseudosounivex at x^*, $\tilde{\phi}_t$ is increasing, and $\tilde{\phi}_t(0) = 0$;

(iii) $\sum_{i\in I_+} u_i^* \bar{\rho}_i(x, x^*) + \sum_{t=1}^{M} \tilde{\rho}_t(x, x^*) \ge 0$ for all $x \in \mathbb{F}$;

(e) *(i) for each $i \in I_{1+} \ne \emptyset$, $\xi \to \Phi_i(\xi, v^*, \lambda^*, \bar{t}, \bar{s})$ is $(\mathcal{F}, \beta, \bar{\phi}_i, \bar{\rho}_i, \theta, m)$-pseudosounivex at x^*, for each $i \in I_{2+}$, $\xi \to \Phi_i(\xi, v^*, \lambda^*, \bar{t}, \bar{s})$ is prestrictly $(\mathcal{F}, \beta, \bar{\phi}_i, \bar{\rho}_i, \theta, m)$-quasisounivex at x^*, and for each $i \in I_+$, $\bar{\phi}_i$ is strictly increasing and $\bar{\phi}_i(0) = 0$, where $\{I_{1+}, I_{2+}\}$ is a partition of I_+;*

(ii) for each $t \in \underline{M}$, $\xi \to \Lambda_t(\xi, v^*, v^*, \bar{t}, \bar{s})$ is $(\mathcal{F}, \beta, \tilde{\phi}_t, \tilde{\rho}_t, \theta, m)$-quasisounivex at x^*, $\tilde{\phi}_t$ is increasing, and $\tilde{\phi}_t(0) = 0$;

(iii) $\sum_{i\in I_+} u_i^* \bar{\rho}_i(x, x^*) + \sum_{t=1}^{M} \tilde{\rho}_t(x, x^*) \ge 0$ for all $x \in \mathbb{F}$;

(f) *(i) for each $i \in I_+$, $\xi \to \Phi_i(\xi, v^*, \lambda^*, \bar{t}, \bar{s})$ is prestrictly $(\mathcal{F}, \beta, \bar{\phi}_i, \bar{\rho}_i, \theta, m)$-quasisounivex at x^*, $\bar{\phi}_i$ is strictly increasing, and $\bar{\phi}_i(0) = 0$;*

(ii) for each $t \in \underline{M}_1 \ne \emptyset$, $\xi \to \Lambda_t(\xi, v^*, \bar{t}, \bar{s})$ is strictly $(\mathcal{F}, \beta, \bar{\phi}_t, \tilde{\rho}_t, \theta, m)$-pseudosounivex at x^*, for each $t \in \underline{M}_2$, $\xi \to \Lambda_t(\xi, v^*, \bar{t}, \bar{s})$ is $(\mathcal{F}, \beta, \tilde{\phi}_t, \tilde{\rho}_t, \theta, m)$-quasisounivex at x^*, and for each $t \in \underline{M}$, $\tilde{\phi}_t$ is increasing and $\tilde{\phi}_t(0) = 0$, where $\{\underline{M}_1, \underline{M}_2\}$ is a partition of $\underline{M}$;

(iii) $\sum_{i\in I_+} u_i^* \bar{\rho}_i(x, x^*) + \sum_{t=1}^{M} \tilde{\rho}_t(x, x^*) \ge 0$ for all $x \in \mathbb{F}$;

(g) *(i) for each $i \in I_{1+}$, $\xi \to \Phi_i(\xi, v^*, \lambda^*, \bar{t}, \bar{s})$ is $(\mathcal{F}, \beta, \bar{\phi}_i, \bar{\rho}_i, \theta, m)$-pseudosounivex at x^*, for each $i \in I_{2+}$, $\xi \to \Phi_i(\xi, v^*, \lambda^*, \bar{t}, \bar{s})$ is prestrictly $(\mathcal{F}, \beta, \bar{\phi}_i, \bar{\rho}_i, \theta, m)$-quasisounivex at x^*, and for each $i \in I_+$, $\bar{\phi}_i$ is strictly increasing and $\bar{\phi}_i(0) = 0$, where $\{I_{1+}, I_{2+}\}$ is a partition of I_+;*

(ii) *for each* $t \in \underline{M}_1$, $\xi \to \Lambda_t(\xi, v^*, \bar{t}, \bar{s})$ *is strictly* $(\mathcal{F}, \beta, \tilde{\phi}_t, \tilde{\rho}_t, \theta, m)$*-pseudosounivex at* x^*, *for each* $t \in \underline{M}_2$, $\xi \to \Lambda_t(\xi, v^*, \bar{t}, \bar{s})$ *is* $(\mathcal{F}, \beta, \tilde{\phi}_t, \tilde{\rho}_t, \theta, m)$*-quasisounivex at* x^*, *and for each* $t \in \underline{M}$, $\tilde{\phi}_t$ *is increasing and* $\tilde{\phi}_t(0) = 0$, *where* $\{\underline{M}_1, \underline{M}_2\}$ *is a partition of* $\underline{M}$;

(iii) $\sum_{i \in I_+} u_i^* \bar{\rho}_i(x, x^*) + \sum_{t=1}^{M} \tilde{\rho}_t(x, x^*) \geq 0$ *for all* $x \in \mathbb{F}$;

(iv) $I_{1+} \neq \emptyset$, $\underline{M_1} \neq \emptyset$, *or* $\sum_{i \in I_+} u_i^* \bar{\rho}_i(x, x^*) + \sum_{t=1}^{M} \tilde{\rho}_t(x, x^*) > 0$.

Then x^* *is an optimal solution of (P).*

Proof

(a): Assume that x^* is not an optimal solution to (P). Then there exists a feasible solution $\bar{x}$ of (P) such that $\varphi(\bar{x}) < \varphi(x^*) = \lambda^*$. Hence, we have

$$f_i(\bar{x}) - \lambda^* g_i(\bar{x}) < 0 \text{ for each } i \in \underline{p}. \tag{2.5}$$

Since $v^* \geq 0$, for each $i \in I_+$, we have

$$\begin{aligned}
\Phi_i(\bar{x}, v^*, \lambda^*, \bar{t}, \bar{s}) &= f_i(\bar{x}) - \lambda^* g_i(\bar{x}) + \sum_{j \in J_0} v_j^* G_j(\bar{x}, t^m) + \sum_{k \in K_0} v_k^* H_k(\bar{x}, s^m) \\
&\leq f_i(\bar{x}) - \lambda^* g_i(\bar{x}) \ \text{(by the feasibility of } \bar{x}) \\
&< 0 \\
&= f_i(x^*) - \lambda^* g_i(x^*) + \sum_{j \in J_0} v_j^* G_j(x^*, t^m) + \sum_{k \in K_0} w_k^* H_k(x^*, s^m) \\
&\text{(by the feasibility of } x^*) \\
&= \Phi_i(x^*, v^*, w^*, \lambda^*, \bar{t}, \bar{s}),
\end{aligned}$$

and so using the properties of the function $\bar{\phi}_i$, we get

$$\bar{\phi}_i\big(\Phi_i(\bar{x}, v^*, \lambda^*) - \Phi_i(x^*, v^*, \lambda^*)\big) < 0,$$

which in view of (i) implies that

$$\begin{aligned}
&\mathcal{F}\Big(x, x^*; \beta(x, x^*)\Big[\nabla f_i(x^*) - \lambda^* \nabla g_i(x^*) + \sum_{m \in J_0} v_m^* \nabla G_{j_m}(x^*, t^m) + \sum_{m \in K_0} v_m^* \nabla H_{k_m}(x^*, s^m)\Big]\Big) + \\
&\quad \frac{1}{2}\Big\langle z^*, \Big[\nabla^2 f_i(x^*) - \lambda^* \nabla^2 g_i(x^*) + \sum_{m \in J_0} v_m^* \nabla^2 G_j(x^*, t^m) + \sum_{m \in K_0} v_m^* \nabla^2 H_{k_m}(x^*, s^m)\Big] z^*\Big\rangle \\
&\qquad < -\bar{\rho}_i(x, x^*) \|\theta(\bar{x}, x^*)\|^m.
\end{aligned}$$

Since $u^* > 0$, $u_i^* = 0$ for each $i \in \underline{p} \backslash I_+$, $\sum_{i=1}^{p} u_i^* = 1$, and $\mathcal{F}(x, x^*, \cdot)$ is sublinear, the above inequalities yield

$$\begin{aligned}
&\mathcal{F}\Big(x, x^*; \beta(x, x^*)\Big\{\sum_{i=1}^{p} u_i^*[\nabla f_i(x^*) - \lambda^* \nabla g_i(x^*)] + \sum_{m \in J_0} v_m^* \nabla G_{j_m}(x^*, t^m) + \sum_{m \in K_0} v_m^* \nabla H_{k_m}(x^*, s^m)\Big\}\Big) + \\
&\quad \frac{1}{2}\Big\langle z^*, \Big\{\sum_{i=1}^{p} u_i^*[\nabla^2 f_i(x^*) - \lambda^* \nabla^2 g_i(x^*)] + \sum_{m \in J_0} v_m^* \nabla^2 G_{j_m}(x^*, t^m) + \sum_{m \in K_0} v_m^* \nabla^2 H_{k_m}(x^*, s^m)\Big\} z^*\Big\rangle \\
&\qquad < -\sum_{i \in I_+} u_i^* \bar{\rho}_i(x, x^*) \|\theta(\bar{x}, x^*)\|^m.
\end{aligned} \tag{2.6}$$

Proceeding as in the proof of Theorem 2.1, we see that our assumptions in (ii) lead to

$$\mathcal{F}\Big(x,x^*;\beta(x,x^*)\sum_{\tau=1}^{M}\Big[\sum_{m\in J_\tau}v_m^*\nabla G_{j_m}(x^*,t^m)+\sum_{m\in K_\tau}v_m^*\nabla H_{k_m}(x^*,s^m)\Big]\Big)$$
$$+\frac{1}{2}\Big\langle z^*,\sum_{\tau=1}^{M}\Big[\sum_{m\in J_\tau}v_m^*\nabla^2 G_j(x^*,t^m)$$
$$+\sum_{m\in K_\tau}v_m^*\nabla^2 H_{k_m}(x^*,s^m)\Big]z^*\Big\rangle\leq-\sum_{\tau=1}^{M}\tilde{\rho}_t(x,x^*)\|\theta(\bar{x},x^*)\|^2,$$

which when combined with (2.1) results in

$$\mathcal{F}\Big(x,x^*;\beta(x,x^*)\Big\{\sum_{i=1}^{p}u_i^*[\nabla f_i(x^*)-\lambda^*\nabla g_i(x^*)]+\sum_{m\in J_0}v_m^*\nabla G_{j_m}(x^*,t^m)$$
$$+\sum_{m\in K_0}v_m^*\nabla H_{k_m}(x^*,s^m)\Big\}\Big)+\frac{1}{2}\Big\langle z^*,\Big\{\sum_{i=1}^{p}u_i^*[\nabla^2 f_i(x^*)-\lambda^*\nabla^2 g_i(x^*)]$$
$$+\sum_{m\in J_0}v_m^*\nabla^2 G_{j_m}(x^*,t^m)+\sum_{m\in K_0}v_m^*\nabla^2 H_{k_m}(x^*,s^m)\Big\}z^*\Big\rangle\geq\sum_{\tau=1}^{M}\tilde{\rho}_\tau(x,x^*)\|\theta(\bar{x},x^*)\|^2.$$

Based on (iii), this inequality contradicts based on our assumption (2.6). Hence, x^* is an optimal solution to (P).
(b) - (g) : The proofs are similar to that of part (a). □

In the following theorem, we construct a different partitioning method which appears to be general for formulating a general duality model for a multiobjective fractional programming problem, and then we present another collection of sufficient optimality results for (P) which are different from those stated in Theorems 2.1 and 2.2. These results are formulated on a partition of $\underline{p}$ in addition to those of $\underline{\nu_0}$ and $\underline{\nu}\backslash\underline{\nu_0}$, and by placing appropriate generalized $(\mathcal{F},\beta,\phi,\rho,\theta)$-univexity requirements on certain combinations of the functions $z\to u_i[f_i(z)-\lambda g_i(z)]$, $z\to G_j(z,t)$, and $z\to H_k(z,s)$.

Let $\{I_0,I_1,\ldots,I_d\}$, $\{J_0,J_1,\ldots,J_e\}$, and $\{K_0,K_1,\ldots,K_e\}$ be partitions of $\underline{p}$, $\underline{\nu_0}$, and $\underline{\nu}\backslash\underline{\nu_0}$, respectively, such that $D=\{0,1,2,\ldots,d\}\subset E=\{0,1,\ldots,e\}$, and let the function $z\to\Pi_\tau(z,\lambda,u,v,\bar{t},\bar{s}):\mathbb{R}^n\to\mathbb{R}$ be defined, for fixed $\lambda,u,v,\bar{t}\equiv(t^1,t^2,\ldots,t^{\nu_0})$, and $\bar{s}\equiv(s^{\nu_0+1},s^{\nu_0+2},\ldots,s^{\nu})$, by

$$\Pi_\tau(z,\lambda,u,v,\bar{t},\bar{s})=\sum_{i\in I_\tau}u_i[f_i(z)-\lambda g_i(z)]+\sum_{m\in J_\tau}v_m G_{j_m}(x,t^m)$$
$$+\sum_{m\in K_\tau}v_m H_{k_m}(z,s^m),\quad \tau\in D.$$

Theorem 2.3. *Let $x^*\in\mathbb{F}$, let $\lambda^*=\varphi(x^*)$, let the functions f_i, g_i, $i\in\underline{p}$, $z\to G_j(z,t)$, and $z\to H_k(z,s)$ be differentiable at x^* for all $t\in T_j$ and $s\in S_k$, $j\in\underline{q}$, $k\in\underline{r}$, and assume that there exist $u^*\in U,u^*>0$, and integers ν_0 and ν, with $0\leq\nu_0\leq\nu\leq n+1$, such that there exist ν_0 indices j_m, with $1\leq j_m\leq q$, together with ν_0 points $t^m\in\hat{T}_{j_m}(x^*)$, $m\in\underline{\nu_0}$, $\nu-\nu_0$ indices k_m, with $1\leq k_m\leq r$, together with $\nu-\nu_0$ points $s^m\in S_{k_m}$, $m\in\underline{\nu}\backslash\underline{\nu_0}$, and ν real numbers v_m^* with $v_m^*>0$ for $m\in\underline{\nu_0}$, such that (1) and (2) hold. Assume, furthermore, that any one of the following seven sets of hypotheses is satisfied:*

(a) (i) for each $\tau\in D$, $z\to\Pi_\tau(z,\lambda^*,u^*,v^*,\bar{t},\bar{s})$ is $(\mathcal{F},\beta,\phi_\tau,\rho_\tau,\theta,m)$ - pseudosounivex at x^*, ϕ_τ is increasing, and $\phi_\tau(0)=0$;

(ii) for each $\tau \in E \setminus D, z \to \Lambda_\tau(z, v^*, \bar{t}, \bar{s})$ *is* $(\mathcal{F}, \beta, \phi_\tau, \rho_\tau, m)-$ *quasisounivex at* x^*, ϕ_τ *is increasing, and* $\phi_\tau(0) = 0$*;*

(iii) $\sum_{\tau \in E} \rho_\tau \geq 0$*;*

(b) *(i) for each* $\tau \in D,\ z \to \Pi_\tau(z, \lambda^*, u^*, v^*, \bar{t}, \bar{s})$ *is prestrictly* $(\mathcal{F}, \beta, \phi_\tau, \rho_\tau, \theta, m)$*- quasisounivex at* x^*, ϕ_τ *is increasing, and* $\phi_\tau(0) = 0$*;*

(ii) for each $\tau \in E \setminus D,\ z \to \Lambda_\tau(z, v^*, \bar{t}, \bar{s})$ *is strictly* $(\mathcal{F}, \beta, \phi_\tau, \rho_\tau, \theta, m)$*- pseudosounivex at* x^*, ϕ_τ *is increasing, and* $\phi_\tau(0) = 0$*;*

(iii) $\sum_{\tau \in E} \rho_\tau \geq 0$*;*

(c) *(i) for each* $\tau \in D,\ z \to \Pi_\tau(z, \lambda^*, u^*, v^*, \bar{t}, \bar{s})$ *is prestrictly* $(\mathcal{F}, \beta, \phi_\tau, \rho_\tau, \theta, m)$*- quasisounivex at* x^*, ϕ_τ *is increasing, and* $\phi_\tau(0) = 0$*;*

(ii) for each $\tau \in E \setminus D,\ z \to \Lambda_\tau(z, v^*, \bar{t}, \bar{s})$ *is* $(\mathcal{F}, \beta, \phi_\tau, \rho_\tau, \theta, m)$*- quasisounivex at* x^*, ϕ_τ *is increasing, and* $\phi_\tau(0) = 0$*;*

(iii) $\sum_{\tau \in E} \rho_\tau > 0$*;*

(d) *(i) for each* $\tau \in D_1,\ z \to \Pi_\tau(z, \lambda^*, u^*, v^*, \bar{t}, \bar{s})$ *is* $(\mathcal{F}, \beta, \phi_\tau, \rho_\tau, \theta, m)$*- pseudosounivex at* x^**, for each* $\tau \in D_2,\ z \to \Pi_\tau(z, \lambda^*, u^*, v^*, \bar{t}, \bar{s})$ *is prestrictly* $(\mathcal{F}, \beta, \phi_\tau, \rho_\tau, \theta, m)$*- quasisounivex at* x^**, and for each* $\tau \in D,\ \phi_\tau$ *is increasing and* $\phi_\tau(0) = 0$*, where* $\{D_1, D_2\}$ *is a partition of D;*

(ii) for each $\tau \in E \setminus D,\ z \to \Lambda_\tau(z, v^*, \bar{t}, \bar{s})$ *is strictly* $(\mathcal{F}, \beta, \phi_\tau, \rho_\tau, \theta, m)$*- pseudosounivex at* x^*, ϕ_τ *is increasing, and* $\phi_\tau(0) = 0$*;*

(iii) $\sum_{\tau \in E} \rho_\tau \geq 0$*;*

(e) *(i) for each* $\tau \in D_1 \neq \emptyset,\ z \to \Pi_\tau(z, \lambda^*, u^*, v^*, \bar{t}, \bar{s})$ *is* $(\mathcal{F}, \beta, \phi_\tau, \rho_\tau, \theta, m)$*- pseudosounivex at* x^**, for each* $\tau \in D_2,\ z \to \Pi_\tau(z, \lambda^*, u^*, v^*, \bar{t}, \bar{s})$ *is prestrictly* $(\mathcal{F}, \beta, \phi_\tau, \rho_\tau, \theta, m)$*- quasisounivex at* x^**, and for each* $\tau \in D,\ \phi_\tau$ *is increasing and* $\phi_\tau(0) = 0$*, where* $\{D_1, D_2\}$ *is a partition of D;*

(ii) for each $\tau \in E \setminus D, z \to \Lambda_\tau(z, v^*, \bar{t}, \bar{s})$ *is* $(\mathcal{F}, \beta, \phi_\tau, \rho_\tau, \theta, m)$*- quasisounivex at* x^*, ϕ_τ *is increasing, and* $\phi_\tau(0) = 0$*;*

(iii) $\sum_{\tau \in E} \rho_\tau \geq 0$*;*

(f) *(i) for each* $\tau \in D,\ z \to \Pi_\tau(z, \lambda^*, u^*, v^*, \bar{t}, \bar{s})$ *is prestrictly* $(\mathcal{F}, \beta, \phi_\tau, \rho_\tau, \theta, m)$*- quasisounivex at* x^*, ϕ_τ *is increasing, and* $\phi_\tau(0) = 0$*;*

(ii) for each $\tau \in (E \setminus D)_1 \neq \emptyset,\ z \to \Lambda_\tau(z, v^*, \bar{t}, \bar{s})$ *is strictly* $(\mathcal{F}, \beta, \phi_\tau, \rho_\tau, \theta, m)$*- pseudosounivex at* x^**, for each* $\tau \in (E \setminus D)_2,\ z \to \Lambda_\tau(z, v^*, \bar{t}, \bar{s})$ *is* $(\mathcal{F}, \beta, \phi_\tau, \rho_\tau, \theta, m)$*- quasisounivex at* x^**, and for each* $\tau \in D,\ \phi_\tau$ *is increasing and* $\phi_\tau(0) = 0$*, where* $\{(E \setminus D)_1, (E \setminus D)_2\}$ *is a partition of* $E \setminus D$*;*

(iii) $\sum_{\tau \in E} \rho_\tau \geq 0$*;*

(g) *(i) for each* $\tau \in D_1,\ z \to \Pi_\tau(z, \lambda^*, u^*, v^*, \bar{t}, \bar{s})$ *is* $(\mathcal{F}, \beta, \phi_\tau, \rho_\tau, \theta, m)$*- pseudosounivex at* x^**, for each* $\tau \in D_2,\ z \to \Pi_\tau(z, \lambda^*, u^*, v^*, \bar{t}, \bar{s})$ *is prestrictly* $(\mathcal{F}, \beta, \phi_\tau, \rho_\tau, \theta, m)$*- quasisounivex at* x^**, and for each* $\tau \in D,\ \phi_\tau$ *is increasing and* $\phi_\tau(0) = 0$*, where* $\{D_1, D_2\}$ *is a partition of D;*

(ii) for each $\tau \in (E \setminus D)_1,\ z \to \Lambda_\tau(z, v^*, \bar{t}, \bar{s})$ *is strictly* $(\mathcal{F}, \beta, \phi_\tau, \rho_\tau, \theta, m)$*- pseudosounivex at* x^**, for each* $\tau \in (E \setminus D)_2,\ z \to \Lambda_\tau(z, v^*, \bar{t}, \bar{s})$ *is* $(\mathcal{F}, \beta, \phi_\tau, \rho_\tau, \theta, m)$*- quasisounivex at* x^**, and for each* $\tau \in D,\ \phi_\tau$ *is increasing and* $\phi_\tau(0) = 0$*, where* $\{(E \setminus D)_1, (E \setminus D)_2\}$ *is a partition of* $E \setminus D$*;*

(iii) $\sum_{\tau \in E} \rho_\tau \geq 0$*;*

(iv) $D_1 \neq \emptyset$, $(E \setminus D)_1 \neq \emptyset$*, or* $\sum_{\tau \in E} \rho_\tau > 0$.

Then x^* *is an optimal solution of (P).*

Proof

(a): Assume that x^* is not an optimal solution to (P). Then based on the proof of Theorem 2.2, we arrive at the inequalities $f_i(\bar{x}) - \lambda^* g_i(\bar{x}) < 0,\ i \in \underline{p}$, for some $\bar{x} \in \mathbb{F}$. Since $u^* > 0$, we see that for each $\tau \in D$,

$$\sum_{i \in I_\tau} u_i^* [f_i(\bar{x}) - \lambda^* g_i(\bar{x})] < 0. \tag{2.7}$$

Now using this inequality, we see that

$$\begin{aligned}
&\Pi_\tau(\bar{x}, \lambda^*, u^*, v^*, \bar{t}, \bar{s}) \\
&= \sum_{i\in I_\tau} u_i^*[f_i(\bar{x}) - \lambda^* g_i(\bar{x})] + \sum_{m\in J_\tau} v_m^* G_{j_m}(\bar{x}, t^m) + \sum_{m\in K_\tau} v_m^* H_{k_m}(\bar{x}, s^m) \\
&\leq \sum_{i\in I_\tau} u_i^*[f_i(\bar{x}) - \lambda^* g_i(\bar{x})] \text{ (by the feasibility of } \bar{x} \text{ and positivity of } v_m^*,\ m \in \underline{\nu_0}) \\
&< 0 \\
&= \sum_{i\in I_\tau} u_i^*[f_i(x^*) - \lambda^* g_i(x^*)] + \sum_{m\in J_\tau} v_m^* G_{j_m}(x^*, t^m) + \sum_{m\in K_\tau} v_m^* H_{k_m}(x^*, s^m) \\
&\quad \text{(since (1.3) holds, } x^* \in \mathbb{F}, \text{ and } t^m \in \hat{T}_{j_m}(x^*),\ m \in \underline{\nu_0}) \\
&= \Pi_\tau(x^*, \lambda^*, u^*, v^*, \bar{t}, \bar{s}),
\end{aligned}$$

and so using the properties of the functions ϕ_τ, $\tau \in D$, we get

$$\phi_\tau\big(\Pi_\tau(\bar{x}, \lambda^*, u^*, v^*, \bar{t}, \bar{s}) - \Pi_\tau(x^*, \lambda^*, u^*, v^*, \bar{t}, \bar{s})\big) < 0,$$

which in view of (i) implies that

$$\begin{aligned}
&\mathcal{F}\Big(x, x^*; \beta(x, x^*)\Big\{\sum_{i\in I_\tau} u_i^*[\nabla f_i(x^*) - \lambda^* \nabla g_i(x^*)] + \sum_{m\in J_\tau} v_m^* \nabla G_{j_m}(x^*, t^m) \\
&\qquad + \sum_{m\in K_\tau} v_m^* \nabla H_{k_m}(x^*, s^m)\Big\}\Big) \\
&\qquad + \frac{1}{2}\Big\langle z^*, \Big\{\sum_{i\in I_t} u_i^*[\nabla^2 f_i(x^*) - \lambda^* \nabla^2 g_i(x^*)] + \sum_{j\in J_t} v_j^* \nabla^2 G_j(x^*, t^m) \\
&\qquad + \sum_{k\in K_t} v_k^* \nabla^2 H_k(x^*, s^m)\Big\} z^*\Big\rangle < -\rho_t(\bar{x}, x^*)\|\theta(\bar{x}, x^*)\|^m.
\end{aligned}$$

Summing over $\tau \in D$ and using the sublinearity of $\mathcal{F}(x, x^*; \cdot)$, we get

$$\begin{aligned}
&\mathcal{F}\Big(x, x^*; \beta(x, x^*)\Big\{\sum_{i=1}^{p} u_i^*[\nabla f_i(x^*) - \lambda \nabla g_i(x^*)] + \sum_{\tau\in D}\Big[\sum_{m\in J_\tau} v_m^* \nabla G_{j_m}(x^*, t^m) \\
&\qquad + \sum_{m\in K_\tau} v_m^* \nabla H_{k_m}(x^*, s^m)\Big]\Big\}\Big) \\
&\qquad + \frac{1}{2}\Big\langle z^*, \Big\{\sum_{i\in I_\tau} u_i^*[\nabla^2 f_i(x^*) - \lambda^* \nabla^2 g_i(x^*)] + \sum_{\tau\in D}\Big[\sum_{m\in J_\tau} v_m^* \nabla^2 G_{j_m}(x^*, t^m) \\
&\qquad + \sum_{m\in K_\tau} v_m^* \nabla^2 H_{k_m}(x^*, s^m)\Big]\Big\} z^*\Big\rangle < -\sum_{\tau\in D} \rho_\tau(\bar{x}, x^*)\|\theta(\bar{x}, x^*)\|^m. \qquad (2.8)
\end{aligned}$$

As shown in the proof of Theorem 2.2, for each $\tau \in E\backslash D$, $\Lambda_\tau(\bar{x}, v^*, \bar{t}, \bar{s}) \leq \Lambda_\tau(x^*, v^*, \bar{t}, \bar{s})$, and so using the properties of ϕ_τ, $\tau \in E\backslash D$, we get the inequality $\phi_\tau\big(\Lambda_\tau(\bar{x}, v^*, \bar{t}, \bar{s}) - \Lambda_\tau(x^*, v^*, \bar{t}, \bar{s})\big) \leq 0$, which in view of (ii) implies that

$$\begin{aligned}
&\mathcal{F}\Big(x, x^*; \beta(x, x^*)\Big[\sum_{m\in J_\tau} v_m^* \nabla G_{j_m}(x^*, t^m) + \sum_{m\in K_\tau} v_m^* \nabla H_{k_m}(x^*, s^m)\Big]\Big) \\
&\qquad + \frac{1}{2}\Big\langle z^*, \Big[\sum_{j\in J_\tau} v_j^* \nabla^2 G_j(x^*, t^m) + \sum_{k\in K_\tau} v_k^* \nabla^2 H_k(x^*, s^m)\Big] z^*\Big\rangle \\
&\qquad\qquad \leq -\rho_\tau(\bar{x}, x^*)\|\theta(\bar{x}, x^*)\|^m.
\end{aligned}$$

Summing over $\tau \in E\backslash D$ and using the sublinearity of $\mathcal{F}(x, x^*; \cdot)$, we obtain

$$\mathcal{F}\Big(x, x^*; \beta(x, x^*) \sum_{\tau\in E\backslash D} \Big[\sum_{m\in J_\tau} v_m^* \nabla G_{j_m}(x^*, t^m) + \sum_{m\in K_\tau} v_m^* \nabla H_{k_m}(x^*, s^m)\Big]\Big)$$
$$+\frac{1}{2}\Big\langle z^*, \sum_{\tau\in E\backslash D} \Big[\sum_{j\in J_\tau} v_j^* \nabla^2 G_j(x^*, t^m) + \sum_{k\in K_\tau} v_k^* \nabla^2 H_k(x^*, s^m)\Big] z^*\Big\rangle$$
$$\leq - \sum_{\tau\in E\backslash D} \rho_\tau(\bar{x}, x^*) \|\theta(\bar{x}, x^*)\|^m. \quad (2.9)$$

Now combining (2.8) and (2.9) and using (iii), we have

$$\mathcal{F}\Big(x, x^*; \beta(x, x^*)\Big\{\sum_{i=1}^{p} u_i^*[\nabla f_i(x^*) - \lambda^* \nabla g_i(x^*)] + \sum_{m=1}^{\nu_0} v_m^* \nabla G_{j_m}(x^*, t^m) +$$
$$\sum_{m=\nu_0+1}^{\nu} v_m^* \nabla H_{k_m}(x^*, s^m)\Big\}\Big)$$
$$+ \langle z^*, \Big[\sum_{i=1}^{p} u_i^*[\nabla^2 f_i(x^*) - \lambda^* \nabla^2 g_i(x^*)] + \sum_{m=1}^{\nu_0^*} v_m^* \nabla^2 G_{j_m}(x^*, t^m) + \sum_{m=\nu_0^*+1}^{\nu^*} v_m^* \nabla^2 H_{k_m}(x^*, s^m)\Big] z^*\rangle$$
$$< - \sum_{\tau\in E} \rho_\tau(x, x^*) \|\theta(x, x^*)\|^m \leq 0,$$

which contradicts (1.1) and (1.2) (since $\beta(x, x^*) > 0$ and $\mathcal{F}(x, x^*; 0) = 0$). Therefore, x^* is an optimal solution of (P).

(b) - (g) : The proofs are similar to that of part (a). □

The modified versions of Theorems 2.2 and 2.3 can be stated in a similar manner. Theorems 2.1 - 2.3, and their modified versions encompass a large family of sufficient optimality conditions for (P) whose members can easily be identified by appropriate choices of the partitioning sets J_μ, K_μ, $\mu \in \underline{m} \cup \{0\}$, and I_t, $t \in \underline{\ell} \cup \{0\}$. To this context, we state explicitly some important special cases of part (a) of Theorem 2.2, for example, the following corollary.

Corollary 1. *Let $x^* \in \mathbb{F}$, let $\lambda^* = \varphi(x^*)$, let the functions f_i, g_i, $i \in \underline{p}$, $z \to G_j(z, t)$, and $z \to H_k(z, s)$ be twice continuously differentiable at x^* for all $t \in T_j$ and $s \in S_k$, $j \in \underline{q}$, $k \in \underline{r}$, and assume that there exist $u^* \in U$ and integers ν_0 and ν, with $0 \leq \nu_0 \leq \nu \leq n+1$, such that there exist ν_0 indices j_m, with $1 \leq j_m \leq q$, together with ν_0 points $t^m \in \hat{T}_{j_m}(x^*)$, $m \in \underline{\nu_0}$, $\nu - \nu_0$ indices k_m, with $1 \leq k_m \leq r$, together with $\nu - \nu_0$ points $s^m \in S_{k_m}$, $m \in \underline{\nu}\backslash\underline{\nu_0}$, and ν real numbers v_m^* with $v_m^* > 0$ for $m \in \underline{\nu_0}$, such that (1.1), (1.2) and (1.3) hold. Assume further that any one of the following ten sets of hypotheses is satisfied:*

(a) (i) for each $i \in I_+ \equiv \{i \in \underline{p} : u_i^* > 0\}$, the function $\xi \to f_i(\xi) - \lambda^* g_i(\xi)$ is $(\mathcal{F}, \beta, \bar{\phi}_i, \bar{\rho}_i, \theta, m)$-pseudosounivex at x^*, $\bar{\phi}_i$ is strictly increasing, and $\bar{\phi}_i(0) = 0$;

(ii) the function $\xi \to \sum_{j=1}^{q} v_j^* G_j(\xi, t) + \sum_{k=1}^{r} v_k^* H_k(\xi, s)$ is $(\mathcal{F}, \beta, \tilde{\phi}, \tilde{\rho}, \theta, m)$-quasisounivex at x^*, $\tilde{\phi}$ is increasing, and $\tilde{\phi}(0) = 0$;

(iii) $\sum_{i\in I_+} u_i^* \bar{\rho}_i(x, x^*) + \tilde{\rho}(x, x^*) \geq 0$ for all $x \in \mathbb{F}$;

(b) (i) for each $i \in I_+$, the function $\xi \to f_i(\xi) - \lambda^* g_i(\xi) + \sum_{j=1}^{q} v_j^* G_j(\xi, t) + \sum_{k=1}^{r} v_k^* H_k(\xi, s)$ is $(\mathcal{F}, \beta, \bar{\phi}_i, \bar{\rho}_i, \theta, m)$-pseudosounivex at x^*, $\bar{\phi}_i$ is strictly increasing, and $\bar{\phi}_i(0) = 0$;

(ii) $\sum_{i\in I_+} u_i^* \bar{\rho}_i(x, x^*) \geq 0$ for all $x \in \mathbb{F}$;

(c) (i) *for each* $i \in I_+$, *the function* $\xi \to f_i(\xi) - \lambda^* g_i(\xi) + \sum_{j \in J_0} v_j^* G_j(\xi, t) + \sum_{k=1}^{r} v_k^* H_k(\xi, s)$ *is* $(\mathcal{F}, \beta, \bar{\phi}_i, \bar{\rho}_i, \theta, m)$*-pseudosounivex at* x^*, $\bar{\phi}_i$ *is strictly increasing, and* $\bar{\phi}_i(0) = 0$;

(ii) *for each* $t \in \underline{M}$, $\xi \to \sum_{j \in J_t} v_j^* G_j(\xi, t)$ *is* $(\mathcal{F}, \beta, \tilde{\phi}_t, \tilde{\rho}_t, \theta, m)$*-quasisounivex at* x^*, $\tilde{\phi}_t$ *is increasing, and* $\tilde{\phi}_t(0) = 0$;

(iii) $\sum_{i \in I_+} u_i^* \bar{\rho}_i(x, x^*) + \sum_{t=1}^{M} \tilde{\rho}_t(x, x^*) \geq 0$ *for all* $x \in \mathbb{F}$;

(d) (i) *for each* $i \in I_+$, *the function* $\xi \to f_i(\xi) - \lambda^* g_i(\xi) + \sum_{j=1}^{q} v_j^* G_j(\xi, t) + \sum_{k \in K_0} v_k^* H_k(\xi, s)$ *is* $(\mathcal{F}, \beta, \bar{\phi}_i, \bar{\rho}_i, \rho, \theta, m)$*-pseudosounivex* x^*, $\bar{\phi}_i$ *is strictly increasing, and* $\bar{\phi}_i(0) = 0$;

(ii) *for each* $t \in \underline{M}$, $\xi \to \sum_{k \in K_t} v_k^* H_k(\xi, s)$ *is* $(\mathcal{F}, \beta, \tilde{\phi}_t, \tilde{\rho}_t, \theta, m)$*-quasisounivex at* x^*, $\tilde{\phi}_t$ *is increasing, and* $\tilde{\phi}_t(0) = 0$;

(iii) $\sum_{i \in I_+} u_i^* \bar{\rho}_i(x, x^*) + \sum_{t=1}^{M} \tilde{\rho}_t(x, x^*) \geq 0$ *for all* $x \in \mathbb{F}$;

(e) (i) *for each* $i \in I_+$, *the function* $\xi \to f_i(\xi) - \lambda^* g_i(\xi) + \sum_{j=1}^{q} v_j^* G_j(\xi, t)$ *is* $(\mathcal{F}, \beta, \bar{\phi}_i, \bar{\rho}_i, \theta, m)$*-pseudosounivex at* x^*, $\bar{\phi}_i$ *is strictly increasing, and* $\bar{\phi}_i(0) = 0$;

(ii) *the function* $\xi \to \sum_{k=1}^{r} v_k^* H_k(\xi, s)$ *is* $(\mathcal{F}, \beta, \tilde{\phi}, \tilde{\rho}, \theta, m)$*-quasisounivex at* x^*, $\tilde{\phi}$ *is increasing, and* $\tilde{\phi}(0) = 0$;

(iii) $\sum_{i \in I_+} u_i^* \bar{\rho}_i(x, x^*) + \tilde{\rho}(x, x^*) \geq 0$ *for all* $x \in \mathbb{F}$;

(f) (i) *for each* $i \in I_+$, *the function* $\xi \to f_i(\xi) - \lambda^* g_i(\xi) + \sum_{k=1}^{r} v_k^* H_k(\xi, s)$ *is* $(\mathcal{F}, \beta, \bar{\phi}_i, \bar{\rho}_i, \theta, m)$*-pseudosounivex at* x^*, $\bar{\phi}_i$ *is strictly increasing, and* $\bar{\phi}_i(0) = 0$;

(ii) *the function* $\xi \to \sum_{j=1}^{q} v_j^* G_j(\xi)$ *is* $(\mathcal{F}, \beta, \tilde{\phi}, \tilde{\rho}, \theta, m)$*-quasisounivex at* x^*, $\tilde{\phi}$ *is increasing, and* $\tilde{\phi}(0) = 0$;

(iii) $\sum_{i \in I_+} u_i^* \bar{\rho}_i(x, x^*) + \tilde{\rho}(x, x^*) \geq 0$ *for all* $x \in \mathbb{F}$;

(g) (i) *for each* $i \in I_+$, *the function* $\xi \to f_i(\xi) - \lambda^* g_i(z) + \sum_{k=1}^{r} v_k^* H_k(\xi, s)$ *is* $(\mathcal{F}, \beta, \bar{\phi}_i, \bar{\rho}_i, \theta, m)$*-pseudosounivex at* x^*, $\bar{\phi}_i$ *is strictly increasing, and* $\bar{\phi}_i(0) = 0$;

(ii) *for each* $j \in \underline{q}$, *the function* $\xi \to v_j^* G_j(\xi, t)$ *is* $(\mathcal{F}, \beta, \tilde{\phi}_j, \tilde{\rho}_j, \theta, m)$*-quasisounivex at* x^*, $\tilde{\phi}_j$ *is increasing, and* $\tilde{\phi}_j(0) = 0$;

(iii) $\sum_{i \in I_+} u_i^* \bar{\rho}_i(x, x^*) + \sum_{j=1}^{q} \tilde{\rho}_j(x, x^*) \geq 0$ *for all* $x \in \mathbb{F}$;

(h) (i) *for each* $i \in I_+$, *the function* $\xi \to f_i(\xi) - \lambda^* g_i(\xi) + \sum_{j=1}^{q} v_j^* G_j(\xi, t)$ *is* $(\mathcal{F}, \beta, \bar{\phi}_i, \bar{\rho}_i, \theta, m)$*-pseudosounivex at* x^*, $\bar{\phi}_i$ *is strictly increasing, and* $\bar{\phi}_i(0) = 0$;

(ii) *for each* $k \in \underline{r}$, $\xi \to w_k^* H_k(\xi, s)$ *is* $(\mathcal{F}, \beta, \tilde{\phi}_k, \tilde{\rho}_k, \theta, m)$*-quasisounivex at* x^* *and* $\tilde{\phi}_k(0) = 0$;

(iii) $\sum_{i \in I_+} u_i^* \bar{\rho}_i(x, x^*) + \sum_{k=1}^{r} \tilde{\rho}_k(x, x^*) \geq 0$ *for all* $x \in \mathbb{F}$;

(i) (i) *for each* $i \in I_+$, *the function* $\xi \to f_i(\xi) - \lambda^* g_i(\xi) + \sum_{j \in J_0} v_j^* G_j(\xi, t) + \sum_{k=1}^{r} v_k^* H_k(\xi, s)$ *is* $(\mathcal{F}, \beta, \bar{\phi}_i, \bar{\rho}_i, \theta, m)$*-pseudosounivex at* x^*, $\bar{\phi}_i$ *is strictly increasing, and* $\bar{\phi}_i(0) = 0$;

(ii) *the function* $\xi \to \sum_{j \in J_1} v_j^* G_j(\xi, t)$ *is* $(\mathcal{F}, \beta, \tilde{\phi}, \tilde{\rho}, \theta, m)$*-quasisounivex at* x^*, $\tilde{\phi}$ *is increasing, and* $\tilde{\phi}(0) = 0$;

(iii) $\sum_{i \in I_+} u_i^* \bar{\rho}_i(x, x^*) + \tilde{\rho}(x, x^*) \geq 0$ *for all* $x \in \mathbb{F}$;

(j) (i) *for each* $i \in I_+$, $\xi \to f_i(\xi) - \lambda^* g_i(\xi) + \sum_{j=1}^{q} v_j^* G_j(\xi, t) + \sum_{k \in K_0} v_k^* H_k(\xi, s)$ *is* $(\mathcal{F}, \beta, \bar{\phi}_i, \bar{\rho}_i, \theta, m)$*-pseudosounivex at* x^*, $\bar{\phi}_i$ *is strictly increasing, and* $\bar{\phi}_i(0) = 0$;

(ii) *the function* $\xi \to \sum_{k \in K_1} v_k^* H_k(\xi, s)$ *is* $(\mathcal{F}, \beta, \tilde{\phi}, \tilde{\rho}, \theta, m)$*-quasisounivex at* x^*, $\tilde{\phi}$ *is increasing, and* $\tilde{\phi}(0) = 0$;

(iii) $\sum_{i \in I_+} u_i^* \bar{\rho}_i(x, x^*) + \tilde{\rho}(x, x^*) \geq 0$ *for all* $x \in \mathbb{F}$.

Then x^* *is an optimal solution of (P).*

Proof

In general proofs follow based on appropriate manipulations, while comparing parts (f) and (g) of the above corollary, we observe that they represent two extreme cases with respect to the $(\mathcal{F}, \beta, \phi, \rho, \theta, m)$-quasisounivexity assumptions in the sense that in (f) all the functions $\xi \to v_j^* G_j(\xi)$, are lumped together, whereas in (g) separate $(\mathcal{F}, \beta, \phi, \rho, \theta, m)$-quasisounivexity conditions are imposed on the individual functions. □

3. Concluding Remarks

We have established several sets of generalized second-order parametric sufficient optimality criteria for a semiinfinite discrete minmax fractional programming problem using a variety of generalized $(\mathcal{F}, \beta, \phi, \rho, \theta, m)$-sounivexity assumptions. These optimality results can be applied to constructing various duality models as well as for developing new algorithms for the numerical solution of minmax fractional programming problems. The results investigated in this communication seem to be applicable to further generalizations and challenging applications to higher order exponential type $(\mathcal{F}, \beta, \phi, \rho, \theta, m)$-sounivexities based on the recently introduced notion of the Hanson-Antczak type invexities by Zalmai [8].

REFERENCES

1. W. Dinkelbach, On nonlinear fractional programming, *Management Sci.* **13** (1967), 492 - 498.
2. M. A. Hanson, Second order invexity and duality in mathematical programming, *Opsearch* **30** (1993), 313 - 320.
3. B. Mond and T. Weir, Generalized concavity and duality, *Generalized Concavity in Optimization and Economics* (S. Schaible and W. T. Ziemba, eds.), Academic Press, New York, 1981, pp. 263 - 279.
4. R. U. Verma and G. J. Zalmai, Second-order parametric optimality conditions in semiinfinite discrete minmax fractional programming based on generalized $(\phi, \eta, \rho, \theta, \tilde{m})$-sonvexities, *Transactions on Mathematical Programming and Applications* **4** (1) (2016), 62 - 84.
5. R. U. Verma and G. J. Zalmai, Generalized higher order $(\phi, \eta, \zeta, \rho, \theta, m)$-invexities and parametric optimality conditions for discrete minmax fractional programming, *Transactions on Mathematical Programming and Applications* **4** (1) (2016), 85 - 101.
6. G. J. Zalmai, Optimality conditions and duality for constrained measurable subset selection problems with minmax objective functions, *Optimization* **20** (1989), 377 - 395.
7. G. J. Zalmai, Generalized second-order $(F, \beta, \phi, \rho, \theta)$-univex functions and parametric duality models in semiinfinite discrete minmax fractional programming, *Adv. Nonlin. Var. Ineq.* **15** (2012), 63 - 91.
8. G. J. Zalmai, Hanson-Antczak-type generalized $(\alpha, \beta, \gamma, \xi, \eta, \rho, \theta)$-V-invex functions in semiinfinite multiobjective fractional programming, Part I: Sufficient efficiency conditions, *Advances in Nonlinear Variational Inequalities* **16** (1) (2013), 91 - 114.
9. G. J. Zalmai and Qinghong Zhang, Generalized $(F, \beta, \phi, \rho, \theta)$-univex functions and global parametric sufficient optimality conditions in semiinfinite discrete minmax fractional programming, *PanAmerican Math. J.* **17** (2007), 1 - 26.

3

Two-step proximal gradient algorithm for low-rank matrix completion

Qiuyu Wang [1,*], Wenjiao Cao [2], Zheng-Fen Jin [3]

[1]*School of Software, Henan University, Kaifeng 475000, China*
[2]*School of Mathematics and Statistics, Henan University, Kaifeng 475000, China*
[3]*School of Mathematics and Statistics, Henan University of Science and Technology, Luoyang 471023, China*

Abstract In this paper, we propose a two-step proximal gradient algorithm to solve nuclear norm regularized least squares for the purpose of recovering low-rank data matrix from sampling of its entries. Each iteration generated by the proposed algorithm is a combination of the latest three points, namely, the previous point, the current iterate, and its proximal gradient point. This algorithm preserves the computational simplicity of classical proximal gradient algorithm where a singular value decomposition in proximal operator is involved. Global convergence is followed directly in the literature. Numerical results are reported to show the efficiency of the algorithm.

Keywords Matrix nuclear norm minimization, Matrix completion, Proximal gradient algorithm, Singular value decomposition

1. Introduction

Over the past few years, finding low-rank matrix subject to some given constraints has attracted significant attentions due to its wide range of engineering applications. The general rank minimization problem can be expressed as

$$\begin{array}{c} \min_{X\in\mathbb{R}^{m\times n}} \ \mathrm{Rank}(X) \\ \text{s.t. } X\in\mathcal{C}, \end{array} \tag{1}$$

where $X \in \mathbb{R}^{m\times n}$ is a decision variable, and $\mathcal{C}$ is a convex set. As a special case, the affine rank minimization problem can be expressed as follows:

$$\begin{array}{c} \min_{X\in\mathbb{R}^{m\times n}} \ \mathrm{Rank}(X) \\ \text{s.t. } \mathcal{A}(X)=b, \end{array} \tag{2}$$

where $\mathcal{A} : \mathbb{R}^{m\times n} \to \mathbb{R}^p$ is a closed linear operator and $\mathcal{A}^*$ is its adjoint, and $b \in \mathbb{R}^p$ is a known observation value. The rank minimization problem is generally known to be computationally intractable (NP-hard) [1, 2].

When the matrix variable is restricted to be diagonal, the problem (2) reduces to the linearly constrained nonsmooth cardinality minimization problem. One common approach is to replace the cardinality term by ℓ_1-norm and then to solve a convex minimization problem. Similarly, to solve (2) via a tractable problem, it is natural

*Correspondence to: Qiuyu Wang (Email: wqy@henu.edu.cn). School of Mathematics and Statistics, Henan University, Kaifeng, 475000, China.

and common to replace "Rank(X)" by the so-called nuclear norm $\|\cdot\|_*$, that is

$$\begin{array}{l} \min_{X\in\mathbb{R}^{m\times n}} \|X\|_* \\ \text{s.t. } \mathcal{A}(X) = b, \end{array} \tag{3}$$

where $\|X\|_*$ is defined as the sum of its singular values. That is, assume that X has r positive singular values of $\sigma_1 \geq \sigma_2 \geq \ldots \geq \sigma_r > 0$, then $\|X\|_* = \sum_{i=1}^{r} \sigma_i$. A frequent alternative to (3) is the following nuclear norm regularized linear least squares problem

$$\min_{X\in\mathbb{R}^{m\times n}} \frac{1}{2}\|\mathcal{A}(X) - b\|_F^2 + \mu\|X\|_*, \tag{4}$$

where $\|\cdot\|_F$ is the Frobenius norm induced by the standard trace inner product, and parameter $\mu > 0$ is used to trade off both terms for minimization. The nuclear norm is alternatively known as the Schatten ℓ_1-norm and the Ky Fan norm [2], and it is the best convex approximation of the rank function over the unit ball of matrices [3].

A particular category of (3) is the so-called matric completion problem: given a random subset of entries of low-rank matrix, it would like to recover all the missing entries. Specially, according to Candès & Recht [4], it can be formulated as

$$\begin{array}{lll} \min_{X\in\mathbb{R}^{m\times n}} & \|X\|_* & \\ \text{s.t.} & X_{i,j} = M_{i,j}, & (i,j) \in \Omega, \end{array} \tag{5}$$

where M is the unknown matrix with some available sampled entries and Ω is a set of index pairs (i, j). Throughout our discussion, the solution set of (5) is assumed to be nonempty. It is known that when the observations are corrupted by Gaussian noise, the following nuclear norm regularized linear least squares problem is preferable,

$$\min_{X\in\mathbb{R}^{m\times n}} \frac{1}{2}\|P_\Omega(X) - P_\Omega(M)\|_F^2 + \mu\|X\|_*, \tag{6}$$

where $P_\Omega(\cdot)$ is defined as

$$P_\Omega(X)_{i,j} = \begin{cases} X_{i,j} & \text{if } (i,j) \in \Omega, \\ 0 & \text{otherwise.} \end{cases} \tag{7}$$

Over the past few years, the numerical approaches for solving the formulations (3) and its different variants have been draw increasingly attention, e.g. [7, 8, 10]. Cai, Candès & Shen [3] proposed a singular value thresholding algorithm and showed that if the parameter goes infinity, then the sequence of optimal solutions converges to the one of (3) with minimum Frobenius norm. Ma, Goldfarb & Chen [11] introduced a fixed point continuation algorithm (FPCA), which is a matrix extension of the fixed point continuation algorithm by Hale et al. [12]. Liu, Sun & Toh [13] developed a proximal point algorithm from the primal, dual, and primal-dual forms, in which the inner sub-problems are solved by gradient projection method or accelerated proximal gradient method. Toh and Yun [5] extended Beck & Teboulle's algorithm [6] to solve matrix completion problems and designed an accelerated proximal gradient algorithm.

Due to its simplicity, the proximal gradient algorithm is popular to solve a closed proper convex optimization problem. However, the classical proximal gradient algorithm has been regarded as a slower until the exciting progress in recent years. The "accelerated" approaches mainly base on an extrapolation step which relies on not only the current point, but also two or more previous computed iterations (e.g., [14, 15]). Other dramatic algorithms can refer to Nesterov [16], Beck & Teboulle [6], Tseng [17], O'Donoghue & Candès [18], and references therein.

The main contribution of the paper is to further investigate the efficiency of the proximal gradient algorithms in solving the matric completion problem (6). The distinguished characters of the proposed algorithm is that each generated iteration is a combination of the previous point, the current point, and its proximal gradient point. As a result, the combination improves the algorithm's flexibility and practicability. The algorithm is also closely related to the well-known two-step iterative shrinkage/thresholding algorithm (TwIST) of the Bioucas-Dias & Figueiredo's algorithm [14] for sparse signal recovering. In other words, the proposed algorithm can be regarded as an extension or application of TwIST to solve nuclear norm regularized least squares for the purpose of testing its efficiency to find low rank matrices. The proposed algorithm preserves the computational simplicity of classical proximal gradient algorithm and mainly involves a singular value decomposition in proximal operator. We give its global convergence property without proof under some appropriate conditions. Finally, we test the algorithm to show its numerical performance.

We organize the rest of this paper as follows. In Section 2, we summarize some preliminaries which are useful for further analysis, and quickly review some closed related approaches in the literature. In Section 3, we develop a two-step proximal gradient algorithm and present its convergence theorem. In Section 4, in order to investigate the benefit of the proposed algorithm, we test the algorithm by a series of numerical experiments. Finally, we conclude our paper in Section 5.

We summarize the notation used in this paper. Matrices are written as uppercase letters. Vectors are written as lowercase letters. For matrix $X \in \mathbb{R}^{m\times n}$, its Frobenius norm is defined as $\|X\|_F = \sqrt{\sum_{i=1}^m \sum_{j=1}^n x_{i,j}^2}$, where $x_{i,j}$ is the (i,j)-th component of X. For any two matrices $X, Y \in \mathbb{R}^{m\times n}$, we define $\langle X, Y\rangle = \mathrm{tr}(X^\top Y)$ (the standard trace inner product), so that $\|X\|_F = \sqrt{\langle X, X\rangle}$. For any $x \in \mathbb{R}^n$, we denote by $\mathrm{Diag}(x)$ the diagonal matrix possessing the components of vector x on the diagonal. For any $X \in \mathbb{R}^{n\times n}$, we use $\lambda_{\max}(X)$ to denote its largest eigenvalue. We define "$\top$" as the transpose of a vector or a matrix. Additional notation will be introduced when it occurs.

2. Preliminaries

In this section, we review some preliminaries on proximal gradient algorithms which are useful in the later analysis. Let $f : \mathbb{R}^n \to \mathbb{R} \cup \{+\infty\}$ be a closed proper convex function. The proximal operator $\mathrm{prox}_{\lambda f} : \mathbb{R}^n \to \mathbb{R}^n$ of the scaled function λf is defined by

$$\mathrm{Prox}_{\lambda f}(y) = \arg\min_{x\in\mathbb{R}^n} \Big(\lambda f(x) + \frac{1}{2}\|x - y\|_2^2\Big), \tag{8}$$

which is also called the proximal operator of f with parameter λ. The definition indicates that $\mathrm{Prox}_{\lambda f}(y)$ is a point that compromises between minimizing f and being near to y. Proximal gradient algorithm is to solve the convex separable minimization problem

$$\min_{x\in\mathbb{R}^n} F(x) := f(x) + g(x), \tag{9}$$

where $f : \mathbb{R}^n \to \mathbb{R}$ and $g : \mathbb{R}^n \to \mathbb{R} \cup \{+\infty\}$ are closed proper convex and f is differentiable. The proximal gradient algorithm is

$$x_{k+1} = \mathrm{Prox}_{\lambda_k g}\Big(x_k - \lambda_k \nabla f(x_k)\Big), \tag{10}$$

where $\lambda_k > 0$ is a step size. It is well-known that when ∇f is Lipschitz continuous with constant L, this method converges with rate $O(1/k)$ when a fixed step size $\lambda_k \in (0, 1/L]$ is used.

The accelerated version of the basic proximal gradient algorithm mainly includes an extrapolation. One popular approach is the well-known fast iterative shrinkage-thresholding algorithm (FISTA) of Beck & Teboulle [6]. Dramatically, the iteration complexity can achieve $O(1/k^2)$ with proper choice on steplength λ_k. Another accelerated approach is the so-called two-step iterative shringkage/thresholding algorithm (TwIST) of Bioucas-Dias & Figueiredo [14]. TwIST is motivated from [19] for solving a linear system in case that the coefficient matrix can be split into two terms and one of them is positive definite and easy to invert. Compared to the standard proximal gradient iteration (10), a general formulation with scalars β and α is formulated as

$$x_{k+1} = \big(1 - \alpha\big)x_{k-1} + \big(\alpha - \beta\big)x_k + \beta\,\mathrm{Prox}_{\lambda_k g}\Big(x_k - \lambda_k \nabla f(x_k)\Big). \tag{11}$$

Clearly, the case of $\alpha = 1$ and $\beta = 1$ reduces to (10). It was shown in [14] that, the iteration (11) converges globally at the case of $f(x) = \|Ax - b\|^2$ and $g(x) = \|x\|_1$ by using some proper values of α and β.

3. Two-step proximal gradient algorithm

Based on the above preliminaries, we are now ready to construct the two-step proximal gradient algorithm to solve problem (6). Comparing the formulations of (6) and (9), we can take

$$f(X) = \frac{1}{2}\|P_\Omega(X) - P_\Omega(M)\|_F^2 \quad \text{and} \quad g(X) = \mu\|X\|_*.$$

It's easy to know that $g : \mathbb{R}^{m\times n} \to \mathbb{R}$ is a nonsmooth and continuous convex function and $f : \mathbb{R}^{m\times n} \to \mathbb{R}$ is a continuously differentiable convex function with Lipschitz continuous gradient

$$\|\nabla f(X) - \nabla f(Y)\|_F \leq L\|X - Y\|_F, \quad \forall\, X, Y \in \mathbb{R}^{m\times n},$$

where L is a Lipschitz constant.

For convenience, we denote $P_\Omega(X) = \mathcal{A}(X)$ and $P_\Omega(M) = b$. Using the notation, we consider the following quadratic approximation of $F(x)$ at a given point X_k

$$Q_k(X) := f(X_k) + \langle X - X_k, \nabla f(X_k)\rangle + \frac{L}{2}\|X - X_k\|^2 + g(X). \tag{12}$$

Its unique minimizer admits the following formulation:

$$\begin{aligned}
& \arg\min_{X\in\mathbb{R}^{m\times n}} Q_k(X) \\
= \;& \arg\min_{X\in\mathbb{R}^{m\times n}} f(X_k) + \langle X - X_k, \nabla f(X_k)\rangle + \tfrac{L}{2}\|X - X_k\|^2 + g(X) \\
= \;& \arg\min_{X\in\mathbb{R}^{m\times n}} g(X) + \tfrac{L}{2}\left\|X - \left(X_k - \tfrac{1}{L}\nabla f(X_k)\right)\right\|_F^2 \\
= \;& \mathrm{Prox}_{g/L}\left(X_k - \tfrac{1}{L}\mathcal{A}^*\left[\mathcal{A}(X_k) - b\right]\right).
\end{aligned}$$

By definition, evaluating a proximal operator involves solving a convex optimization problem generally. In some cases, exploiting special structure in the problem like sparsity may produce the simplest algorithm, even derive analytical solutions. Let $Y_k = X_k - \frac{1}{L}\mathcal{A}^*\left[\mathcal{A}(X_k) - b\right]$, and suppose that the rank of Y_k is r. The reduced singular value decomposition of Y_k can be expressed as

$$Y_k = U_k\Sigma_k V_k^\top, \quad \Sigma_k = \mathrm{Diag}(\{\sigma_i\}_{1\leq i\leq r}),$$

where U_k and V_k are respectively $m \times r$ and $n \times r$ matrix with orthogonal columns, $\sigma \in \mathbb{R}^r$ is the vector of positive singular values arranged in descending order $\sigma_1 \geq \sigma_2 \geq \cdots \geq \sigma_r > 0$. It follows [3] that the proximal operator can be expressed as

$$\begin{aligned}
& \mathrm{Prox}_{g/L}\left(X_k - \tfrac{1}{L}\mathcal{A}^*\left[\mathcal{A}(X_k) - b\right]\right) \\
= \;& \mathrm{Prox}_{g/L}\left(Y_k\right) \\
= \;& U_k\mathrm{Diag}\left(\left[\sigma - \tfrac{\mu}{L}\right]_+\right)V_k^\top,
\end{aligned}$$

where $[\cdot]_+ = \max\{\cdot, 0\}$ and "$\max$" is interpreted as componentwise.

Based on the above analysis, the standard proximal gradient algorithm for solving (4) reduces to

$$X_{k+1} = \mathrm{Prox}_{g/L}(Y_k),$$

which means that the next iteration is only determined by the current X_k. For deriving an algorithm with more general form, as formula (11), we choose two positive scalars α and β, and set

$$X_{k+1} = (1 - \alpha)X_{k-1} + (\alpha - \beta)X_k + \beta\,\mathrm{Prox}_{g/L}(Y_k).$$

It is clear that the new iteration is a combination of the previous point X_{k+1}, the current point X_k, and its proximal gradient point $\mathrm{Prox}_{g/L}(Y_k)$.

For the general nuclear norm regularized linear least squares problem (4), the Lipschitz constant L which depends on the maximum eigenvalue of $\mathcal{A}^*\mathcal{A}$ is not always easily computable. Specially, in this case of the matrix completion problem (6), the linear operator $\mathcal{A}$ always satisfies $\mathcal{A}^*\mathcal{A} = I$ (i.e, the identity matrix), and then $L = 1$ from the fact that $L = \lambda_{\max}(\mathcal{A}^*\mathcal{A})$. Based on the above analysis, the standard version of the two-step proximal gradient algorithm (abbr. Ts_PGA) for solving separable convex minimization (6) can be outlined as follows.

Algorithm 1
(Ts_PGA)

Initialization. Choose positive constants α and β. Choose an initial point X_0 and set $X_{-1} = 0$. Let $k := 0$.
Step 1. Compute

$$\widetilde{X}_k = \text{Prox}_g\Big(X_k - \mathcal{A}^*\big[\mathcal{A}(X_k) - b\big]\Big)$$

where

$$X_{k+1} = \big(1-\alpha\big)X_{k-1} + \big(\alpha - \beta\big)X_k + \beta\widetilde{X}_k. \tag{13}$$

Step 2. If termination criterion is not met, let $k := k + 1$. Go to Step 1.

For being easily understood, we make a remark on the close relationship between Ts_PGA and the solver TwIST of Bioucas-Dias & Figueiredo [14], and the solver FPCA of Ma, Goldfarb, and Chen [11].

Remark 1
It is well-known that TwIST is designed for solving ℓ_1-norm regularized minimization problems in compressive sensing. It is also easy to derive that Ts_PGA is equivalent to TwIST when it used to recover large sparse signal in compressive sensing. Generally speaking, the proposed algorithm Ts_PGA can be considered as an extension of TwIST to solve the problems of recovering corrupted low rank matrix. At the special case of $\alpha = \beta = 1$, the iteration (13) reduces to the classical proximal gradient algorithm, also named the fixed point continuation algorithm when it used to solve (4). In other words, Ts_PGA can also be regarded as a generalized variant of FPAC.

Given that Ts_PGA is an application of the well-known TwIST to solve nuclear norm least squares for matric completion, hence, its global convergence can be followed directly. For completeness of this paper, we list the convergence theorem without proof at the end of this section. For the proof of the theorem, one can refer to [14, Appendix II].

Theorem 1
Let ξ be a real number such that $0 < \xi \leq \lambda_i(\mathcal{A}^*\mathcal{A}) \leq 1$ where $\lambda_i(\cdot)$ is the i-th eigenvalue and let $\hat{\rho} = \frac{1-\sqrt{\xi}}{1+\sqrt{\xi}}$. Let $\bar{X}$ be the minimizer of (6) and define

$$E_k = X_k - \bar{X}_k \quad \& \quad W_k = [E_{k+1},\, E_k]^\top.$$

(1) There exists a matrix Q_k such that $W_{k+1} = Q_k W_k$. If $0 < \alpha < 2$ and $0 < \beta < 2\alpha$, then $\rho(Q_k) < 1$, where $\rho(Q_k)$ is the spectral radius of $Q(k)$.
(2) Setting $\alpha = \hat{\rho}^2 + 1$ and $\beta = \frac{2\alpha}{1+\xi}$ guarantees that $\rho(Q_k) < 1$.
(3) If $0 < \alpha \leq 1$, and $0 < \beta < 2\alpha$, then $\lim_{k\to\infty} W_k = 0$

4. Numerical Experiments

In this section, to illustrate the efficiency of Ts_PGA, we test it to recover some missing entries of a low-rank matrix at the case of different rank, sample radios, and noisy levels. At the meantime, we compare it with the closed related algorithm FPCA [11] for performance comparison. In the first place, we summary some useful parameters to make our experiments more easy to follow:

m:	the row number of matrix;
n:	the column number of matrix;
r:	the rank of original matrix, which is far less than $\min\{m, n\}$;
sr:	the sample ratio;
p:	the number of measurements, which is set to be $p = sr \times m \times n$;
dr:	the number of degree of freedom for a rank r matrix, which is defined as $dr = r(m + n - r)$;
Fr:	$Fr = r(m + nr)/p$ and $Fr \in (0, 1)$ is important to successfully recover M;
M:	a real low rank matrix to be recovered, which is generated by $M = M_L M_R^\top$, where matrix $M_L \in \mathbb{R}^{m\times r}$ and $M_R \in \mathbb{R}^{n\times r}$ are generated via the Matlab script "$randn(m, r)$";
Ω:	the index set of known elements, which are selected randomly;
b:	the given measurement vector, which is $b = \mathcal{A}(X) + \omega$;
ω:	Gaussian noise with mean zero and standard deviation σ generated by $\sigma \times randn(p, 1)$;
μ:	a penalty parameter, which is updated by continuation technique.
X^*:	an optimal solution.

All experiments are performed under Window 7 premium and MATLAB v7.8(2009a) running on a Lenovo laptop with an Intel core CPU at 2.4 GHz and 2 GB memory. The iterative process is terminated when the optimal solution X^* satisfies the following criterion:

$$RelErr = \frac{\|X^* - M\|_F}{\|M\|_F} \leq tol. \tag{14}$$

The relative error (RelErr) measures the quality of X^* to the original M. As usual [3, 20], we say that M is recovered successfully if $RelErr$ is less than $tol = 10^{-3}$. By the way, the terminated condition (14) is also used in [11]. On the other hand, for comparing in a relatively fair way, we also use the Matlab package as in FPCA for matrix singular value decomposition (SVD).

Our tests are partitioned into four parts. In the first three parts, we test Ts_PGA and FPCA to solve different cases of matrix completion problems. In the forth part, we use two typical images to visibly show the efficiency of both tested algorithms. In each test, we use the same technique as in [14, (26) and (27)] to choose the parameters α and β for the purpose of guarantee the algorithm's convergence. The numerical results generated by Ts_PGA and FPCA for solving matrix completion problems with different sr and r are reported in Table 1, where "Time" denotes the CPU time required by the algorithm. As can be seen from the top part of the Table 1, both algorithms required more computing time as Fr is greater than 0.38. For each cases with different Fr, we can clearly see that Ts_PGA performs better than FPCA in terms of running time and relative error. The worse thing is that FPCA failed to obtain high accuracy solutions at the cases of $sr = 0.8, 0.9$. From the bottom part of this table, we observe that all the problems are successfully solved with different r when $Fr = 0.38$, and see that Ts_PGA is a winner in this case. From the simple test, it is concluded that Ts_PGA is more efficient than FPCA.

In the second test, we apply Ts_PGA and FPCA to solve some easy matrix completion problems with different dimension. The numerical results are listed in Table 2 where Fr restricts into $(0.01, 0.2)$ and the m and n vary from 100 to 2000. From this table, we see that both algorithms produce satisfied optimal solutions within almost equivalent running time. But the accuracy of the solutions obtained by FPCA looks slightly higher, which means that the performance of both algorithms are competitive.

In the third test, we use Ts_PGA and FPCA to recover low rank matrix which corrupted by Gaussian noise. The noisy level is set as $\sigma = 1e - 3$. The numerical results are reported in Table 3. From this table, it is clear to see that Ts_PGA requires less running time to get solutions with similar accuracy than FPCA. This test shows that Ts_PGA runs faster to solve noisy matrix completion problem.

To visibly illustrate the efficiency of Ts_PGA, in the last test, we use both algorithms to recover a couple of randomly corrupted images which widely used in the literature. In this test, we choose $sr = 0.5$ and $r = 40$. The original low-rank images, the corrupted images, and recovered images by each algorithm are presented in Figure

Table 1. Numerical results of Ts_PGA and FPCA

				Ts_PGA		FPCA	
(m, n)	r	sr	Fr	Time	RelErr	Time	RelErr
(100,100)	10	0.4	0.47	3.68	5.10e-006	4.35	2.66e-006
(100,100)	10	0.5	0.38	4.35	2.66e-006	9.06	1.37e-006
(100,100)	10	0.6	0.32	0.18	8.57e-005	0.26	4.70e-005
(100,100)	10	0.7	0.27	0.20	7.32e-005	0.85	4.91e-005
(100,100)	10	0.8	0.24	0.20	6.10e-005	1.35	3.36e-002
(100,100)	10	0.9	0.21	0.25	5.67e-005	1.32	2.27e-001
(100,100)	10	0.5	0.38	4.35	2.66e-006	9.06	1.37e-006
(200,200)	20	0.5	0.38	20.40	2.37e-006	44.27	1.21e-006
(300,300)	30	0.5	0.38	60.00	2.43e-006	132.96	1.02e-006
(400,400)	40	0.5	0.38	131.56	2.21e-006	284.10	8.81e-007
(500,500)	50	0.5	0.38	239.57	2.24e-006	510.55	9.21e-007

Table 2. Comparison of Ts_PGA and FPCA for easy matrix completion problems

				Ts_PGA		FPCA	
(m, n)	r	sr	Fr	Time	RelErr	Time	RelErr
(100,100)	5	0.5	0.20	0.20	1.00e-004	0.21	2.69e-005
(200,200)	5	0.5	0.10	0.70	8.63e-005	0.83	6.95e-006
(300,300)	5	0.5	0.07	1.94	8.96e-005	2.19	3.64e-006
(400,400)	5	0.5	0.05	4.17	8.60e-005	5.06	1.83e-006
(500,500)	5	0.5	0.04	6.30	9.75e-005	6.94	1.39e-006
(600,600)	5	0.5	0.03	12.15	8.17e-005	11.77	9.12e-007
(700,700)	5	0.5	0.03	18.09	9.35e-005	19.68	7.75e-007
(800,800)	5	0.5	0.02	27.31	9.69e-005	29.83	3.21e-007
(900,900)	5	0.5	0.02	37.48	9.74e-005	41.24	2.88e-007
(1000,1000)	5	0.5	0.02	39.64	9.30e-005	44.53	2.52e-007
(1500,1500)	5	0.5	0.01	232.00	9.42e-005	262.56	1.62e-007
(2000,2000)	5	0.5	0.01	554.60	9.33e-005	554.60	9.33e-005

Table 3. Numerical Results of Ts_PGA and FPCA for (6) with Gaussian noise

				Ts_PGA		FPCA	
(m, n)	r	sr	Fr	Time	RelErr	Time	RelErr
(200,200)	10	0.5	0.10	5.81	1.86e-004	6.18	2.07e-004
(400,400)	10	0.5	0.05	3.95	1.31e-004	43.87	1.43e-004
(600,600)	10	0.5	0.03	16.77	1.19e-004	141.90	1.13e-004
(800,800)	10	0.5	0.02	34.86	1.15e-004	62.10	1.00e-004
(1000,1000)	10	0.5	0.02	55.01	1.04e-004	70.21	8.81e-005

1. From this figure, we see that Ts_PGA successfully recovered the both corrupted images, and the quality of each recovered image is competitive with the one derived by FPCA.

Taking the above four cases together, it illustrates that Ts_PGA provides an alternative approach to low rank matrix completion problems, and its performance is competitive with or slightly better than the widely used solver FPCA.

5. Conclusions

In the field of machine learning, one often meets the problem of exploiting the low-rank matrix from its given noisy partial entries. It has been shown that the task can be characterized as a matrix nuclear-norm minimization problem. However, the non-smoothness of the nuclear norm make the problem is challenging to solve. Due to the simplicity and effectiveness of proximal gradient algorithm for a wide range of convex minimization problems, the

Figure 1. (first column): the original image with size 512×512 and rank $r = 40$; (second column): randomly masked image with $sr = 50\%$; (third column): recovered images by Ts_PGA with "RelErr" $6.88e - 002$ (peppers) and $7.12e - 002$ (man); (forth column): recovered images by FPCA, with "RelErr" $6.23e - 002$ (peppers) and $7.30e - 002$ (man).

algorithm and its accelerated variants have been intensively studied in the past few years. Particularly, the type of this algorithm was successfully used to solve matrix completion problem by Ma, Goldfarb, and Chen [11]. On the one hand, the proposed algorithm is actually an extension or generalization of the proximal gradient algorithm, in which the generated iteration is a combination of the previous point, the current point, and its proximal gradient point. On the other hand, our idea also comes from the effectiveness of the TwIST algorithm of Bioucas-Dias and Figueiredo [14] to recover large and sparse signal from the limited measurement. Hence, our algorithm can also be regarded as an extension of TwIST to solving matrix completion problems. Although both motivations are simple, the numerical experiments illustrated that the proposed algorithm performs well and is competitive with the well-known solver FPCA, which in turn showed the superiority of the proposed algorithm. Surely, this is the contribution of this paper. The paper was paid more attention on solving matrix completion problem where the Lipschitz constant of the gradient of the least square term is 1, which is essential to the convergence of the algorithm. However, its efficiency in solving general linear constrained nuclear norm minimization problems has not been studied. This should be our further task to investigate.

Acknowledgements

The work of Q. Wang was supported by the National Natural Science Foundation of China (Grant No. 11471101). The work of Z.-F. Jin was supported by the National Natural Science Foundation of China (Grant No. 11471102) and the Key Basic Research Foundation of the Higher Education Institutions of Henan Province(No.16A110012).

REFERENCES

1. L. Vandenberghe and S. Boyd, *Semidefinite programming*, SIAM Review, vol.38,pp.49–95,1996.
2. B. Recht, M. Fazel, and PA. Parrilo, *Guaranteed minimum rank solutions of matrix equations via nuclear norm minimization*, SIAM Review,vol.52,pp.471–501,2010.
3. J.F. Cai, E.J. Candès, and Z. Shen, *A singular value thresholding algorithm for matrix completion*, preprint, SIAM J. Optim., vol.20, pp.1956–1982,2010.
4. E.J. Candès and B. Recht, *Exact matrix completion via convex optimization*, Foundations of Computational Mathematics, vol.9, pp.717–772,2009.
5. K.C. Toh, and S.W. Yun, *An accelerated proximal gradient algorithm for nuclear norm regularized least squares problems*, Pacific J. Optim., vol.6, pp.615–640, 2010.
6. A. Beck and M. Teboulle, *A fast iterative shrinkage-thresholding algorithm for linear inverse problems*, SIAM J. Imag. Sci., vol.2, pp.183–202, 2009.
7. R.-P. Wen, X.-H. Yan, *A new gradient projection method for matrix completion*, App. Math. Comput., vol.258, pp.537–544, 2015.

8. H. Zhang, L.Z. Cheng, *Projected Landweber iteration for matrix completion*, J. Comput. Appl. Math., vol.235, pp.593–601,2010.
9. Y.-F. Li, Y.-J. Zhang, and Z.-H. Huang, *A reweighted nuclear norm minimization algorithm for low rank matrix recovery*, J. Comput. Appl. Math., vol.263, pp.338–350, 2014.
10. Y. Xiao, S.-Y. Wu, D.-H. Li, *Splitting and linearizing augmented Lagrangian algorithm for subspace recovery from corrupted observations*, Adv. Comput. Math., vol.38 , pp.837–858, 2013.
11. S. Ma, D. Goldfarb, and L. Chen, *Fixed point and Bregman iterative methods for matrix rank minimization*, Math. Program., vol.128, pp.321–353, (2011).
12. E.T. Hale, W. Yin, and Y. Zhang, *Fixed-point continuation for ℓ_1-minimization: methodology and convergence*, SIAM J. Optim., vol.19, pp.1107–1130, 2008.
13. Y.J. Liu, D. Sun, and K.C. Toh, *An implementable proximal point algorithmic frameword for nuclear norm minimization*, Math. Program., vol.133 , pp.399–436,2012.
14. J.M. Bioucas-Dias and M. Figueiredo, *A new TwIST: Two-step iterative shrinkage/thresholding algorithms for image restoratin*, IEEE Trans. Image. Proces., vol.16, pp.2992–3004, 2007.
15. M. Elad, B. Matalon, and M. Zibulevsky, *Subspace optimization methods for linear least squares with non-quadratic regularization*, Appl. Comput. Harmon. Anal., vol.23, pp.346–367, 2007.
16. Y. Nesterov, *Gradient methods for minimizing composite objective function*, CORE Discussion Paper, Catholic University of Louvain, vol.76, 2007.
17. P. Tseng, *On accelerated proximal gradient methods for convex-concave optimization*, SIAM J. Optim., available at http://pages.cs.wisc.edu/ brecht/cs726docs/Tseng.APG.pdf
18. B. O'Donoghue and E. Candès, *Adaptive restart for accelerated gradient schemes*, Found. Comput. Math., vol.15, pp.715–732, 2015.
19. O. Axelsson, *Iterative Solution Methods*, Cambridge University Press, New York, 1996.
20. Z.-F. Jin, Z. Wan, Y. Jiao, and X. Lu, *An alternating direction method with continuation for nonconvex low rank minimization*, J. Sci. Comput., doi: 10.1007/s10915-015-0045-0.

4

A comparison of compressed sensing and sparse recovery algorithms applied to simulation data

Ya Ju Fan *, Chandrika Kamath

Lawrence Livermore National Laboratory, CA, USA

Abstract The move toward exascale computing for scientific simulations is placing new demands on compression techniques. It is expected that the I/O system will not be able to support the volume of data that is expected to be written out. To enable quantitative analysis and scientific discovery, we are interested in techniques that compress high-dimensional simulation data and can provide perfect or near-perfect reconstruction. In this paper, we explore the use of compressed sensing (CS) techniques to reduce the size of the data before they are written out. Using large-scale simulation data, we investigate how the sufficient sparsity condition and the contrast in the data affect the quality of reconstruction and the degree of compression. We provide suggestions for the practical implementation of CS techniques and compare them with other sparse recovery methods. Our results show that despite longer times for reconstruction, compressed sensing techniques can provide near perfect reconstruction over a range of data with varying sparsity.

Keywords Compressed Sensing, Sparse Recovery, Simulation Data, Algorithms

1. Background and Motivation

The analysis of data from scientific simulations is typically done by writing out the variables of interest at each time step and analyzing them at a later time. For simulations run on massively parallel machines, the volume of data written out frequently approaches terabytes and beyond. In some problems, writing out the data could be more time consuming than the computations done at each time step. As we move to exascale computing, it is expected that the I/O subsystem will not be able to provide the needs of the simulations [8]. Instead of writing out the data for analysis, it has been proposed that the analysis algorithms be moved *in situ*, and only the analysis results, which are hopefully much smaller, be written out. This is a viable option that works when we know the analysis algorithm we want to use and its parameters. However, in many cases, the choice of analysis algorithms and their parameters may depend on what we find as we analyze the simulation output. This is especially true when the motivation for the analysis is scientific understanding and discovery. In such cases, the analysis cannot be done *in situ* and alternate approaches have to be considered to address this problem of limited I/O bandwidth.

One such approach is to reduce the size of the simulation output by compressing the data before they are written out. There are several ways of compressing data, ranging from the traditional compression schemes, such as JPEG, that are used for images, to others that are based on clustering techniques from data mining. For the data set and problem considered in this paper, we require loss-less compression techniques that can be applied to data that are irregularly spaced in two or three dimensions and are distributed across multiple processors (or files). We propose

*Correspondence to: Lawrence Livermore National Laboratory, 7000 East Avenue, Livermore, CA 94551, USA.
E-mail:fan4,kamath2@llnl.gov

addressing this problem of compressing simulation output using compressed sensing (CS) and sparse recovery algorithms. We have three goals. First, we want to understand how we would apply compressed sensing techniques in practice to simulation data, given the irregular spacing of points and their distribution across multiple files. Second, we are interested in how well these techniques preserve the features of interest in the data so that we can perform our analysis on data that were compressed and reconstructed, while obtaining results identical or close to what we would have obtained using the original, uncompressed data. Finally, we want to compare compressed sensing with other sparse recovery algorithms and evaluate their suitability for our data sets using the metrics of reconstruction accuracy, compression time, and reconstruction time.

We note that our interest in compression is for the purpose of reducing the amount of data generated for quantitative analysis, not for visualization. In other words, the fact that the original and the compressed, then reconstructed, data are visually the same is not sufficient; we also want the quantitative results of the analysis to be the same for the two data sets. Therefore, we need perfect or near-perfect reconstruction of the compressed data.

This paper is organized as follows: First, in Section 2, we discuss compression techniques currently in use to reduce the size of scientific data sets, especially those from simulations. Next, we describe the specific characteristics of our data set in Section 3 to provide the background on the constraints that influence our choice of compression algorithm. We describe the algorithms considered in Section 4 and the pre-processing of the data in Section 5, followed by the experimental results in Section 6. We first identify a performance metric to use to evaluate the effectiveness of compression, followed by an evaluation of various preprocessing steps and the use of sparse vs. dense measurement matrices. We then compare compressed sensing with other sparse recovery algorithms and identify which algorithms best meet the needs of our problem. We conclude with a summary in Section 7.

2. Related work

Data compression techniques, such as JPEG and MPEG [28], have long been used to reduce the size of images and video to make it easier to store and transmit them. These techniques are lossy, that is, the data compressed using these techniques cannot be reconstructed exactly. However, if the compression preserves key aspects of the images such that the image or video after reconstruction is visually the same as the original, such compression may be acceptable in some applications while, at the same time, enabling a large reduction in the volume of the data.

However, for scientific data, visual quality is often not the only factor influencing the choice of a compression technique. For example, the wavelet-based Hcompress software [16] used in compressing astronomy images has both a lossy and a loss-less capability; the former is used to generate pre-view images, where fine details may be irrelevant, and the latter is used in scientific analysis where such details are important. Similarly, for medical images, loss-less compression is used to preserve the information in the data, though a hybrid approach that uses lossy compression in regions that are not of interest has been proposed [15] to reduce the size.

Our interest is in scientific data generated not as images, but as a result of scientific simulations run on high-performance computers. These simulations typically generate multi-terabytes of data in the form of double-precision variable values that are specified at grid or mesh points in a two- or three-dimensional spatial domain. The grid could be regular or irregular and could be either fixed or change over time. Since the volume of these data are large, the visualization community has developed several compression schemes to reduce the costs of data transfer and storage. Many of these schemes are lossy, and range from simple approaches, such as storing a double precision number in single precision, to more complex techniques involving sophisticated encoding [14, 24]. These lossy techniques are often sufficient to meet the end goal of qualitative analysis through visual display of the data, especially as the data sizes are much larger than can be accommodated on the displays used to visualize the data. However, to understand the effects of loss of information, efforts have been made to evaluate how various physically-motivated metrics change as a result of compression [22]. A different approach to lossy compression of simulation data borrows ideas from graph-based clustering techniques. It represents the mesh data as a graph, and identifies vertex-sets that can be approximated by a constant value within a user-specified error tolerance [20].

Compressed sensing techniques, the subject of this paper, can also be implemented in a lossy manner by including quantization in the process [29, 26], or by combining it with wavelets [27].

As mentioned earlier, our interest is in loss-less compression of simulation data using compressed sensing and sparse recovery methods. There has been relatively little work in this area. The visualization community has investigated, albeit to a limited extent, the loss-less compression of the meshes used to represent three-dimensional geometric models [19] as well as the floating-point data themselves [25]. The work in this paper is an extension of our early work that was performed to understand if compressed sensing could be applied to unstructured mesh data [12].

3. Description of the Data

We conduct our investigation into compressed sensing and sparse recovery algorithms using data from a fusion simulation [23]. These data describe how the plasma within a tokamak evolves over time. The computational domain is a three-dimensional toroid, representing the tokamak. It is composed of 32 poloidal planes, equi-spaced along the toroid (see schematic in Figure 1(a)). Each poloidal plane has 591,745 grid points; these are arranged in concentric circles around a center that represents the magnetic axis that runs around the toroid (see Figure 1(c)). To correctly simulate the physical processes in fusion without unnecessarily increasing the number of grid points, and hence the computational cost of the simulation, the distances between poloidal planes is larger than the distances between grid points within a poloidal plane. This means that the physics of interest occurs within each poloidal plane and we can analyze each plane independently. The grid points are not regularly spaced, so their spatial coordinates are required for analysis. This also means that compression techniques that assume the data to be available in a two-dimensional regular grid, as in an image, are not applicable to our data set. To reduce the size of the simulation output, the data are written out every five time steps.

Since the computational domain is so large, with nearly 19,200,000 grid points across the 32 poloidal planes, the simulation is run on massively parallel machines. The points in each poloidal plane are sub-divided into smaller domains, as shown in Figure 1(b), with roughly equal number of points in each, and each domain is assigned to one processor. As we shall see later in this section, this distribution of points across multiple processors, causes an additional complication in applying the compression algorithms to our data.

At each time step in the simulation, there are several variables that are written out for each grid point; each variable represents a different physical quantity. In our work, we consider just one of these variables; compressing all the data at a time step can be done by processing each variable separately. The grid points in our problem remain fixed over time, so their coordinates, in three-dimensions, are written out only once.

In our evaluation of compressed sensing and sparse recovery algorithms, we want to ensure that the analysis results obtained using the re-constructed compressed data are very close to the results using the original, uncompressed data. Specifically, we focus on the analysis of coherent structures in the data. Coherent structures are essentially groups of grid points that behave as a coherent whole. The grid points are spatially near each other, have similar values of a variable, and evolve in a similar way over time. Identifying coherent structures and characterizing their behavior over time is a common analysis task in many simulations, so our work is broadly applicable in many domains.

Figure 1(d) shows the values of the variable of interest at the grid points in plane 00 at time step 500. Since each plane has a large number of grid points (nearly 600,000), we progressively zoom-in into the details of this plane in Figure 2, where panel (a) shows a pie-shaped wedge, panel (b) shows the distinct grid points illustrating the coherent structures, and panel (c) shows a region with no structures. The coherent structures can be seen in panel (b) as the groups of grid points with values of the variable higher than the surrounding grid points, where the values are nearly zero. At this time step, there are very few coherent structures on the left half of the poloidal plane and near the center (which is the location of the magnetic axis of the tokamak). As a result, when the plane is sub-divided into domains as shown in Figure 1(b), there are some domains that have a large number of coherent structures, while others have none. Consequently, each domain has a different range of values of a variable, with

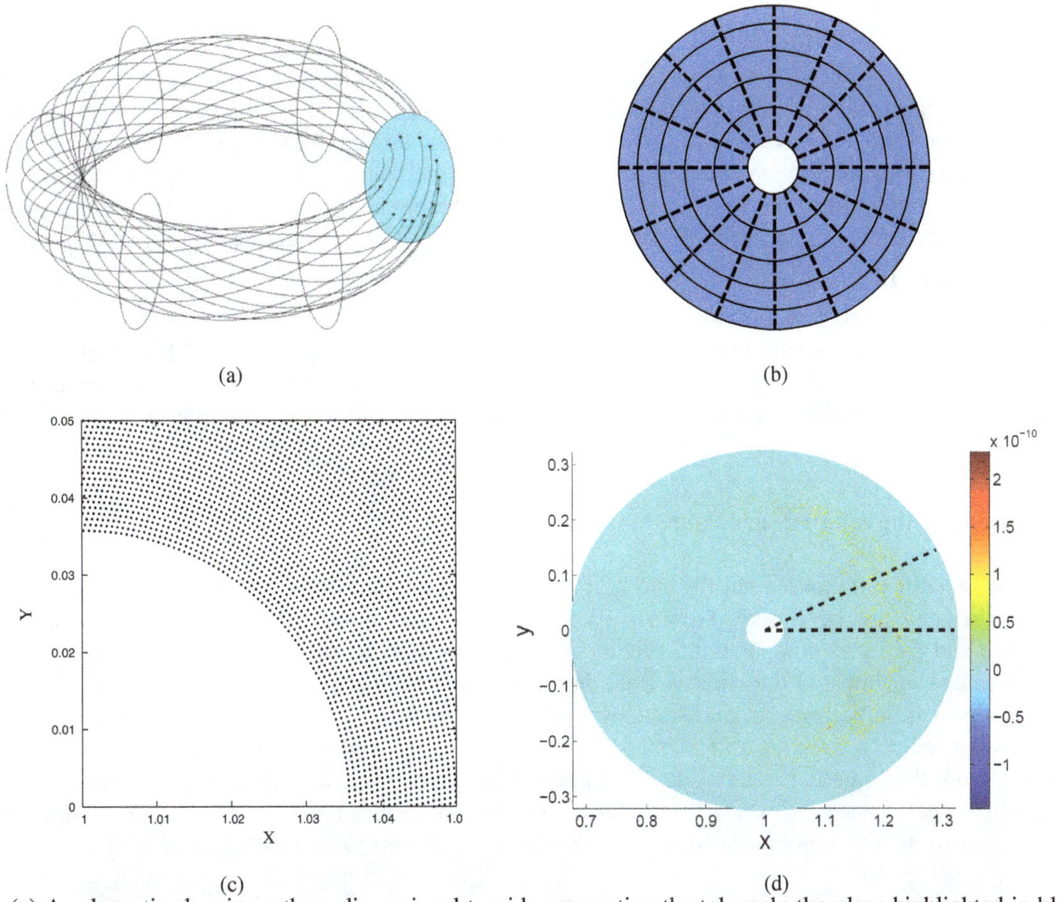

Figure 1. (a) A schematic showing a three-dimensional toroid representing the tokamak; the plane highlighted in blue is a poloidal plane. (b) Black lines show the division of a poloidal plane into regions with roughly equal areas; each region is assigned to a different processor. The center of the circle is the magnetic axis of the toroid. (c) A zoomed-in view showing the locations of the points near the center of the poloidal plane. (d) An example of the data in a poloidal plane at time step 500. The dotted lines indicate a slice of the data that contains $\frac{1}{16}$ of the grid points in this plane.

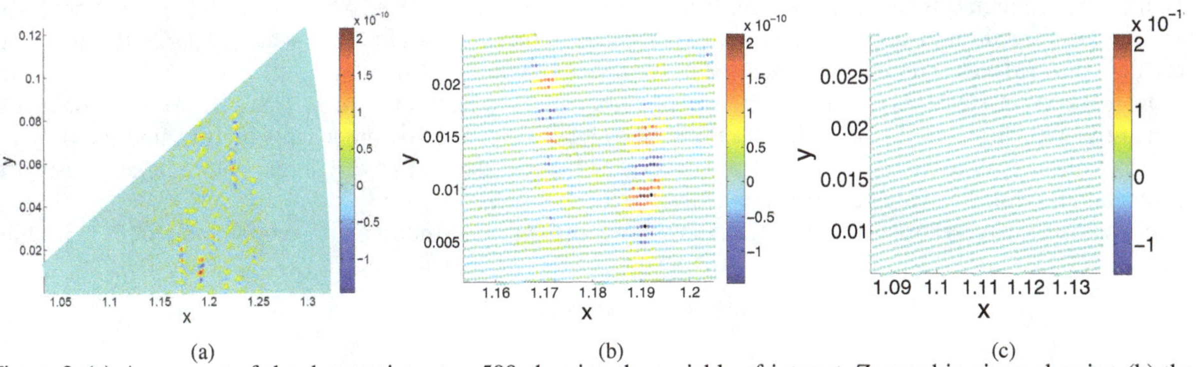

Figure 2. (a) A segment of the data at time step 500 showing the variable of interest. Zoomed-in views showing (b) the patterns of interest and the grid points and (c) the background region with no structures.

domains with many coherent structures having a larger variation. Given this variation in the data, we expect that determining CS parameters, such as the size of the compressed data, will be a challenge.

3.1. Dividing the data into sub-domains

To mimic the decomposition of each poloidal plane among processors, we divided the data on a poloidal plane into parts, or sub-domains, of equal area so that each sub-domain has almost the same number of data points, n. Since the data are in the form of a circular disk, we use the radius r and angle θ for the split. The range of the angle θ is $[-\pi, \pi]$. We make 16 cuts into the circle from the center out. Each part has the same angle $\theta = \frac{\pi}{8}$ in radians. We then divide each part into five pieces of equal area. Let r_0 and r be the closest and farthest distance from the data to the center of the circle, respectively. We compute the radii, $r_1, r_2, \ldots, r_4$, that equally separate the part into five based on r_0 and r. That is $r_i = \sqrt{\frac{1}{5}(i \times r^2 + (5-i) \times r_0^2)}$. Figure 1(b) shows how the data are divided into 80 sub-domains that are almost equal in size.

In practice, the number of sub-domains is determined by the size of the problem and the number of processors used to run the simulation. However, the division into sub-domains shown in Figure 1(b) is typical as it minimizes the communication costs associated with sharing data across sub-domains in neighboring processors.

4. Description of the compression algorithms

We next describe the compressed sensing and sparse reconstruction algorithms used in our work. The ideas behind compressed sensing have been studied extensively [11]. In essence, the method begins by sampling a sparse signal at a low rate and then uses computational techniques to obtain its exact reconstruction. This makes compressing data a simple task and most of the effort is spent on reconstructing the original signal.

Given a signal $\mathbf{u}$, in the form of an n-dimensional vector, we want to compress the signal to a smaller length $m(< n)$ so that we can reduce the cost of moving or transmitting this signal, which in our case is the data from the simulation, from the computer to the disk. We make m measurements, each amounting to a projection of $\mathbf{u}$ on a known vector. These measurements with a linear transformation result in a vector $\mathbf{f}$ of length m, that is, $\mathbf{f} = \mathbf{A}\mathbf{u}$, where the known matrix $\mathbf{A}$ is of size $m \times n$. The vector $\mathbf{f}$ is referred to as a sketch of $\mathbf{u}$ in theoretical computer science [17].

Once the m values of the vector $\mathbf{f}$, which is smaller than the original signal $\mathbf{u}$, have been transmitted, we need to uncompress the vector and recover the original signal. In general, this is a difficult task. Compressed sensing works only when signal $\mathbf{u}$ is sufficiently sparse, with k number of non-zero elements, where $k \ll n$ [7, 5]. The matrix $\mathbf{A}$ contains randomly generated numbers, which, in the conventional CS approach, are drawn independently from a Gaussian distribution. From a computational viewpoint, we do not need to explicitly store the matrix $\mathbf{A}$. Instead, we can store the random seed used to create the matrix and generate the matrix entries as needed, saving on memory requirements for the algorithm. In our work, we focus on the case where the non-zero components are real numbers and the measurements are linearly independent, that is, A is a full-rank matrix. The CS reconstruction problem then reduces to finding a $\mathbf{u}$ that satisfies the linear system $\mathbf{f} = \mathbf{A}\mathbf{u}$.

To make exact signal reconstruction possible in principle, we need at least $m \geq k+1$. We can conduct an exhaustive search on all $\binom{n}{k}$ possible selections of searches on non-zero components of $\mathbf{u}$ to find the only one possible linear system that can be inverted. However, when our data set is very large, with a large n, using the enumeration method quickly becomes impractical.

In practice, the reconstruction step in CS is solved by forming an L_1-norm optimization problem [6, 11] defined as

$$\min_{\hat{\mathbf{u}}} \|\hat{\mathbf{u}}\|_1 \quad \text{subject to} \quad \mathbf{A}\hat{\mathbf{u}} = \mathbf{f} \tag{1}$$

where $\|\hat{\mathbf{u}}\|_1 = \sum_{i=1}^{n} |\hat{\mathbf{u}}_i|$. Equation (1) is also known as the basis pursuit problem, whose solution is the vector with smallest L_1 norm that describes the sketch $\mathbf{f}$.

CS is an example of a sparse recovery algorithm that approximates a sparse signal $\mathbf{u}$ from its lower-dimensional sketch $\mathbf{Au} \in \Re^m$. However, unlike the conventional CS approach, sparse recovery algorithms do not require a solution of the optimization problem in Equation (1), nor do they require the matrix $\mathbf{A}$ to be dense or have entries drawn from a Gaussian distribution.

In this paper, we compare the performance of CS and sparse recovery algorithms for the compression of the data set described in Section 3. We consider six different methods: three of them - L_1 magic [4], Gradient Projection for Sparse Reconstruction (GPSR) [13] and YALL1 [10] - are conventional CS methods where the matrix $\mathbf{A}$ is dense, with values drawn from a Gaussian distribution. The other three methods - Sparse Matching Pursuit (SMP) [2], Sequential Sparse Matching Pursuit (SSMP) [1], and Count-Min sketch [9] - have a sparse measurement matrix $\mathbf{A}$. Numerous methods have been proposed for sparse recovery by different research communities, including signal processing, theoretical computer science, and applied mathematics. These methods vary in the accuracy of reconstruction, computational time, and the length of the sketch. The six methods we selected are illustrative of different approaches. We briefly describe them in the following; more details are available in the references.

- **Compressed sensing using L_1 magic:** L_1 magic [4] is a public domain solver that obtains solutions of Equation (1) using efficient linear programming techniques. In our work, we use the routine `l1eq_pd`, which is a primal-dual algorithm for linear programming. A system of nonlinear equations is formulated using the Karush-Kuhn-Tucker conditions corresponding to Equation (1). The system is linearized and solved using the classical Newton's method. This algorithm starts at an interior point and applies the maximum step size that just keeps the point in the interior. The algorithm iterates and identifies the solution point as close to being optimal when the slackness condition is small and the surrogate duality gap has decreased below a given tolerance. In the conventional CS method, the measurement matrix $\mathbf{A}$ is dense.
- **Compressed sensing using Gradient Projection for Sparse Reconstruction (GPSR):** Gradient projection for sparse reconstruction (GPSR) [13] solves the general optimization problem formulated as

$$\min_{\hat{\mathbf{u}}} \frac{1}{2}\|\mathbf{f} - \mathbf{A}\hat{\mathbf{u}}\|_2^2 + \tau\|\hat{\mathbf{u}}\|_1, \tag{2}$$

  which is a convex unconstrained optimization problem with a quadratic term. τ is a nonnegative parameter and $\|\mathbf{v}\|_2$ denotes the Euclidean norm of a vector $\mathbf{v}$. Equation (2) falls in the category of a sparse recovery problem since the use of the L_1 norm on $\hat{\mathbf{u}}$ tends to reconstruct the signal $\mathbf{u}$ to be as sparse as possible, given the sketch $\mathbf{f}$ and the measurement matrix $\mathbf{A}$. GPSR is a gradient projection algorithm in which the search path from each iterate is obtained by projecting the negative-gradient onto the feasible set from each iteration. GPSR has been shown to perform well in a range of applications and is significantly faster than competing methods [13]. The measurement matrix $\mathbf{A}$ is typically dense.
- **Group sparse reconstruction using alternating direction method (YALL1):** YALL1 (Your ALgorithm for L1) [10] is a solver for group sparse reconstruction using alternating direction method (ADM) [31, 3]. The algorithm makes use of the group sparsity structure in the data. Components within a group are likely to be either all zeros or all nonzeros. Formulating the group sparsity reduces the degrees of freedom in the solution, and hence improves reconstruction performance. Let groups $\{\mathbf{u}_{g_i} \in \Re^{n_i} : i = 1, \ldots, s\}$ be a partition of $\mathbf{u}$, where $g_i \subseteq \{1, 2, \ldots, n\}$ is a set of instance indices contained in the i-th group, and $\mathbf{u}_{g_i}$ are the n_i instances of $\mathbf{u}$ that are in the i-th group. YALL1 solves the following basis pursuit model:

$$\min_{\hat{\mathbf{u}}} \|\hat{\mathbf{u}}\|_{w,2,1} \quad \text{subject to} \quad \mathbf{A}\hat{\mathbf{u}} = \mathbf{f} \tag{3}$$

  where $\|\hat{\mathbf{u}}\|_{w,2,1} := \sum_{i=1}^{s} w_i \|\hat{\mathbf{u}}_{g_i}\|_2$ is the weighted $l_{2,1}$-norm and $w_i \geq 0$ for $i = 1, \ldots, s$ are weights associated with each group. YALL1 applies the classic ADM to both the primal and dual forms of the $l_{w,2,1}$-problem (3) whose objective functions are separable, and constructs an efficient first-order algorithm using augmented Lagrangian that updates both the primal and dual variables at each iteration.
- **Sparse matching pursuit (SMP):** Sparse matching pursuit (SMP) is a method designed for practical, near-optimal, sparse recovery in the L_1 norm [2]. Instead of using a dense matrix generated from a Gaussian distribution in the L_1 norm, as in Equation (1), SMP uses a sparse measurement matrix $\mathbf{A}$ with binary elements, that is, the matrix elements take the values 0 or 1. It is derived from a very fast recovery algorithm, called Expander Matching Pursuit (EMP) [18], which, in turn, is derived from the Orthogonal Matching Pursuit (OMP) [30]. Like these methods, SMP is a greedy iterative algorithm that estimates the vector $\mathbf{y} = (\mathbf{u}^t - \mathbf{u})$, which is the difference between the current approximation $\mathbf{u}^t$ and the signal $\mathbf{u}$ from the sketch

$\mathbf{Ay} = \mathbf{Au}^t - \mathbf{Au}$. The matrix $\mathbf{A}$ can be interpreted as an adjacency matrix of a bipartite graph G, with left vertex set = $\{1, 2, \ldots, n\}$ and right vertex set = $\{1, 2, \ldots, m\}$. Each column of $\mathbf{A}$ contains d elements that are one, that is, every vertex in the left part of the graph has exactly d neighbors in the right part. Each coordinate $\mathbf{y}_i^*$ in the estimation $\mathbf{y}^*$ of $\mathbf{y}$ is the median of the coordinates of the sketch vector $\mathbf{Ay}$ that are neighbors of i in the graph G. The approximated signal $\mathbf{u}^t$ is updated by $\mathbf{y}^*$ and the process is repeated. The idea behind the algorithm is to select the neighbors of i using a voting-like mechanism. Compared to EMP, SMP is slightly slower, but performs successful recovery from a significantly smaller number of measurements, and is therefore more suitable for our application in reducing the cost of data movement.

- **Sequential sparse matching pursuit (SSMP):** Sequential sparse matching pursuit (SSMP) [1] is derived from the SMP method. Both SSMP and SMP use a sparse binary matrix for measurement, and voting-like mechanism in iterative search for signal approximation. The difference is that SSMP updates the approximation signals sequentially while SMP updates them in parallel. SMP works only when the input parameters, the sparsity k and the number of measurements m, are within the theoretically guaranteed region. SSMP does not have such convergence problems and is designed for arbitrary input signals.
- **Count-Min and Count-Median sketch:** The Count-Min sketch algorithm [9] also utilizes a sparse random matrix $\mathbf{A}$ with binary elements. It only works under the assumption that all elements in the signal are greater or equal to zero. That is, the elements of $\mathbf{u}$, $\mathbf{u}_i \geq 0$ for all i. The method divides matrix $\mathbf{A}$ into d segments, $\mathbf{A}(h_1), \mathbf{A}(h_2), \ldots, \mathbf{A}(h_d)$. Each segment of $\mathbf{A}$ corresponds to a function h, from the set H of all functions $h : \{1, 2, \ldots, n\} \rightarrow \{1, 2, \ldots, w\}$. We independently and uniformly choose these d random functions. Each such function generates a binary matrix $\mathbf{A}(h)$ of size $w \times n$ by setting $(\mathbf{A}(h))_{j,i} = 1$ if $j = h(i)$ and $(\mathbf{A}(h))_{j,i} = 0$ otherwise. The one-to-one mapping makes each column have exactly one 1. Combining the d segments, the algorithm constructs the matrix $\mathbf{A}$ whose number of rows is equal to $m = wd$. To approximate $\mathbf{u}^*$ from $\mathbf{Au}$, the algorithm makes use of the construction of matrix $\mathbf{A}$. Observe that for any signal $\mathbf{u}$,

$$(\mathbf{Au})_{(l-1)w+j} = (\mathbf{A}(h_l)\mathbf{u})_j = \sum_{i:h_l(i)=j} \mathbf{u}_i$$

for $l = 1, 2, \ldots, d$ and $j = 1, 2, \ldots, w$. Using this property, computing the approximation of signal $\mathbf{u}$ is fairly simple. It requires only pairwise independent hash functions. Under the assumption of $\mathbf{u} \geq \mathbf{0}$, we define elements of the approximation of $\mathbf{u}^*$ to be:

$$\mathbf{u}_i^* = \min_l \quad (\mathbf{A}(h_l)\mathbf{u})_{h_l(i)} = \min_l \quad \sum_{i':h_l(i')=h_l(i)} \mathbf{u}_{i'}.$$

We cannot directly apply Count-Min to our data sets as they contain negative values. Instead we use the extended Count-Min algorithm, which is also called the Count-Median algorithm, for our problem. This method replaces $\mathbf{u}_i^*$ by

$$(\mathbf{u}_{med}^*)_i = \operatorname*{median}_l \quad (\mathbf{A}(h_l)\mathbf{u})_{h_l(i)}.$$

Removing minimization constraints allows negative estimators. The algorithm then works for general signals with both positive and negative values.

Count-Min and Count-Median sketches are algorithms designed by the database community to handle fast queries on data streams [9]. Therefore, their recovery time is expected to be very fast, but as only a summary of the data is stored, the algorithms are not intended for use in exact recovery. However, we have included them in our study to evaluate how much the accuracy degrades in the context of our problem.

5. Preprocessing the data

Before we could apply the six compression algorithms to our simulation data, we found that the data required some preprocessing. Our data set contains a large number of values that are almost zero as seen in the background region in Figure 2. When we first applied CS to the data, we were not successful in recovering the original signal even though the algorithm had converged (see Figure 5). On examining the results, we found that a majority of the data components have to be exactly zero in order to satisfy the sufficient sparsity condition of CS. This suggests that we need to threshold the data first and set all values below the threshold to zero. From the analysis viewpoint, this is acceptable because one of the first steps in the analysis of coherent structures in the data is to perform such a thresholding as it is the grid points with the larger values that are of interest [21].

The use of thresholding as a pre-processing step is also motivated by a constraint on k, the number of non-zeros, and m, the number of measurements, that must be satisfied for reconstructing the signal. As the results by Candes et al. [5] show, if $k \leq \alpha \cdot m \cdot (\log n)^{-1}$, where $\alpha > 0$ is a constant, then the L_1 problem (1) will find the unique solution that is exactly equal to $\mathbf{u}$ with very high probability. Their empirical study showed that for n between a few hundred and a few thousand, with $k \leq m/8$, the recovery rate is above 90%. These results also suggest that $m^* = k \log n$ is an appropriate choice for the number of rows of $\mathbf{A}$. However, there is still slight chance that we will not get the exact $\mathbf{u}$ from solving the L_1 problem. Since we are looking for exact reconstruction of the patterns in the data, in addition to thresholding values that are very near zero, we need to investigate values around m^* and find a better condition for CS for our simulation data.

Another issue we need to address is the contrast in our data set, that is the range of intensity values at the grid points in the simulations. The data over an entire poloidal plane at early time steps have low contrast, with a relatively small range of intensities. Not only are the majority of the data values almost zero, forming the background, but the rest of the data values, which are part of the coherent structures of interest, are also close to zero. For example, at time step 500, Figure 1 indicates that the absolute values in the poloidal plane are less than 3×10^{-10}. The corresponding histogram of values shown in Figure 3 indicates that most of the values are near zero (the background), with a few grid points (the coherent structures) having slightly larger values, as shown in Figure 3(b). At later time steps in the simulation, the absolute values of the variable in a coherent structure within a poloidal plane can go as high as 10^{-5}, while the background remains at 10^{-10} or smaller. Further, some sub-domains, such as the ones near the center or on the left side of the poloidal plane, continue to have a small range of values as they are composed of mainly the background. The results at later time steps of the simulation allow us to compare the methods for higher dynamic range data with varying sparsity.

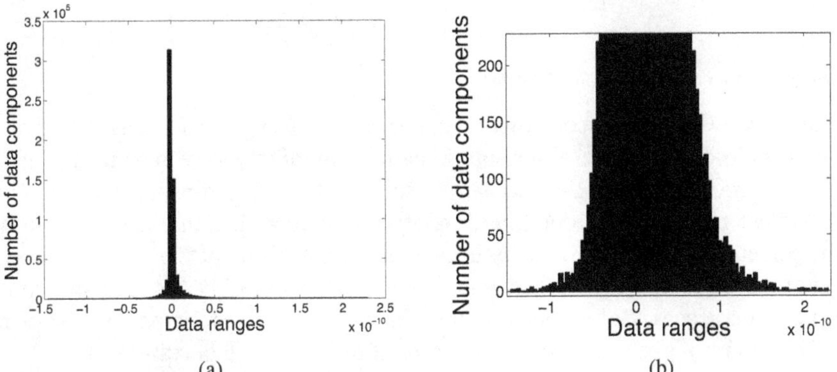

Figure 3. (a) Histogram of the data at time step 500 showing majority of the data values very close to zero. (b) Zoomed-in view of the histogram.

To address this problem of sub-domains in a poloidal plane having a large range of values either within a sub-domain or across sub-domains, we consider the option of scaling the data within a poloidal plane. Scaling just the values in each sub-domain to [0.0,1.0] will result in amplifying the noise in the sub-domains that have all grid points with values very near zero. To scale across the entire poloidal plane, we first use a gather/scatter operation to find the minimum and maximum values across all sub-domains that form a poloidal plane. Then, each sub-domain scales its grid points so that the values in the entire poloidal plane are in the [0.0, 1.0] range. After reconstruction, we only need the two limits to scale the data back to their original values without any loss of information.

Increasing the contrast of the data helps the performance of CS and the sparse recovery algorithms in many ways. First, it improves our ability to control the convergence of these methods. Recall that L_1 magic, GPSR and YALL1 are iterative optimization algorithms and all have convergence conditions with a given tolerance. In practice, the tolerance is set to be some very small number. However, when the contrast or the values of the data are relatively small, the algorithm could terminate prematurely due to the convergence conditions being satisfied, even though the optimal solution has not been obtained. While the user can adjust the tolerance for the algorithm to terminate at an acceptable optimal, it is difficult to determine such a value when the data set contains very small numbers and has a small range of intensity values. Therefore, scaling the data to [0.0,1.0] helps standardize the input data for convergence conditions. This scaling, by improving the contrast in the data values, also helps the sparse recovery algorithms that are iterative in nature. Finally, when multiple algorithms are being compared, as in our study, scaling the data helps us to evaluate the performance of the algorithms more objectively.

6. Experimental Results

We next describe the experiments we conducted to understand how best to apply compressed sensing and sparse recovery algorithms to our data set. Given a data set of n components containing k non-zero entries, we have two main goals: The first is to determine if CS can be applied successfully to our data and if our ideas for scaling the data and selecting the size of the compressed data result in sufficient compression and near-perfect reconstruction. In particular, we want to understand how well the features of interest in our data are preserved during reconstruction. Second, we want to compare CS with other compression methods described in Section 4 in terms of reconstruction accuracy, compression time and reconstruction time to identify those suitable for our problem.

We conduct our experiments in two parts. Our main focus is compressed sensing as it allows perfect reconstruction of the data. Therefore, in the first set of experiments we evaluate different performance metrics and pre-processing ideas with CS using L_1 magic as the compression algorithm. In the second set of experiments, we compare CS with the sparse recovery algorithms. Since the SMP, SSMP and Count-Min methods require a sparse measurement matrix, for a fair comparison, we consider both a dense matrix and a sparse matrix in L_1 magic, GPSR and YALL1. Recall that these three methods were originally proposed with a dense measurement matrix.

6.1. Reconstruction accuracy of compressed sensing

The first evaluated how well CS could reconstruct compressed data using the L_1 magic software. Recall that we divided the data on each poloidal plane into equal sub-domains to mimic the way in which the data are decomposed into domains for parallel processing. As observed earlier, some of these domains have a large number of coherent structures, while others are mainly background, with relatively small values that are likely to be ignored during thresholding. So, the percentage of non-zeros, k/n, in each domain is different. To evaluate the performance of CS on each sub-domain of the data, we vary the percentage of compression, m/n for each sub-domain, and evaluate a metric that determines the accuracy of reconstruction. The ratio k/n gives the percentage of non-zeros in each sub-domain, while the ratio m/n gives the reduction in size of the data for that sub-domain.

We considered three metrics to measure the accuracy of reconstruction: the L_2 norm, the peak signal-to-noise ratio (PSNR), and R^2. The L_2 norm measures the Euclidean distance between the original data, $\mathbf{u}_i, i = 1, \ldots, n$, and the data, $\hat{\mathbf{u}}_i, i = 1, \ldots, n$, after CS and reconstruction. PSNR is derived from the mean squared error (MSE) of the two data sets; it is often used in image compression where removing noise is the desired outcome. Let

$$\text{MSE} = \frac{1}{n}\sum_{i=1}^{n}(\mathbf{u}_i - \hat{\mathbf{u}}_i)^2.$$

Then, PSNR is defined as:

$$\text{PSNR} = 10 \cdot \log_{10}\left(\frac{\max(\{\mathbf{u}_i\}_{i=1,\ldots,n})}{\text{MSE}}\right).$$

The coefficient of determination, denoted by R^2, is widely used in statistics in comparing two data sets and is defined as:

$$R^2 = 1 - \frac{\sum_{i=1}^{n}(\mathbf{u}_i - \hat{\mathbf{u}}_i)^2}{\sum_{i=1}^{n}(\mathbf{u}_i - \frac{\sum_{i=1}^{n}\mathbf{u}_i}{n})^2}.$$

Since we are looking for exact reconstruction, we prefer smaller L_2, larger PSNR and an R^2 value that is close to 1.

6.1.1. Evaluating performance metrics We compared the three performance metrics, the L_2 norm, PSNR, and R^2, by evaluating their effectiveness using the data in all sub-domains from time step 500 in the simulation. First, we scaled the data in each sub-domain using the maximum and the minimum values from the 00 poloidal plane that spans all sub-domains. Then, based on the histogram of the data, we set components whose absolute values are less than 0.3 to zero as a threshold that separates the coherent structures of interest from the background.

Table 1. Data size n and number of non-zero components k for the 11 sub-domains (out of a total of 80) that are non-sparse at time step 500. CS fails to reconstruct these sub-domains using $m < n$.

Pie #	Slice #	n	k
1	2	7454	4171
1	3	7508	4553
2	2	7381	3866
2	3	7448	4914
3	3	7458	4505
4	3	7459	3587
13	3	7468	3933
14	3	7458	4898
15	2	7380	3756
15	3	7459	4435
16	3	7451	4752

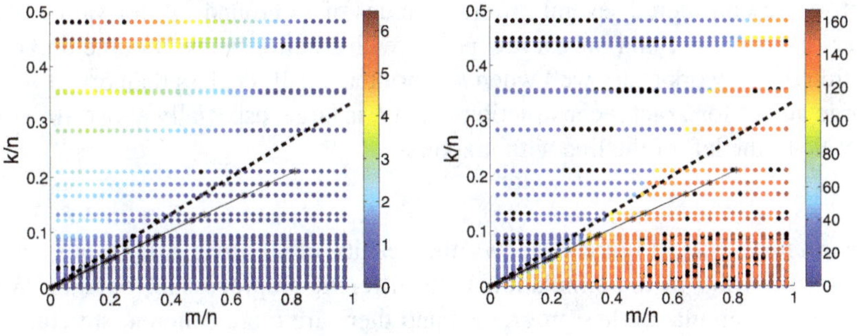

(a) L_2 (b) PSNR (c) R^2

Figure 4. Performance of L_1 magic with a sparse measurement matrix on data from time step 500. The values of the three metrics - L_2, PSNR and R^2 - are shown in color for different levels of compression (m/n) for sub-domains with different levels of sparsity (k/n). The dotted black line represents the slope of $\frac{1}{3}$. The straight line with star markers is the theoretically suggested selection of m^* with respect to k.

When the data in a sub-domain are not sufficiently sparse, CS performs poorly. Table 1 shows the numbers of non-zero components and the sizes of the data for the 11 sub-domains that are not sparse enough and are excluded in our work. The results indicate that in general when k is more than 50% of n, CS fails in reconstruction using $m < n$.

Figure 4 shows the performance of L_1 magic with a sparse measurement matrix on each sub-domain at time step 500 using the three different performance metrics. This measurement matrix was generated as described in the Count-Min sketch algorithm and in Section 6.1.4. There are 80 sub-domains on each poloidal plane (as shown in

Figure 1(c)), each with roughly the same number of grid points, but a different number, k, of non-zero values. We ignore the low sparsity sub-domains listed in Table 1. Before applying CS, the data were first scaled so the poloidal plane had values between [0.0, 1.0], then thresholded to identify the background points, which were set to zero. The dotted black line starting from origin represents the slope of $\frac{1}{3}$. Points below that line could be compressed better by writing out just the x–y coordinates and values of the grid points that remain after the thresholding. The solid straight line with star markers is the theoretically suggested selection of m^* with respect to k.

Each horizontal line in the figure represents the performance of CS on one sub-domain. Taking Figure 4(c) as an example, as we vary m, the number of measurements, for a sub-domain that has k non-zero values, the performance of CS evaluated using R^2 is represented by colors on a horizontal dotted line. The change in color clearly shows that the R^2 increases as the size of compressed data (m/n) increases. Recall that we prefer smaller L_2, larger PSNR and an R^2 value that is close to 1.

Among the three metrics, R^2 gives the most gradual and distinct difference in the performance as we vary m. Regardless of the data used, R^2 is always in the range $[0, 1]$, unlike the L_2 norm and PSNR metrics, whose maximum will vary with the data, making it difficult to compare compression results across data from different time steps or poloidal planes. A value of $R^2 = 1$ obtained using data before and after compression indicates that they are identical. Therefore, we propose using R^2 as our metric of choice in measuring CS for exact reconstruction. We will only show R^2 values in the rest of this paper.

6.1.2. Evaluating the effectiveness of compression In Figure 4, we include two lines to help evaluate the effectiveness of the compression using CS. The thick dotted black line starting from the origin has a slope of $\frac{1}{3}$. Points on the line represent $m = 3k$, indicating that carrying the three values - x–y coordinates of a grid point and its value - for all non-zero points that remain after thresholding requires the same size of storage as the compressed data. Points below the line represent $m > 3k$, where the compressed data have a larger size than the size of k non-zero values and their corresponding x–y locations. Therefore, the sub-domains that benefit from CS are the ones that have $R^2 = 1$ and lie above the dotted line.

The straight line with markers connects the values when $m = m^*$. As mentioned earlier, theory suggests a value of $m^* = k \log n$ for reconstruction using the L_1 norm. To test how accurate this number m^* works in practice, especially in the context of the R^2 metric, we considered several different values of m around m^* for various k. Most points below the line with markers have $R^2 = 1$ and only a few points with small k below the line do not. This indicates that the theoretical selection of m^* works very well when k is not too small, i.e. less than 5% of n in Figure 4(c). Also, we do not need as large an m^* for exact reconstruction when k is large, especially above 10% of n as indicated by the $R^2 = 1$ points that lie to the left of the line with markers.

6.1.3. Evaluating the performance of preprocessing Figure 5 shows the results of L_1 magic with a sparse measurement matrix on data from time step 1000 in the simulation with different preprocessing options. We considered a later time step as by this time the simulation has progressed and there are more coherent structures, leading to data that allows a better evaluation of the preprocessing options. Recall that we need exact reconstruction, so a value of $R^2 = 1$ is desired.

Figure 5(a) shows the results with no preprocessing, followed by the results with only thresholding, only scaling, and both thresholding and scaling, in Figures 5(b)-(d), respectively. To make it easier to compare the plots, each horizontal line in Figures 5(a) and (c) corresponds to the same sub-domain as the corresponding line in (d); however, the number of non-zero components is higher when no thresholding is used. We make several observations. The results with no preprocessing and only thresholding appear very similar. Neither has any sub-domains with perfect reconstruction. The maximum R^2 from the data without any preprocessing is 0.9982, while the maximum R^2 from only thresholding is 0.9977. The use of both scaling and thresholding results the most sub-domains with $R^2 = 1$, while using just scaling, but no thresholding, results in a few sub-domains with $R^2 = 1$. However, all these sub-domains in Figure 5(c) are below the black dotted line, the one at $\frac{1}{3}$ slope, indicating that compression is not preferred. From these observations, we conclude that scaling has very significant effect on the performance of CS, and thresholding improves the performance only when the data are scaled.

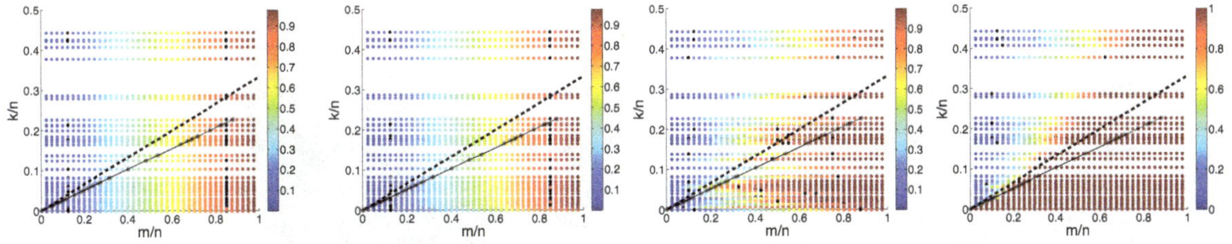

(a) Without preprocessing (b) Thresholding only (c) Scaling only (d) Scaling and thresholding

Figure 5. Performance of L_1 magic with a sparse measurement matrix on data at time step 1000 with different preprocessing options. (a) Results with no scaling or thresholding of the data. (b) Results with thresholding, no scaling. (c) Results with scaling, but no thresholding of the data. (d) Results using both scaling and thresholding. For ease of comparison, each horizontal line in the four plots corresponds to the same sub-domain as indicated in (d), though the sub-domain has a higher number of non-zeros when thresholding is not used.

We next considered the performance of L_1 magic at other time steps. We know that in the simulation, the number of coherent structures changes over time. At early time steps, there is a small number of coherent structures, which grow larger in size over time, until they split into multiple structures. Thus the sparsity of the data reduces over time, and we would expect that CS techniques would start performing poorly at late time. Figure 6 shows the performance of L_1 magic with a sparse measurement matrix evaluated using R^2 on data at time steps 2000, 3000 and 4000 using scaling and thresholding. These results suggest that compression using CS is an option when the sparsity is more than 15% and less than 50%. For sparsity lower than the lower limit, it is preferable to just store the non-zero values and their locations in three dimensions. For sparsity greater than the upper limit, CS does not give perfect reconstruction.

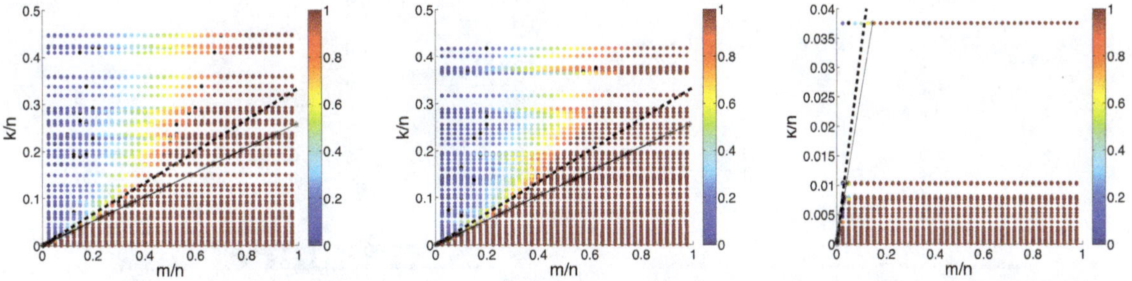

(a) Time step 2000 (b) Time step 3000 (c) Time step 4000

Figure 6. Performance of L_1 magic with a sparse measurement matrix at time steps 2000, 3000 and 4000 evaluated using R^2.

Finally, we evaluated the effectiveness of CS in preserving small but unusual structures in the data. Our earlier work had shown that, at time step 3000, there are some small structures, composed of a few points, with large negative values of the variable, unlike other structures in the data that have positive values. When the simulations are run for scientific discovery, it is important that any compression preserve such structures so they can be analyzed further. Figure 7 and Table 2 show that this is possible using CS, even with a compression ratio of 72%.

Table 2. Performance of compressed sensing on sub-domain data at time steps 3000.

k/n	0.38	0.38	0.38	0.38	0.38	**0.38**	0.38	0.38	0.38	0.38	0.38	0.38	0.38	0.38
m/n	0.60	0.63	0.65	0.68	0.70	**0.73**	0.75	0.78	0.80	0.83	0.85	0.90	0.95	0.98
R^2	0.87	0.89	0.91	0.93	0.95	**1.00**	1.00	1.00	1.00	1.00	1.00	1.00	1.00	1.00

6.1.4. Comparing the use of dense vs. sparse measurement matrix Next, we compare the use of a dense matrix vs. a sparse matrix in L_1 magic, GPSR and YALL1, using the metrics of compression time, recovery time and reconstruction accuracy. The dense matrix is generated using random numbers drawn independently from a

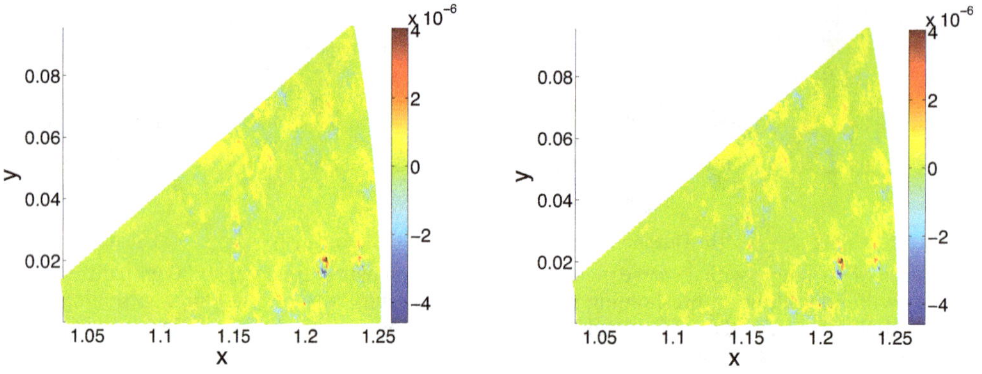

(a) Original data (b) Reconstructed data using $m/n = 72\%$, giving $R^2 = 1$

Figure 7. Sub-domain data at time steps 3000 before and after applying preprocessing and compressed sensing. Using 72% compression rate, CS reconstructed the patterns perfectly with $R^2 = 1$ after the background grid points were set to zero. The small structures near the lower right with large positive and negative values next to each other are of scientific interest.

Gaussian distribution. The sparse matrix is generated using $d = 8$, which divides each column of $\mathbf{A}$ into d segments, randomly assigns one element in each division to be 1, and sets the rest to 0. This sparse matrix satisfies the requirements for SMP, SSMP and Count-Min algorithms to work and therefore, allows us to compare all methods on an equal footing.

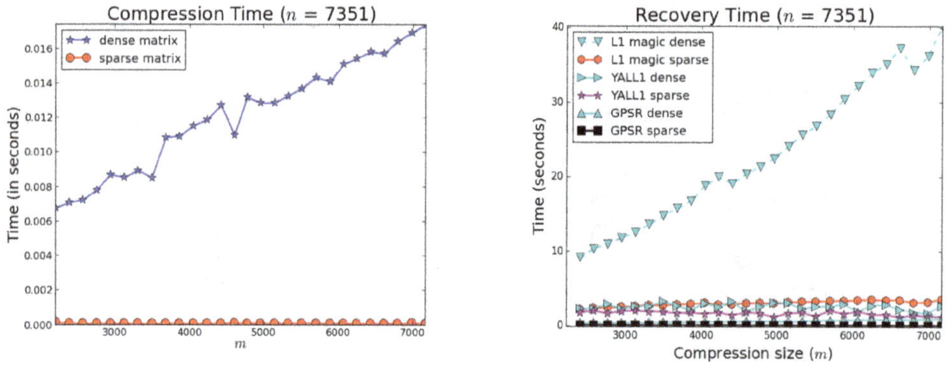

(a) (b)

Figure 8. Comparison of (a) compression and (b) reconstruction times using L_1 magic, GPSR and YALL1 with a dense and a sparse matrix on a subdomain data of length $n = 7351$ and sparsity $k = 26$ at time step 2000.

Since we compress data $\mathbf{u}$ using a matrix $\mathbf{A}$, the compression time is the time for the calculation of $\mathbf{Au}$, and is the same for all methods. Figure 8(a) compares the compression time using a dense matrix and a sparse matrix on a subdomain of length $n = 7351$ and sparsity $k = 26$ at time step 2000, as the size of compression m is varied. The compression time using a dense matrix obviously increases linearly as m increases; in contrast, using a sparse matrix takes a smaller amount of time, which is almost constant with increasing m. The corresponding recovery times for L_1 magic, GPSR and YALL1 with both dense and sparse matrices on the same subdomain data at time step 2000 are shown in Figure 8(b), with additional details shown in Figure 13. Similarly using a dense matrix takes dramatically more time for the recovery, especially in the case of L_1 magic. In general, the recovery times increase as the compression size m increases, except for YALL1.

Figure 9(a) and Figure 9(c) display the reconstruction accuracy of L_1 magic using a dense matrix and a sparse matrix, respectively, for time step 2000. Figures 9(b) and (d) show just the points where $R^2 = 1$, that is, the points when L_1 magic gives perfect reconstruction. Focusing on the points with perfect reconstruction, we observe that the performance of a dense matrix is very similar to that of a sparse matrix. There are only a few points that are

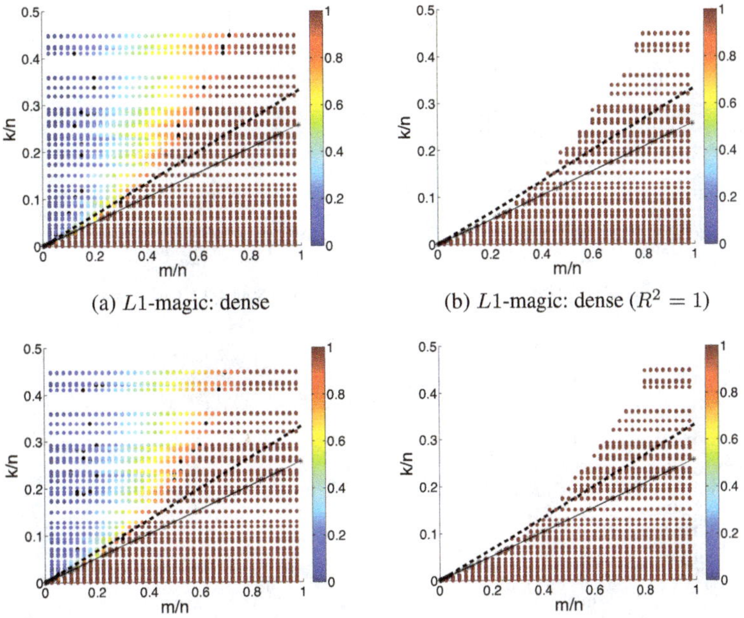

(a) $L1$-magic: dense (b) $L1$-magic: dense ($R^2 = 1$)

(c) $L1$-magic: sparse (d) $L1$-magic: sparse ($R^2 = 1$)

Figure 9. Comparison of reconstruction accuracy of $L1$-magic using (a) a dense matrix and (c) a sparse matrix. (b) and (d) display just the points that satisfy $R^2 = 1$.

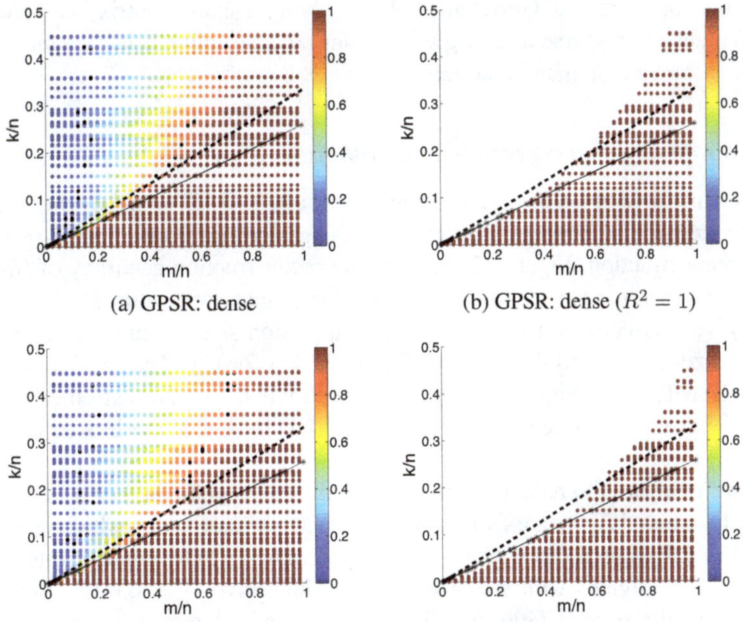

(a) GPSR: dense (b) GPSR: dense ($R^2 = 1$)

(c) GPSR: sparse (d) GPSR: sparse ($R^2 = 1$)

Figure 10. Comparison of reconstruction accuracy of GPSR using (a) a dense matrix and (c) a sparse matrix. (b) and (d) display just the points that satisfy $R^2 = 1$.

different. A similar result is observed for GPSR and YALL1 with dense and sparse matrices as shown in Figure 10 and Figure 11.

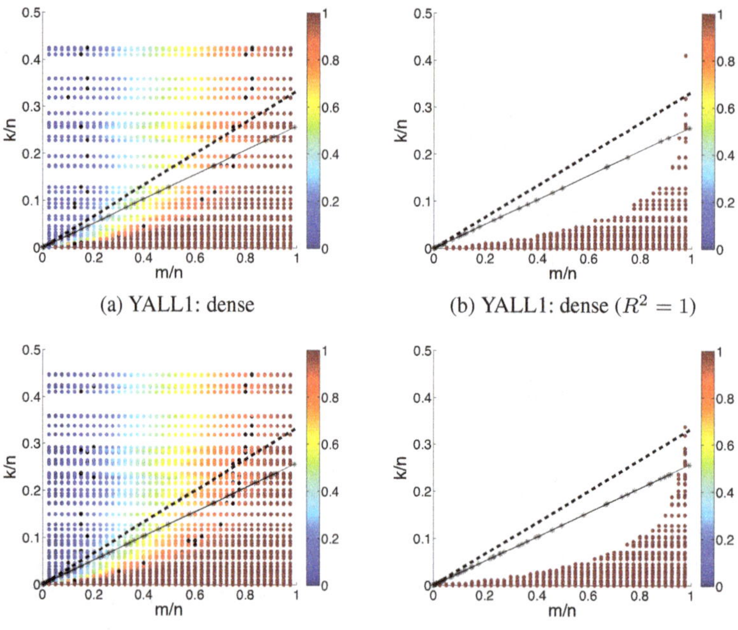

(a) YALL1: dense (b) YALL1: dense ($R^2 = 1$)

(c) YALL1: sparse (d) YALL1: sparse ($R^2 = 1$)

Figure 11. Comparison of reconstruction accuracy of YALL1 using (a) a dense matrix and (c) a sparse matrix. (b) and (d) display just the points that satisfy $R^2 = 1$.

Our results suggest that for L_1 magic, GPSR and YALL1, using a sparse matrix is a better choice compared to using a dense matrix. The use of a sparse matrix gives similar reconstruction accuracy and requires a significantly smaller amount of time in compression and recovery.

6.2. Comparison of CS with other sparse recovery algorithms

We next compare the performance of CS using L_1 magic, GPSR and YALL1 with the other sparse recovery methods described in Section 4. Recall that, unlike compressed sensing algorithms, the sparse recovery algorithms have no guarantee of exact reconstruction. Figure 12 displays the reconstruction accuracy of SMP, SSMP and Count-Min using a sparse matrix for time step 2000, along with the points with value $R^2 = 1$ where we have perfect reconstruction. SSMP gives perfect reconstruction with compression size m almost aligned with the theoretically suggested compression size m^* (the straight line in Figure 12(b)) for CS. The results also show that SMP does not perform as well as SSMP, and Count-Min only works well when data are extremely sparse. All three sparse recovery methods and the group sparse recovery method (YALL1) have fewer points with $R^2 = 1$ when compared with L_1 magic and GPSR.

Next, in Figure 13, we compare the recovery time for the different methods. For a fair comparison, we use the same subdomain as in the case of L_1 magic from time step 2000, the results for which are shown in Figure 9. We also consider a range of m values such that we have full recovery ($R^2 = 1$) for some points for all the methods. The subdomain contains $n = 7351$ signals with sparsity $k = 26$, which gives $\frac{k}{n} = 0.0035$. Since all our sub-domains have roughly the same signal size $n \approx 7400$, we did not vary n as a factor in the recovery time. These results show that L_1 magic with a sparse matrix, and GPSR with a dense matrix, require more time for the recovery step, with the time increasing as the compression size m increases. Recovery times for YALL1 with both dense and sparse matrices are mostly in between the times for L_1 magic and GPSR. The remaining methods in decreasing order of recovery times are SSMP, GPSR sparse, SMP and Count-Min; all methods are almost indifferent to the compression size m.

Based on the performance of all methods using compression time, recovery time and accuracy, we make the following observations: GPSR is a good method if a shorter recovery time is desired; Count-Min is the fastest

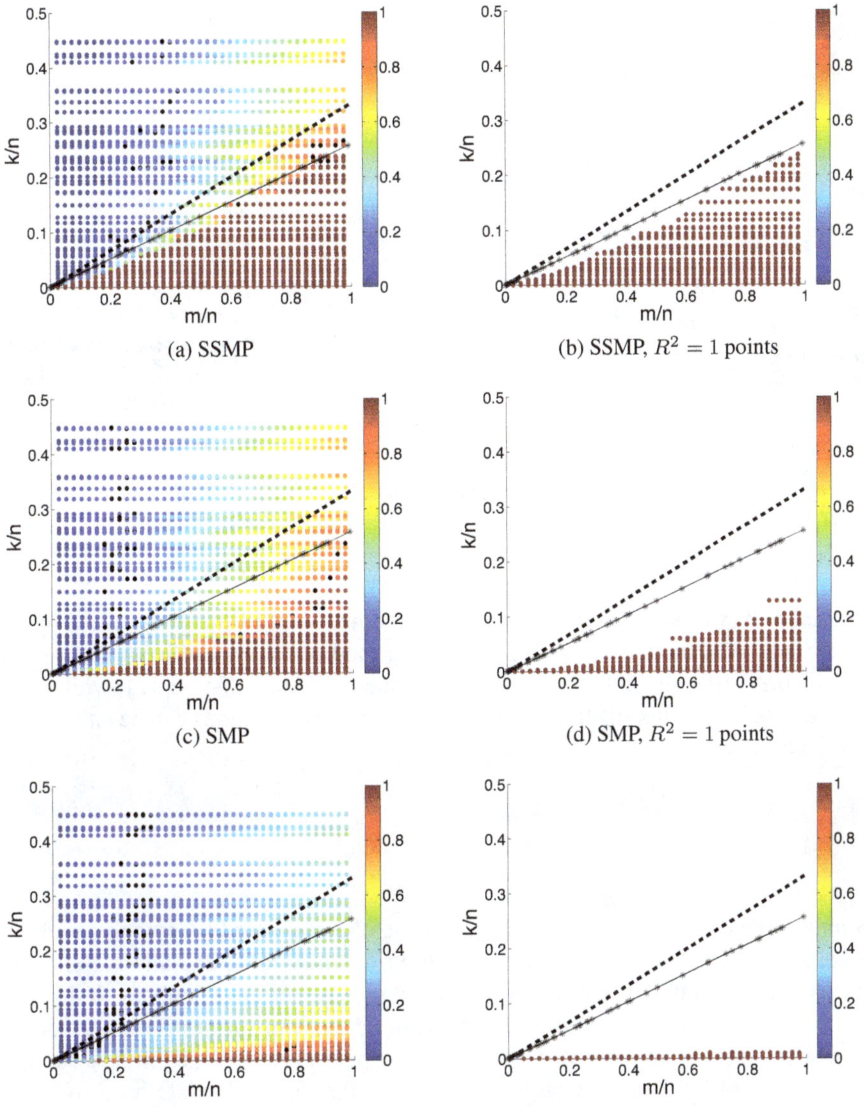

(a) SSMP

(b) SSMP, $R^2 = 1$ points

(c) SMP

(d) SMP, $R^2 = 1$ points

(e) Count-Min

(f) Count-Min, $R^2 = 1$ points

Figure 12. Performance comparison of SSMP, SMP and Count-Min using sparse matrix for time step 2000, showing just the points with $R^2 = 1$ in the right column.

method but only works well when data are extremely sparse; and SSMP is particularly better than SMP in reconstruction accuracy but takes longer time in recovery. L_1 magic gives the best performance in accuracy but takes a much longer time than the other methods. It also takes longer to compress and recover the signal using a dense matrix than using a sparse matrix for L_1 magic, GPSR and YALL1. We suggest using L_1 magic if the goal is to obtain perfect reconstruction as it has more points with $R^2 = 1$ that meet the criterion of using less storage than the non-zero values and their locations.

Finally, we compare how well different methods perform in preserving the small structures of interest in time step 3000 (see Figure 7). In Figure 14, we show the region around the structures after thresholding and the difference between this and the reconstruction by each of the methods. The reconstructed data by each method appears visually similar to the thresholded data, so we present the difference plot to clearly show the quality of reconstruction for

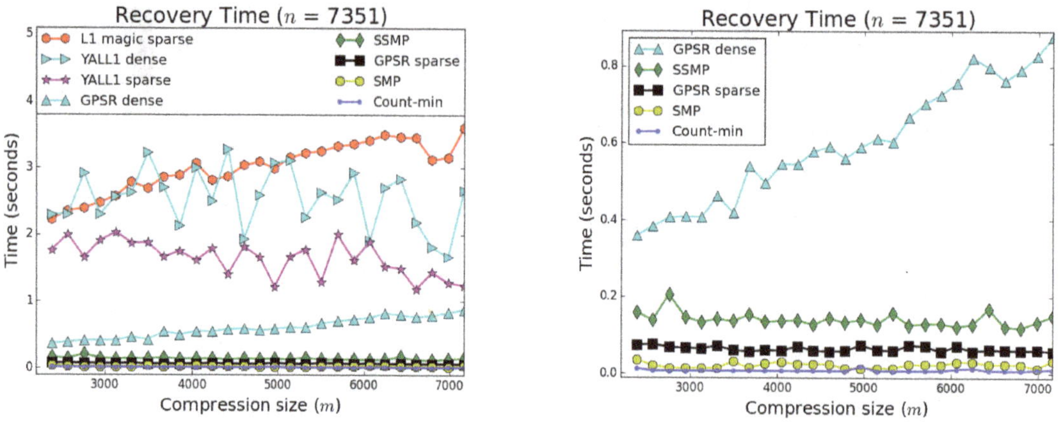

(a) Sparse recovery (b) A zoomed-in view of (a)

Figure 13. Comparisons of recovery times for the different methods. The values of m are chosen such that all six methods give full recovery ($R^2 = 1$). Results for L_1 magic with a dense matrix are shown in Figure 8.

each method. For each plot, we list the minimum and maximum values in the thresholded data or the difference data sets. We observe that even though the worst performing method appears to have a large difference from the thresholded data in Figure 14(h), this difference is in the range of [-7.6e-12, 5.1e-12], while the thresholded data are in the range [-4.6e-6, 4.1e-6]. This suggests that the sparse recovery methods can be an alternative when perfect reconstruction is not required.

7. Conclusions

In this paper, we considered practical issues in the application of compressed sensing to simulation data. We were motivated by the need to reduce the size of the data being written out from simulations run on massively-parallel processors, while enabling accurate reconstruction of the data for quantitative analysis. Using data from a plasma physics simulation, we showed how we could scale and threshold the data that are distributed across processors and determine the amount of compression that would enable near perfect reconstruction. We found that a successful application of CS is bounded by the percentage of sparsity in the data - the data have to be sparse enough for compression using CS, but not so sparse that it is more cost effective to just write out the locations and values of the non-zero data points.

Our experiments, conducted using the R^2 metric, show that scaling the data is very helpful and thresholding helps both with compression and the coherent structure analysis that is performed on the data. After comparing the performance of CS and sparse recovery algorithms, we suggest using L_1 magic with a sparse measurement matrix for CS if the goal is to obtain near perfect reconstruction, even though the method takes longer time for the recovery task. Our future work will focus on determining to what extent we can relax the perfect reconstruction requirement and still obtain results close to what we would get by analyzing the original data.

Acknowledgment

LLNL-JRNL-681293. We thank Zhihong Lin, UC Irvine, for providing access to the data that was generated as part of the GSEP SciDAC project. This work was done as part of the Exa-DM project, funded by Dr. Lucy Nowell, program manager, ASCR, US Department of Energy. This work performed under the auspices of the U.S. Department of Energy by Lawrence Livermore National Laboratory under Contract DE-AC52-07NA27344.

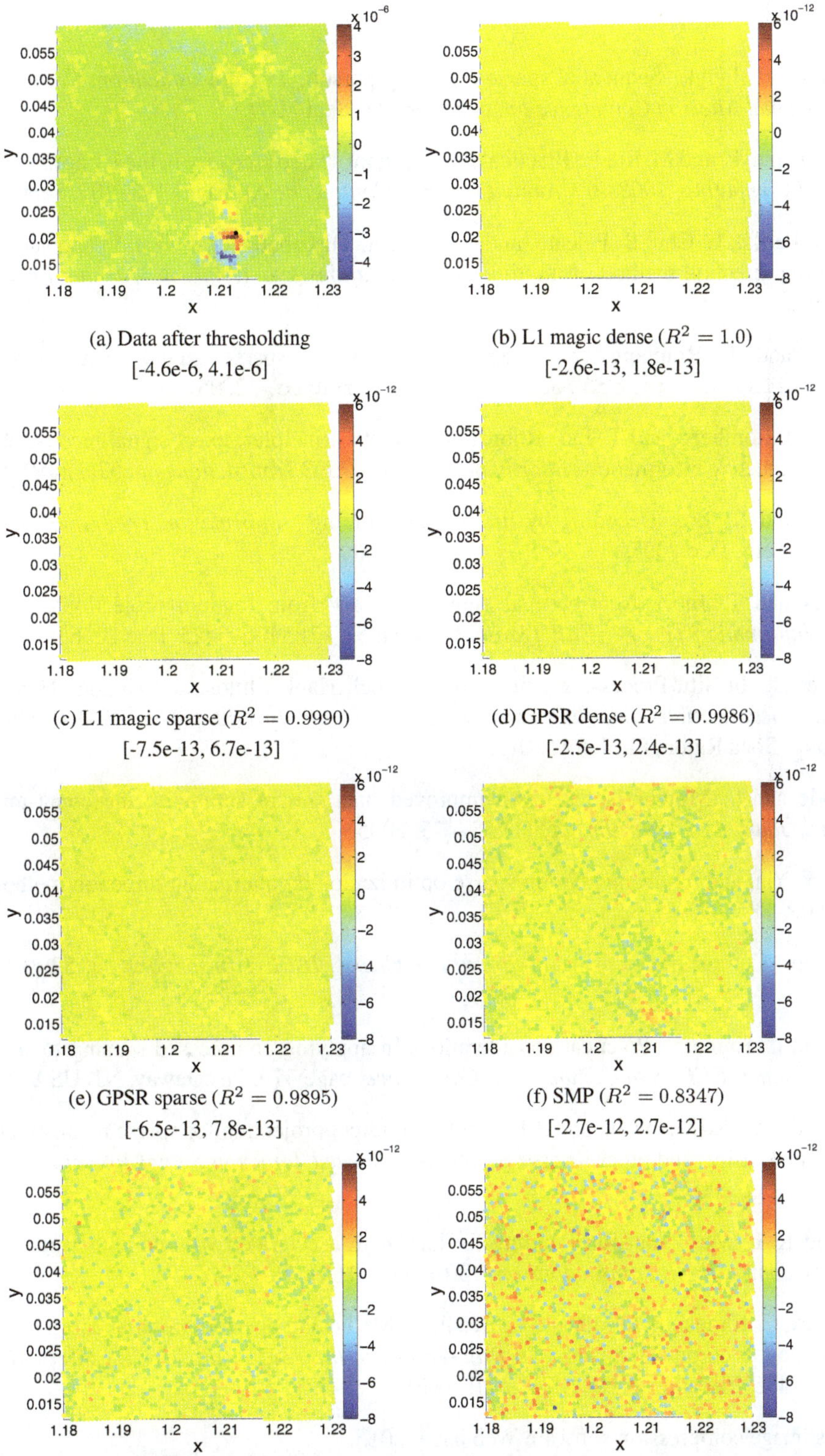

(a) Data after thresholding
[-4.6e-6, 4.1e-6]

(b) L1 magic dense ($R^2 = 1.0$)
[-2.6e-13, 1.8e-13]

(c) L1 magic sparse ($R^2 = 0.9990$)
[-7.5e-13, 6.7e-13]

(d) GPSR dense ($R^2 = 0.9986$)
[-2.5e-13, 2.4e-13]

(e) GPSR sparse ($R^2 = 0.9895$)
[-6.5e-13, 7.8e-13]

(f) SMP ($R^2 = 0.8347$)
[-2.7e-12, 2.7e-12]

(g) SSMP ($R^2 = 0.8260$)
[-2.7e-12, 2.7e-12]

(h) Count-min ($R^2 = 0.5898$)
[-7.6e-12, 5.1e-12]

Figure 14. A slice from sub-domain data at time step 3000 before and after applying compression methods. (a) displays the thresholded data. (b)-(h) show the difference between the thresholded data and the reconstructed data for each method. Numbers in $[\min, \max]$ are the minimum and maximum values in each plot.

REFERENCES

[1] R. Berinde and P. Indyk. Sequential sparse matching pursuit. In *Communication, Control, and Computing, 2009 47th Annual Allerton Conference on*, pages 36–43, Sept 2009.

[2] R. Berinde, P. Indyk, and M. Ruzic. Practical near-optimal sparse recovery in the L1 norm. In *Communication, Control, and Computing, 2008 46th Annual Allerton Conference on*, pages 198–205, Sept 2008.

[3] S. Boyd, N. Parikh, E. Chu, B. Peleato, and J. Eckstein. Distributed optimization and statistical learning via the alternating direction method of multipliers. *Foundations and Trends in Machine Learning*, 3(1):1–122, January 2010.

[4] E. Candes and J. Romberg. l-magic: Recovery of sparse signals via convex programming. `http://users.ece.gatech.edu/~justin/l1magic/`, 2005.

[5] E.J. Candes, J. Romberg, and T. Tao. Robust uncertainty principles: exact signal reconstruction from highly incomplete frequency information. *Information Theory, IEEE Transactions on*, 52(2):489–509, Feb 2006.

[6] E.J. Candes and T. Tao. Decoding by linear programming. *Information Theory, IEEE Transactions on*, 51(12):4203–4215, Dec 2005.

[7] E.J. Candes and T. Tao. Near-optimal signal recovery from random projections: Universal encoding strategies? *Information Theory, IEEE Transactions on*, 52(12):5406–5425, Dec 2006.

[8] H. Childs et al. In Situ Processing. In E. Wes Bethel, Hank Childs, and Charles Hansen, editors, *High Performance Visualization—Enabling Extreme-Scale Scientific Insight*, pages 171–198. CRC Press/Francis–Taylor Group, Boca Raton, FL, USA, 2012.

[9] G. Cormode and S. Muthukrishnan. An improved data stream summary: the count-min sketch and its applications. *Journal of Algorithms*, 55(1):58 – 75, 2005.

[10] W. Deng, W. Yin, and Y. Zhang. Group sparse optimization by alternating direction method. volume 8858, pages 88580R–88580R–15, September 2013.

[11] D.L. Donoho. Compressed sensing. *Information Theory, IEEE Transactions on*, 52(4):1289–1306, April 2006.

[12] Y. J. Fan and C. Kamath. Practical considerations in applying compressed sensing to simulation data. In *Proceedings of the IEEE Data Compression Conference*, page 479, Piscataway, NJ, USA, 2015.

[13] M.A.T. Figueiredo, R.D. Nowak, and S.J. Wright. Gradient projection for sparse reconstruction: Application to compressed sensing and other inverse problems. *Selected Topics in Signal Processing, IEEE Journal of*, 1(4):586–597, Dec 2007.

[14] N. Fout and K.-L. Ma. Transform coding for hardware-accelerated volume rendering. *Visualization and Computer Graphics, IEEE Transactions on*, 13(6):1600–1607, Nov 2007.

[15] S. B. Gokturk, C. Tomasi, B. Girod, and C. Beaulieu. Medical image compression based on region of interest, with application to colon CT images. In *Proceedings of the 23rd Annual EMBS International Conference*, pages 2453–2456, Piscataway, NJ, USA, 2001. IEEE.

[16] Hcompress image compression software Web page, 2015. `http://www.stsci.edu/software/hcompress.html`.

[17] P. Indyk, N. Koudas, and S. Muthukrishnan. Identifying representative trends in massive time series data sets using sketches. In *Proceedings, Conference on Very Large Data Bases*, pages 363–372, 2000.

[18] P. Indyk and M. Ruzic. Near-optimal sparse recovery in the L1 norm. In *Proceedings, 49th Symposium on Foundations of Computer Science*, pages 199–207, October 2008.

[19] M. Isenburg, P. Lindstrom, and J. Snoeyink. Lossless compression of predicted floating-point geometry. *Comput. Aided Des.*, 37(8):869–877, July 2005.

[20] J. Iverson, C. Kamath, and G. Karypis. Fast and effective lossy compression algorithms for scientific datasets. In *Proceedings of the 18th International Conference on Parallel Processing*, Euro-Par'12, pages 843–856, Berlin, Heidelberg, 2012.

[21] C. Kamath, J. Iverson, R. Kirk, and G. Karypis. Detection of coherent structures in extreme-scale simulations. In *Proceedings of the Exascale Research Conference*, April 2012.

[22] D. Laney, S. Langer, C. Weber, P. Lindstrom, and A. Wegener. Assessing the effects of data compression in simulations using physically motivated metrics. In *Proceedings of the International Conference on High Performance Computing, Networking, Storage and Analysis*, SC '13, pages 76:1–76:12, New York, NY, USA, 2013. ACM.

[23] Z. Lin, T. S. Hahm, W. W. Lee, W. M. Tang, and R. B. White. Turbulent transport reduction by zonal flows: Massively parallel simulations. *Science*, (1835):281, 1998.

[24] P. Lindstrom. Fixed-rate compressed floating-point arrays. *Visualization and Computer Graphics, IEEE Transactions on*, 20(12):2674–2683, Dec 2014.

[25] P. Lindstrom and M. Isenburg. Fast and efficient compression of floating-point data. *Visualization and Computer Graphics, IEEE Transactions on*, 12(5):1245–1250, Sep-Oct 2006.

[26] R. Saab, R. Wang, and O. Yilmaz. Near-optimal compression for compressed sensing. In *Data Compression Conference (DCC), 2015*, pages 113–122, April 2015.

[27] M. Salloum, N. Fabian, D. M. Hensinger, and J. A. Templeton. Compressed sensing and reconstruction of unstructured mesh datasets. `arXiv:1508.06314`, August 2015.

[28] K. Sayood. *Introduction to Data Compression (3rd Ed.)*. Morgan Kaufmann Publishers Inc., San Francisco, CA, USA, 2005.

[29] R. Soundararajan and S. Vishwanath. Quantization using compressive sensing. In *Information Theory and Applications Workshop (ITA), 2011*, pages 1–6, Feb 2011.

[30] J.A. Tropp and A.C. Gilbert. Signal recovery from random measurements via orthogonal matching pursuit. *Information Theory, IEEE Transactions on*, 53(12):4655–4666, Dec 2007.

[31] J. Yang and Y. Zhang. Alternating direction algorithms for l_1-problems in compressive sensing. *SIAM Journal on Scientific Computing*, 33(1):250–278, 2011.

5

A criterion for testing hypothesis about impulse response function

Yu. V. Kozachenko [1], I.V. Rozora [2,*]

[1]*Department of Probability Theory, Statistics and Actuarial Mathematics, Faculty of Mathematics and Mechanics, Taras Shevchenko National University of Kyiv, Ukraine, yvk@univ.kiev.ua.*

[2]*Department of Applied Statistics, Faculty of Cybernetics, Taras Shevchenko National University of Kyiv, Ukraine, irozora@bigmir.net.*

Abstract In this paper a time-invariant continuous linear system with a real-valued impulse response function is considered. A new method for the estimator construction of the impulse response function is proposed. Two criteria on the shape of the impulse response function are given.

Keywords Impulse response function, cross-correlogram, Hermite polynomials, large deviation probability, rate of convergence, criterion for testing hypothesis.

1. Introduction

The problem of estimation of a stochastic linear system has been a matter of active research for the last five decades. One of the simplest models considers a black box?with some input and giving a certain output. The input may be single or multiple and there is the same choice for the output. This generates a great amount of models that can be considered. The sphere of applications of these models is vary extensive, ranging from signal processing and automatic control to econometrics (errors-in-variables models). For more details, see [15], [22] and [23].

We are interested in the estimation of the so-called impulse function from observations of responses of a SISO (single-input single-output) system to certain input signals. This problem can be considered both for linear and non-linear systems. To solve this problem, different statistical approaches were used as well as various deterministic methods that are based on a perturbation of the system by stationary stochastic processes and the further analysis of some characteristics of both input and output processes. Let us mention two monographs on this problem by Bendat and Piersol [3] and Schetzen [22]. Akaike [1] studied a MISO (multiple-input single-output) linear system and obtained estimates of the Fourier transform of the response function in each component. He considered later a scenario involving non-Gaussian processes [2].

Some methods for estimation of unknown impulse response function of linear system and the study of properties of corresponding estimators were considered in the works of Buldygin and his followers. These methods are based on constructing a sample cross-correlogram between the input stochastic process and the response of the system.

Conditions for asymptotic unbiasedness, consistency, asymptotic normality of the integral-type zero-mean cross-correlogram estimators for the response function in the space of continuous functions were investigated in the papers [9], [10], [4], [5], [6].

*Correspondence to: 60 Volodymyrska str., Department of Applied Statistics, Faculty of Cybernetics, Taras Shevchenko National University of Kyiv 01601,Ukraine (irozora@bigmir.net).

The discrete-time sample inputoutput cross-correlogram as estimator of the response function was considered by Buldygin, Utzet, Kurochka and Zaiats [8], [11]. Both asymptotic normality of finite-dimensional distributions of the estimates and their asymptotic normality in spaces of continuous functions were studied. An inequality of the distribution for supremum of estimation error in the space of continuous functions in the case of integral-type cross-correlogram estimator was obtained in [19].

In this paper a time-invariant continuous linear system with a real-valued impulse response function is considered. A new method for the construction of estimator of the impulse response function is proposed.

The paper consists of 7 sections.

In the second Section we describe the main definitions and general properties of the estimator. The input signal process is supposed to be a zero mean Gaussian stochastic process which is represented as a treamed sum with respect to orthonormal basis in $L_2(\mathbf{R})$.

In section 3 properties of the Hermite polynomials are described. Since the family of the Hermite functions forms an orthonormal basis in $L_2(\mathbf{R})$, then the input process of the system can be represented as a series with respect to the Hermite polynomials. Estimates of mathematical expectation, variance and variance of the increments for the estimator of impulse function are found.

Section 4 deals with square Gaussian random variables and processes. Inequalities for the $C(T)$ norm as well as for $L_p(T)$ norm of a square Gaussian stochastic process are shown.

In the fifth section the convergence rate for the estimator of unknown impulse response function in the space of continuous functions and in the space $L_p([0, A])$ is investigated.

In the sixth section two criteria are developed on the shape of the impulse response function.

2. The estimator of an impulse response function and its properties

Consider a time-invariant continuous linear system with a real-valued square integrable impulse response function $L(\tau)$, $\tau \in \mathbf{R}$. This means that the response of the system to an input signal $X(t)$, $t \in \mathbf{R}$, has the following form:

$$Y(t) = \int_0^{+\infty} L(\tau)X(t-\tau)d\tau \tag{1}$$

and $L \in L_2(\mathbf{R})$. In practice the system is often supposed to be a causal linear system, that is, $L(\tau) = 0$ as $\tau < 0$. Hereinafter we will consider only such feasible system. Then the system (1) can be written as

$$Y(t) = \int_{-\infty}^{+\infty} L(\tau)X(t-\tau)d\tau.$$

One of the problems arising in the theory of linear systems is to estimate the function L from observations of responses of the system to certain input signals.

Consider a real-valued Gaussian zero mean stochastic process $X_N = (X_{N,t}(u),\ u \in \mathbf{R})$, that can be presented as

$$X_N(u) = X_{t,N}(u) = \sum_{k=1}^{N} \xi_k \varphi_k(t-u), \tag{2}$$

where a fixed value $t > 0$, the system of functions $\{\varphi_k(t),\ k = \overline{1,\infty}\}$ is an orthonormal basis (ONB) in $L_2(\mathbf{R})$ and random variables ξ_k, $k \geq 1$, are independent with $\mathbf{E}\xi_k = 0$, $\mathbf{E}\xi_k\xi_l = \delta_{kl}$, where δ_{kl} is a Kronecker symbol.

Let us denote

$$a_k = \int_0^{+\infty} L(\tau)\varphi_k(\tau)d\tau. \tag{3}$$

If the system (1) is perturbed by the stochastic process X_N, then for the output process we obtain

$$
\begin{aligned}
Y_N(t) &= \int_0^{+\infty} L(\tau)X_N(t-\tau)d\tau \\
&= \sum_{k=1}^{N} \xi_k \int_0^{+\infty} L(\tau)\varphi_k(\tau)d\tau = \sum_{k=1}^{N} \xi_k a_k.
\end{aligned} \tag{4}
$$

Consider the sequence of independent copies $\{X_{N,i}(u),\ i = 1, ..., n\}$ of the Gaussian process (2), that perturb the system (1). That is,

$$X_{N,i}(u) = \sum_{k=1}^{N} \xi_{k,i}\varphi_k(t-u), \tag{5}$$

where $\xi_{k,i}$ are independent normal distributed random variables, $\mathbf{E}\xi_{k,i} = 0$ and $E\xi_{k,i}\xi_{l,j} = \delta_{kl}\delta_{ij}, k, l = \overline{1, N},\ i, j = \overline{1, n}$.

By $\{Y_{N,i}(t),\ i = 1, ..., n\}$ denote the reactions of the system on input signals $\{X_{N,i}(u)\}$.

An estimator for impulse function L at the point τ, $\tau > 0$, is defined by

$$\hat{L}_{N,n}(\tau) = \frac{1}{n}\sum_{i=1}^{n} Y_{N,i}(\tau)X_{N,i}(t-\tau),\ \tau > 0. \tag{6}$$

Since $L \in L_2(\mathbf{R})$, then the following remarks hold true.

Remark 2.1. The integral in (1) is considered as the mean-square Riemann integral.
The integral in (1) exists if and only if there exists the Riemann integral (see [14], p. 278)

$$\int_0^{\infty}\int_0^{\infty} L(\tau)K_{X_N}(s, \tau)L(s)dsd\tau. \tag{7}$$

Since the covariance function of the process X_N is

$$K_{X_N}(s, \tau) = \sum_{k=1}^{N} \varphi_k(t-s)\varphi_k(t-\tau)$$

and $\{\varphi_k(t),\ k \geq 1\}$ is an orthonormal basis in $L_2(\mathbf{R})$ and $L \in L_2(\mathbf{R})$, then the integral (7) exists. Therefore, there exists also the integral in (1).

Remark 2.2. The process $X_{N,i}(t-\tau)$ in (6) depends only on τ and doesn't depend on t. It follows from the definition of the process in (5).

Lemma 2.1
The following relations hold true:

$$\mathbf{E}\hat{L}_{N,n}(\tau) = \sum_{k=1}^{N} \varphi_k(\tau)a_k, \quad \tau > 0, \tag{8}$$

and

$$L(\tau) - \mathbf{E}\hat{L}_{N,n}(\tau) = \sum_{k=N+1}^{\infty} \varphi_k(\tau)a_k, \quad \tau > 0. \tag{9}$$

Proof
The joint covariance function the processes $X_{N,i}$ and $Y_{N,i}$ equals

$$\begin{aligned} \mathbf{E}\left(Y_{N,i}(\tau)X_{N,i}(t-\tau)\right) &= \mathbf{E}\sum_{k=1}^{N}\sum_{l=1}^{N}\varphi_k(\tau)a_l\xi_{k,i}\xi_{l,i} \\ &= \sum_{k=1}^{N}\varphi_k(\tau)a_k. \end{aligned} \tag{10}$$

From the equality above it follows that

$$\mathbf{E}\hat{L}_{N,n}(\tau) = \frac{1}{n}\sum_{i=1}^{n}\mathbf{E}\left(Y_{N,i}(\tau)X_{N,i}(t-\tau)\right) = \sum_{k=1}^{N}\varphi_k(\tau)a_k.$$

Therefore, relation (8) is proved.

Since $L \in L_2(\mathbf{R})$, then the function L can be expanded into the series by orthonormal basis $\{\varphi_k(t),\ k \geq 1\}$. We obtain

$$L(\tau) = \sum_{k=1}^{\infty}\varphi_k(\tau)\cdot\int_0^{+\infty}L(\tau)\varphi_k(\tau)d\tau = \sum_{k=1}^{\infty}\varphi_k(\tau)\cdot a_k, \tag{11}$$

where a_k is from (3).

Remark 2.3. The series (11) converges in the mean square sense.

Equalities (11) and (8) imply that

$$L(\tau) - \mathbf{E}\hat{L}_{N,n}(\tau) = \sum_{k=N+1}^{\infty}\varphi_k(\tau)\cdot a_k, \quad \tau > 0.$$

□

We prove now the following auxiliary Lemma.

Lemma 2.2
The joint moments of $\hat{L}_{T,N}$ is equal to

$$\begin{aligned} \mathbf{E}\hat{L}_{N,n}(\tau)\hat{L}_{N,n}(\theta) &= \frac{1}{n}\sum_{k=1}^{N}\sum_{l=1}^{N}\left(na_ka_l\varphi_k(\tau)\varphi_l(\theta) + a_k^2\varphi_l(\tau)\varphi_l(\theta)\right. \\ &+ \left. a_ka_l\varphi_l(\tau)\varphi_k(\theta)\right), \end{aligned} \tag{12}$$

where the coefficients a_k are defined in (3).

Proof
By definition of estimator $\hat{L}_{T,N}$ (6) and the values of the processes $X_{N,i}$ from (5) and $Y_{N,i}$ from (4) we have

$$\begin{aligned} \mathbf{E}\hat{L}_{N,n}(\tau)\hat{L}_{N,n}(\theta) &= \frac{1}{n^2}\sum_{i=1}^{n}\sum_{j=1}^{n}\mathbf{E}Y_{N,i}(\tau)X_{N,i}(t-\tau)Y_{N,j}(\theta)X_{N,j}(t-\theta) \\ &= \frac{1}{n^2}\sum_{i=1}^{n}\sum_{j=1}^{n}\sum_{k,l,u,v=1}^{N}a_ka_u\varphi_l(\tau)\varphi_v(\theta)\mathbf{E}\xi_{k,i}\xi_{l,i}\xi_{u,j}\xi_{v,j} \end{aligned} \tag{13}$$

To prove the assertion it's enough to find $\mathbf{E}\xi_{k,i}\xi_{l,i}\xi_{u,j}\xi_{v,j}$. Using the Isserlis?formula for the centered Gaussian random variables [7] we obtain:

$$\begin{aligned}\mathbf{E}\xi_{k,i}\xi_{l,i}\xi_{u,j}\xi_{v,j} &= \mathbf{E}\xi_{k,i}\xi_{l,i}\mathbf{E}\xi_{u,j}\xi_{v,j} \\ &+ \mathbf{E}\xi_{k,i}\xi_{u,j}\mathbf{E}\xi_{l,i}\xi_{v,j} \\ &+ \mathbf{E}\xi_{k,i}\xi_{v,j}\mathbf{E}\xi_{l,i}\xi_{u,j}.\end{aligned} \tag{14}$$

By the definition of the processes $X_{N,i}$ and $Y_{N,i}$, we have

$$\begin{aligned}\mathbf{E}\xi_{k,i}\xi_{l,i} &= \delta_{kl}, & \mathbf{E}\xi_{u,j}\xi_{v,j} &= \delta_{uv}, \\ \mathbf{E}\xi_{k,i}\xi_{u,j} &= \delta_{ku}\delta_{ij}, & \mathbf{E}\xi_{l,i}\xi_{v,j} &= \delta_{lv}\delta_{ij}, \\ \mathbf{E}\xi_{k,i}\xi_{v,j} &= \delta_{kv}\delta_{ij}, & \mathbf{E}\xi_{l,i}\xi_{u,j} &= \delta_{lu}\delta_{ij},\end{aligned}$$

(12) will be completely proved if the equalities above and (14) will be substituted in (13). □

The next Corollary follows from Lemma 2.2.

Corollary 2.1
The variance of the estimator $\hat{L}_{N,n}$ is equal to

$$\begin{aligned}Var\hat{L}_{N,n}(\tau) &= \mathbf{E}\hat{L}^2_{N,n}(\tau) - (\mathbf{E}\hat{L}_{N,n}(\tau))^2 \\ &= \frac{1}{n}\sum_{l=1}^{N}\sum_{k=1}^{N}\left(a_k^2\varphi_l^2(\tau) + a_k a_l \varphi_k(\tau)\varphi_l(\tau)\right).\end{aligned} \tag{15}$$

$$Var(\hat{L}_{N,n}(\tau) - \hat{L}_{N,n}(\theta)) = Var\hat{L}_{N,n}(\tau) + Var\hat{L}_{N,n}(\theta) - 2cov(\hat{L}_{N,n}(\tau), \hat{L}_{N,n}(\theta)) \tag{16}$$

Lemma 2.3
Suppose that the Lipschitz condition with a rate $\alpha \in (0, 1]$ for the functions $\varphi_k(t)$ holds on a segment $[0, A]$. It means that there exist constants $c_{k,\varphi}$ such that

$$|\varphi_k(t) - \varphi_k(s)| \leq c_{k,\varphi}|t - s|^\alpha \quad \text{for all} \quad t, s \in [0, A] \quad \text{and} \quad k \geq 1. \tag{17}$$

In this case

$$Var(\hat{L}_{N,n}(\tau) - \hat{L}_{N,n}(\theta)) \leq C(N, n)|\tau - \theta|^\alpha, \quad \alpha \in (0, 1],\ \tau, \theta \in [0, A],$$

where

$$C(N, n) = \frac{1}{n}\sum_{k=1}^{N}\sum_{l=1}^{N}\left(a_k^2(|\varphi_l(\tau)| + |\varphi_l(\theta)|)c_{l,\varphi} + a_k a_l(|\varphi_l(\tau)|c_{k,\varphi} + |\varphi_k(\theta)|c_{l,\varphi})\right) \tag{18}$$

Proof
From relations (8) and (12) it follows that

$$\begin{aligned}cov(\hat{L}_{N,n}(\tau), \hat{L}_{N,n}(\theta)) &= \mathbf{E}\hat{L}_{N,n}(\tau)\hat{L}_{N,n}(\theta) - \mathbf{E}\hat{L}_{N,n}(\tau)\mathbf{E}\hat{L}_{N,n}(\theta) \\ &= \frac{1}{n}\sum_{k=1}^{N}\sum_{l=1}^{N}\left(a_k^2\varphi_l(\tau)\varphi_l(\theta) + a_k a_l \varphi_l(\tau)\varphi_k(\theta)\right),\end{aligned} \tag{19}$$

Therefore, by (15), (16) and (19), making elementary reduction we obtain

$$\begin{aligned}Var(\hat{L}_{N,T}(\tau) - \hat{L}_{N,T}(\theta)) &= \frac{1}{n}\sum_{k,l=1}^{N}\left(a_k^2\left(\varphi_l(\tau)(\varphi_l(\tau) - \varphi_l(\theta)) + \varphi_l(\theta)(\varphi_l(\theta) - \varphi_l(\tau))\right)\right. \\ &+ \left. a_k a_l\left(\varphi_l(\tau)(\varphi_k(\tau) - \varphi_k(\theta)) + \varphi_k(\theta)(\varphi_l(\theta) - \varphi_l(\tau))\right)\right)\end{aligned} \tag{20}$$

The Lemma will be fully proved if in (20) the Lipschitz condition (17) for the function $\varphi_k(\tau)$, $\tau \in [0, A]$, will be used. □

3. The use of Hermite polynomials for the presentation of the estimator

Consider the Hermite polynomials of the degree $n \geq 1$:

$$H_n(x) = (-1)^n e^{\frac{x2}{2}} \frac{d^n}{dx^n}(e^{-\frac{x^2}{2}}).$$

It is shown in the book [13] that

$$\frac{d^k H_n(x)}{dx^k} = n(n-1)(n-2)\cdots(n-k+1)H_{n-k}(x).$$

In particular case $k = 1$ we have

$$\frac{dH_n(x)}{dx} = nH_{n-1}(x). \tag{21}$$

It is known that the system of the Hermite functions

$$g_n(x) = \frac{e^{-\frac{x^2}{4}} H_n(x)}{\sqrt{n!}(2\pi)^{1/4}} = \frac{(-1)^n e^{\frac{x^2}{4}}}{\sqrt{n!}(2\pi)^{1/4}} \frac{d^n}{dx^n}(e^{-\frac{x^2}{2}}), \quad n \geq 0, \tag{22}$$

is an orthonormal basis (ONB) in $L_2(\mathbf{R})$ (see, for example, [13]).

Suppose now that the input signal processes of the system (1) are zero mean Gaussian stochastic processes that are formed by the Hermite ONB (22). This means that the processes $X_{N,i}(u)$, $i = \overline{1,n}$, are represented in such way:

$$X_{N,i}(u) = \sum_{k=1}^{N} \xi_{k,i} g_k(t-u) \quad \text{for}\, t, u \in \mathbf{R}. \tag{23}$$

It follows from (1) that the output processes $Y_{N,i}(t)$, $t \in \mathbf{R}$, equal

$$Y_{N,i}(t) = \int_0^{+\infty} L(\tau) X_{N,i}(t-\tau) d\tau, \quad \text{for} \quad i = \overline{1,n}. \tag{24}$$

The estimator for impulse function L in the point τ, $\tau > 0$, is defined by (6).

Consider the following conditions:

Condition A. There exists an integral

$$\int_{-\infty}^{+\infty} Z_L(z) dz < \infty,$$

where the function $Z_L(z)$ is equal to

$$Z_L(z) = \frac{d^3 L(z)}{dz^3} - \frac{3z}{2}\frac{d^2 L(z)}{dz^2} + \frac{3z^2 - 6}{4}\frac{dL(z)}{dz} + \frac{6z - z^3}{8} L(z). \tag{25}$$

Denote

$$I_L = \left| \int_{-\infty}^{+\infty} Z_L(z) dz \right|.$$

Condition B. The function $L(z)$ increases on z no faster than $e^{z^2/4}$. It means that there exists a constant $c \in (0, \frac{1}{4})$ that

$$L(z) e^{-cz^2} \to 0, \ z \to +\infty.$$

The following Lemma gives an estimate for $|L(\tau) - \mathbf{E}\hat{L}_{N,n}(\tau)|$ and for the variance $Var(\hat{L}_{T,N}(\tau))$.

Lemma 3.1
Assume that the conditions **A** and **B** are satisfied. Then

$$|L(\tau) - \mathbf{E}\hat{L}_{N,n}(\tau)| \leq \frac{2K^2}{3\sqrt{N+1}} I_L, \tag{26}$$

$$Var\hat{Z}_{N,n}(\tau) < \frac{K^4 I_L^2}{n}\left(N\frac{N^2-1}{2N^2} + \frac{(2\sqrt{N}-2)^2}{N}\right), \tag{27}$$

where the number $K = 1,086435$.

Proof
From (9) it follows that

$$L(\tau) - \mathbf{E}\hat{L}_{N,n}(\tau) = \sum_{k=N+1}^{\infty} g_k(\tau) a_k \tag{28}$$

Calculate the required values a_k by using equality (21) and partial integration:

$$\begin{aligned}
a_k &= \int_{-\infty}^{+\infty} L(z) g_k(z) dz \\
&= \int_{-\infty}^{+\infty} \frac{L(z) e^{-\frac{z^2}{4}} H_k(z)}{\sqrt{k!}(2\pi)^{1/4}} dz \\
&= \frac{1}{\sqrt{k+1}} \int_{-\infty}^{+\infty} \frac{L(z) e^{-\frac{z^2}{4}}}{\sqrt{(k+1)!}(2\pi)^{1/4}} \frac{dH_{k+1}(z)}{dz} dz \\
&= \frac{1}{\sqrt{k+1}} \left(\frac{L(z) e^{-\frac{z^2}{4}} H_{k+1}(z)}{\sqrt{(k+1)!}(2\pi)^{1/4}} \Big|_{-\infty}^{+\infty} - \int_{-\infty}^{+\infty} \frac{d\left(L(z)e^{-\frac{z^2}{4}}\right)}{dz} \frac{H_{k+1}(z)}{\sqrt{(k+1)!}(2\pi)^{1/4}} dz \right)
\end{aligned}$$

Since $H_k(z)e^{-z^2/4}$ tends to zero as $z \to \pm\infty$, and by condition A the function $L(z)$ increases on z no faster than $e^{z^2/4}$, then

$$a_k = -\frac{1}{\sqrt{k+1}} \int_{-\infty}^{+\infty} \frac{d\left(L(z)e^{-\frac{z^2}{4}}\right)}{dz} \frac{H_{k+1}(z)}{\sqrt{(k+1)!}(2\pi)^{1/4}} dz.$$

Let's apply the integration by parts two times more. We obtain

$$\begin{aligned}
a_k &= -\frac{1}{\sqrt{(k+1)(k+2)}} \int_{-\infty}^{+\infty} \frac{d\left(L(z)e^{-\frac{z^2}{4}}\right)}{dz} \frac{dH_{k+2}(z)}{dz} \frac{1}{\sqrt{(k+2)!}(2\pi)^{1/4}} dz \\
&= -\frac{1}{\sqrt{(k+1)(k+2)}} \frac{d\left(L(z)e^{-\frac{z^2}{4}}\right)}{dz} \frac{H_{k+2}(z)}{\sqrt{(k+2)!}(2\pi)^{1/4}} \Big|_{-\infty}^{+\infty} \\
&+ \frac{1}{\sqrt{(k+1)(k+2)}} \int_{-\infty}^{+\infty} \frac{d^2\left(L(z)e^{-\frac{z^2}{4}}\right)}{dz^2} \frac{H_{k+2}(z)}{\sqrt{(k+2)!}(2\pi)^{1/4}} dz \\
&= \frac{1}{\sqrt{(k+1)(k+2)}} \int_{-\infty}^{+\infty} \frac{d^2\left(L(z)e^{-\frac{z^2}{4}}\right)}{dz^2} \frac{H_{k+2}(z)}{\sqrt{(k+2)!}(2\pi)^{1/4}} dz \\
&= \frac{1}{\sqrt{(k+1)(k+2)(k+3)}} \int_{-\infty}^{+\infty} \frac{d^3\left(L(z)e^{-\frac{z^2}{4}}\right)}{dz^3} \frac{H_{k+3}(z)}{\sqrt{(k+3)!}(2\pi)^{1/4}} dz
\end{aligned}$$

$$= \frac{1}{\sqrt{(k+1)(k+2)(k+3)}} \int_{-\infty}^{+\infty} Z_L(z) g_{k+3}(z) dz, \tag{29}$$

where $g_{k+3}(z)$ is the Hermite function of and the function $Z_L(z)$ is from (25).

By using (28) and (29) we obtain that

$$\begin{aligned} L(\tau) - \mathbf{E}\hat{L}_{N,n}(\tau) &= \sum_{k=N+1}^{\infty} g_k(\tau) a_k \\ &= \sum_{k=N+1}^{\infty} \frac{g_k(\tau)}{\sqrt{(k+1)(k+2)(k+3)}} \int_{-\infty}^{+\infty} Z_L(z) g_{k+3}(z) dz \end{aligned} \tag{30}$$

From the Cramér inequality [13] it follows that for all $z \in \mathbf{R}$ $|g_k(z)| < K$, where $K = 1,086435$. Therefore, the coefficients a_k can be estimated as follows

$$|a_k| \leq \frac{K}{\sqrt{(k+1)(k+2)(k+3)}} \left| \int_{-\infty}^{+\infty} Z_L(z) dz \right|. \tag{31}$$

From relations (30) and (31) it follows that

$$|L(\tau) - \mathbf{E}\hat{L}_{N,n}(\tau)| \leq K^2 I_L \sum_{k=N+1}^{\infty} \frac{1}{\sqrt{(k+1)(k+2)(k+3)}} \tag{32}$$

Evaluate now the sum in (32).

$$\begin{aligned} \sum_{k=N+1}^{\infty} \frac{1}{\sqrt{(k+1)(k+2)(k+3)}} &< \sum_{k=N+1}^{\infty} \frac{1}{\sqrt{(k+1)^3}} \\ &= \sum_{k=N+1}^{\infty} \int_k^{k+1} \frac{1}{\sqrt{(k+1)^3}} dx < \sum_{k=N+1}^{\infty} \int_k^{k+1} \frac{1}{x^{3/2}} dx \\ &= \int_{N+1}^{\infty} \frac{1}{x^{3/2}} dx = \frac{2}{3\sqrt{N+1}}. \end{aligned} \tag{33}$$

Hence, inequalities (32) and (33) imply (26).

Find the estimate for the variance of the process $\hat{Z}_{N,n}(\tau)$. From (15) follows that

$$Var\hat{Z}_{N,n}(\tau) = \frac{1}{n} \sum_{l=1}^{N} \sum_{k=1}^{N} \left(a_k^2 g_l^2(\tau) + a_k a_l g_k(\tau) g_l(\tau) \right).$$

Since $|g_k(\tau)| < K$ and using inequality (31) for a_k, we obtain that

$$\begin{aligned} Var\hat{Z}_{N,n}(\tau) &< \frac{K^4 I_L^2}{n} \sum_{l=1}^{N} \sum_{k=1}^{N} \left(\frac{1}{(k+1)(k+2)(k+3)} \right. \\ &+ \left. \frac{1}{\sqrt{(k+1)(k+2)(k+3)}\sqrt{(l+1)(l+2)(l+3)}} \right) \\ &= \frac{K^4 I_L^2}{n} \left(N \sum_{k=1}^{N} \frac{1}{(k+1)(k+2)(k+3)} + \left(\sum_{k=1}^{N} \frac{1}{\sqrt{(k+1)(k+2)(k+3)}} \right)^2 \right). \end{aligned} \tag{34}$$

Similarly to (33) the sums above can be evaluated in the following way

$$\begin{aligned}\sum_{k=1}^{N}\frac{1}{(k+1)(k+2)(k+3)} &< \sum_{k=1}^{N}\frac{1}{(k+1)^3}\\ &= \sum_{k=1}^{N-1}\int_k^{k+1}\frac{1}{(k+1)^3}dx < \sum_{k=1}^{N-1}\int_k^{k+1}\frac{1}{x^3}dx\\ &= \int_1^N\frac{1}{x^3}dx = \frac{N^2-1}{2N^2};\end{aligned} \tag{35}$$

$$\sum_{k=1}^{N}\frac{1}{\sqrt{(k+1)(k+2)(k+3)}} < \int_1^N\frac{1}{x^{3/2}}dx = \frac{2\sqrt{N}-2}{\sqrt{N}}. \tag{36}$$

If substitute (35), (36) into (34), then we obtain (27). □

The following Lemma gives an estimate for the difference of the Hermite functions $|g_k(x)-g_k(y)|$ as $x, y \in [0, A]$.

Lemma 3.2

Let $g_k(x)$, $k \geq 1$, be the Hermite functions from (22). Assume that the conditions of Lemma 3.1 are satisfied. Then for any $x, y \in [0, A]$, where $A > 0$ is some constant, the inequality

$$|g_k(x)-g_k(y)| \leq c_{k,g}|x-y|^{\alpha}, \quad \alpha \in (0,1], \tag{37}$$

holds true, where the value $c_{k,g}$ is equal to

$$\begin{aligned}c_{k,g} &= \frac{1}{\sqrt{k!}(2\pi)^{1/4}}\left(\frac{2^{\frac{k-\alpha}{2}+1}e^{\frac{A^2}{4}}}{\sqrt{\pi}}\Gamma\left(\frac{k+\alpha+1}{2}\right)+\frac{2^{\frac{k}{2}-\alpha}A^{\alpha}}{\sqrt{\pi}}\Gamma(\frac{k+1}{2})e^{\frac{A^2}{2}}\right)\\ &= \frac{2^{\frac{k-\alpha}{2}-\frac{1}{4}}e^{\frac{A^2}{2}}}{\sqrt{k!}\pi^{3/4}}\left(2\Gamma\left(\frac{k+\alpha+1}{2}\right)+2^{-\frac{\alpha}{2}}A^{\alpha}\Gamma(\frac{k+1}{2})\right),\end{aligned} \tag{38}$$

where $\Gamma(r) = \int_0^{\infty} t^{r-1}e^{-t}dt$ is an Euler gamma function.

Proof

Without limiting the generality assume that $x > y \geq 0$. By the definition of the Hermite function:

$$\begin{aligned}|g_k(x)-g_k(y)| &= \left|\frac{e^{\frac{-x^2}{4}}}{\sqrt{k!}(2\pi)^{1/4}}H_k(x)-\frac{e^{\frac{-y^2}{4}}}{\sqrt{k!}(2\pi)^{1/4}}H_k(y)\right|\\ &= \frac{1}{\sqrt{k!}(2\pi)^{1/4}}\left|e^{\frac{x^2}{4}}\frac{d^k}{dx^k}(e^{-\frac{x^2}{2}}) \pm e^{\frac{x^2}{4}}\frac{d^k}{dy^k}(e^{-\frac{y^2}{2}})-e^{\frac{y^2}{4}}\frac{d^k}{dy^k}(e^{-\frac{y^2}{2}})\right|\\ &\leq \frac{1}{\sqrt{k!}(2\pi)^{1/4}}\left(e^{\frac{x^2}{4}}\left|\frac{d^k}{dx^k}(e^{-\frac{x^2}{2}})-\frac{d^k}{dy^k}(e^{-\frac{y^2}{2}})\right|+|\frac{d^k}{dy^k}(e^{-\frac{y^2}{2}})|\left|e^{\frac{x^2}{4}}-e^{\frac{y^2}{4}}\right|\right)\\ &= \frac{1}{\sqrt{k!}(2\pi)^{1/4}}(I_1+I_2).\end{aligned} \tag{39}$$

To estimate the values I_1 and I_2 let's use the next formula

$$e^{-\frac{x^2}{2}} = \frac{1}{\sqrt{2\pi}}\int_{-\infty}^{+\infty}e^{iux-\frac{u^2}{2}}du, \tag{40}$$

that is, actually, a characteristic function for a standard Gaussian random variable. By (40) we have

$$\frac{d^k e^{-\frac{x^2}{2}}}{dx^k} = \frac{1}{\sqrt{2\pi}} \int_{-\infty}^{+\infty} (iu)^k e^{iux - \frac{u^2}{2}} du. \tag{41}$$

From (41) follows that the the quantity I_1 in the case of $x, y \in [0, A]$ is bounded in a such way:

$$\begin{aligned} I_1 &\le e^{\frac{A^2}{4}} \left| \frac{1}{\sqrt{2\pi}} \int_{-\infty}^{+\infty} (iu)^k e^{iux - \frac{u^2}{2}} du - \frac{1}{\sqrt{2\pi}} \int_{-\infty}^{+\infty} (iu)^k e^{iuy - \frac{u^2}{2}} du \right| \\ &\le \frac{e^{\frac{A^2}{4}}}{\sqrt{2\pi}} \int_{-\infty}^{+\infty} |u|^k |e^{iux} - e^{iuy}| e^{-\frac{u^2}{2}} du. \end{aligned}$$

Since

$$|e^{iux} - e^{iuy}| = 2|\sin(\frac{u}{2}(x-y))|,$$

then

$$I_1 \le \frac{2e^{\frac{A^2}{4}}}{\sqrt{2\pi}} \int_{-\infty}^{+\infty} |u|^k |\sin(\frac{u}{2}(x-y))| e^{-\frac{u^2}{2}} du. \tag{42}$$

For $u \ge 0$ and $v > 0$ the inequality holds true (see, for example, [18])

$$\left|\sin \frac{u}{v}\right| \le \frac{u^\alpha}{v^\alpha}, \quad \alpha \in (0, 1]. \tag{43}$$

If (43) will be used for the value (42), then we obtain that

$$\begin{aligned} I_1 &\le \frac{2e^{\frac{A^2}{4}}}{2^\alpha \sqrt{2\pi}} |x-y|^\alpha \int_{-\infty}^{+\infty} |u|^{k+\alpha} e^{-\frac{u^2}{2}} du \\ &= \frac{2^{\frac{k-\alpha}{2}+1} e^{\frac{A^2}{4}}}{\sqrt{\pi}} \Gamma\left(\frac{k+\alpha+1}{2}\right) |x-y|^\alpha, \quad \alpha \in (0, 1]. \end{aligned} \tag{44}$$

Remind that $\Gamma(r) = \int_0^\infty t^{r-1} e^{-t} dt$ is an Euler gamma function.

From (41) follows that

$$\begin{aligned} \left|\frac{d^k e^{-\frac{y^2}{2}}}{dy^k}\right| &\le \frac{1}{\sqrt{2\pi}} \int_{-\infty}^{+\infty} |u|^k e^{-\frac{u^2}{2}} du \\ &= \frac{2^{\frac{k}{2}}}{\sqrt{\pi}} \Gamma(\frac{k+1}{2}). \end{aligned} \tag{45}$$

Estimate now the value I_2 from (39). By (45) we have

$$\begin{aligned} I_2 &= \left|\frac{d^k e^{-\frac{y^2}{2}}}{dy^k}\right| \left| e^{\frac{x^2}{4}} - e^{\frac{y^2}{4}} \right| \\ &\le \frac{2^{\frac{k}{2}}}{\sqrt{\pi}} \Gamma(\frac{k+1}{2}) e^{\frac{x^2}{2}} |1 - e^{-\frac{x^2-y^2}{4}}|. \end{aligned} \tag{46}$$

Since for $t > 0$

$$1 - e^{-t} \le t^\alpha, \quad \alpha \in (0, 1],$$

then for $x > y \geq 0$, $x, y \in [0, A]$, we obtain

$$\begin{aligned} I_2 &\leq \frac{2^{\frac{k}{2}}}{\sqrt{\pi}}\Gamma(\frac{k+1}{2})e^{\frac{x^2}{2}}\left|\frac{x^2-y^2}{4}\right|^{\alpha} \\ &\leq \frac{2^{\frac{k}{2}-\alpha}A^{\alpha}}{\sqrt{\pi}}\Gamma(\frac{k+1}{2})e^{\frac{A^2}{2}}|x-y|^{\alpha}. \end{aligned} \tag{47}$$

If we substitute (44) and (47) into (39), we will have that (37) completely proved. □

Corollary 3.1
From conditions of Lemma 3.1 it follows that

$$Var(\hat{Z}_{N,n}(\tau) - \hat{Z}_{N,n}(\theta)) \leq C_g(N,n)|\tau-\theta|^{\alpha}, \quad \alpha \in (0,1],\ \tau, \theta \in [0, A], \tag{48}$$

where

$$C_g(N,n) = \frac{K^3 I_L^2}{n}\left(2\frac{N^2-1}{2N^2}\sum_{l=1}^{N} c_{l,g} + \sum_{k=1}^{N}\frac{c_{k,g}}{\sqrt{(k+1)(k+2)(k+3)}}\right). \tag{49}$$

Proof
The assertion (48) follows from Lemmas 2.3 and 3.2. □

4. Square Gaussian random variables and processes

In this section the definition and some properties of square Gaussian random variables and processes are presented. Let (Ω, L, P) be a probability space and let $(\mathbf{T}, \rho)$ be a compact metric space with metric ρ.

Definition 4.1. [7] Let $\Xi = \{\xi_t, t \in \mathbf{T}\}$ be a family of joint Gaussian random variables for which $E\xi_t = 0$ (e.g., ξ_t, $t \in \mathbf{T}$, is a Gaussian stochastic process).

The space $SG_{\Xi}(\Omega)$ is the space of square Gaussian random variables if any element $\eta \in SG_{\Xi}(\Omega)$ can be presented as

$$\eta = \bar{\xi}^{\top} A\bar{\xi} - \mathbf{E}\bar{\xi}^{\top} A\bar{\xi}, \tag{50}$$

where $\bar{\xi}^{\top} = (\xi_1, \xi_2, \ldots, \xi_n)$, $\xi_k \in \Xi$, $k = 1, \ldots, n$, A is a real-valued matrix or the element $\eta \in SG_{\Xi}(\Omega)$ is the square mean limit of the sequence (50)

$$\eta = l.i.m._{n\to\infty}(\bar{\xi}_n^{\top} A\bar{\xi}_n - \mathbf{E}\bar{\xi}_n^{\top} A\bar{\xi}_n).$$

Definition 4.2. [7] A stochastic process $\xi(t) = \{\xi(t), t \in \mathbf{T}\}$ is square Gaussian if for any $t \in \mathbf{T}$ a random variable $\xi(t)$ belongs to the space $SG_{\Xi}(\Omega)$.

There are shown in the book by Buldygin and Kozachenko [7] that

- $SG_{\Xi}(\Omega)$ is a Banach space with respect to the norm $||\zeta|| = \sqrt{\mathbf{E}\zeta^2}$;
- $SG_{\Xi}(\Omega)$ is a subspace of the Orlicz space $L_U(\Omega)$ generated by the function

$$U(x) = \exp|x| - 1;$$

- the norm $||\zeta||_{L_U(\Omega)}$ on $SG_{\Xi}(\Omega)$ is equivalent to the norm $||\zeta||$.

Example 4.1. Consider a family of Gaussian centered stochastic processes $\xi_1(t), \xi_2(t), \ldots, \xi_n(t)$, $t \in \mathbf{T}$. Let the matrix $A(t)$ be symmetric. Then

$$X(t) = \bar{\xi}^{\top}(t)A(t)\bar{\xi}(t) - \mathbf{E}\bar{\xi}^{\top}(t)A(t)\bar{\xi}(t),$$

where $\bar{\xi}^{\top}(t) = (\xi_1(t), \xi_2(t), \ldots, \xi_n(t))$, is a square Gaussian stochastic process.

For more information about properties of square Gaussian random processes see [16], [17], [18], [7], [20].

4.1. An estimate for the $C(T)$ norm of a square Gaussian stochastic process

Denote by $N(u)$ the metric massiveness, that is the least number of closed balls of radius u, covering the set $\mathbf{T}$ with respect to the metric ρ. Let $\xi(t) = \{\xi(t),\, t \in \mathbf{T}\}$ be a square Gaussian stochastic process. Assume that there exists a monotonically increasing continuous function $\sigma(h),\ h > 0$, such that $\sigma(h) \to 0$ as $h \to 0$, and the inequality

$$\sup_{\rho(t,s)\leq h} (Var(\xi(t) - \xi(s)))^{\frac{1}{2}} \leq \sigma(h)$$

holds true.

Define now the following values:

$$\varepsilon_0 = \inf_{t\in\mathbf{T}} \sup_{s\in\mathbf{T}} \rho(t,s), \qquad t_0 = \sigma(\varepsilon_0),$$

$$\gamma_0 = \sup_{t\in\mathbf{T}} (Var\,\xi(t))^{1/2},$$

Let C be a maximum of t_0 and γ_0, $C = \max\{t_0, \gamma_0\}$. The next theorem gives an estimate for the large deviation probability of square Gaussian process in the norm of continuous function. The proof of the Theorem can be found in the article [19].

Theorem 4.1
Let $\xi(t) = \{\xi(t), t \in \mathbf{T}\}$ be a separable square Gaussian stochastic process. Suppose that there exists an increasing function $r(u) \geq 0, u \geq 1$, with the properties: $r(u) \to \infty$ and $u \to \infty$ and let the function $r(\exp\{t\})$ be convex. Assume that the following integral

$$\int_0^{t_0} r\left(N\left(\sigma^{(-1)}(u)\right)\right) du$$

is convergent. Then for all $x > 0$

$$\mathbf{P}\left\{\sup_{t\in\mathbf{T}} |\xi(t)| > x\right\} \leq 2 \inf_{0<p<1} \left\{ r^{(-1)}\left(\tfrac{1}{t_0 p}\int_0^{t_0 p} r\left(N\left(\sigma^{(-1)}(\nu)\right)\right) d\nu\right) \right.$$
$$\left. \times \left(1 + \tfrac{\sqrt{2}x(1-p)}{C}\right)^{\frac{1}{2}} \exp\left\{-\tfrac{x(1-p)}{\sqrt{2}\gamma_0}\right\}\right\}. \tag{51}$$

4.2. An estimate for the $L_p(T)$ norm of a square Gaussian stochastic process

The following theorem can be found in article [21].

Theorem 4.2
[21] Let $\{T, \mathfrak{A}, \mu\}$ be a measurable space, where $\mathbf{T}$ is a parametric set, and let $\xi = \{\xi(t), t \in \mathbf{T}\}$ be a measurable square Gaussian stochastic process. Assume that the Lebesgue integral $\int_{\mathbf{T}} (\mathbf{E}\xi^2(t))^{\frac{p}{2}} d\mu(t)$ is well defined for $p \geq 1$. Then the integral $\int_{\mathbf{T}} (\mathbf{E}\xi^2(t))^p d\mu(t)$ exists with probability 1, and

$$P\left\{\int_{\mathbf{T}} |\xi(t)|^p d\mu(t) > x\right\} \leq 2\sqrt{1 + \frac{x^{1/p}\sqrt{2}}{C_p^{\frac{1}{p}}}} \exp\left\{-\frac{x^{1/p}}{\sqrt{2}C_p^{\frac{1}{p}}}\right\}, \tag{52}$$

for all $x \geq (\frac{p}{\sqrt{2}} + \sqrt{(\frac{p}{2}+1)p})^p C_p$, where $C_p = \int_{\mathbf{T}} (\mathbf{E}\xi^2(t))^{\frac{p}{2}} d\mu(t)$.

5. On the rate of convergence of the estimator of impulse response function

This section is devoted to the investigation of the rate of convergence of estimators of unknown impulse response function in the space of continuous functions and in the space $L_2([0, A])$.

Suppose that $X_{N,i} = (X_{N,i}(u),\ u \in \mathbb{R})$, $i = \overline{1,n}$, are real-valued independent centered Gaussian processes from (23). Assume that they perturb a time-invariant casual continuous linear system (1).

Consider estimator (6) for the impulse response function L. The output process $Y(t)$ is defined by (24).

The next Lemma is clear.

Lemma 5.1

Stochastic process $\hat{Z}_{N,n}(\tau) = \hat{L}_{N,n}(\tau) - E\hat{L}_{N,n}(\tau)$, $\tau > 0$, is a square Gaussian one.

Consider a difference of the estimator $\hat{L}_{N,n}(\tau)$ and the impulse function $L(\tau)$

$$L(\tau) - \hat{L}_{N,n}(\tau), \quad \tau > 0.$$

At first we study the distribution of supremum for this difference on the domain $[0, A]$, where A is a fixed positive number,

$$P\{ \sup_{\tau\in[0,A]} |\hat{L}_{N,n}(\tau) - L(\tau)| \geq \varepsilon\}, \quad \varepsilon > 0.$$

Denote

$$l^*_{N,n} = \frac{2K^2}{3\sqrt{N+1}} I_L.$$

Then from (26) it follows that

$$|E\hat{L}_{N,n}(\tau) - L(\tau)| \leq l^*_{N,n}, \quad \tau \in [0, A].$$

Put

$$\gamma_0(N,n) = \gamma_0 = \frac{K^4 I_L^2}{n} \left(N\frac{N^2-1}{2N^2} + \frac{(2\sqrt{N}-2)^2}{N} \right). \tag{53}$$

From (34) we have that

$$\sup_{\tau\in[0,A]} Var\hat{Z}_{N,n}(\tau) \leq \gamma_0.$$

Let

$$M_\alpha = 2^{2-\frac{1}{2\alpha}} e^{1/\alpha} \gamma_0^{-\frac{1}{2}-\frac{1}{\alpha}} \alpha^{1/\alpha - 1/2}.$$

The following theorem gives the rate of convergence of the impulse function estimator in the space of continuous function.

Theorem 5.1

Suppose that the conditions **A** , **B** are fulfilled. Then the inequality

$$\begin{aligned} P & \left\{ \sup_{\tau\in[0,A]} |L(\tau) - \hat{L}_{N,n}(\tau)| > \varepsilon \right\} \leq M_\alpha (\varepsilon - l^*_{N,n})^{\frac{1}{\alpha}} \\ \times & \left(C\alpha + \sqrt{2}\alpha(\varepsilon - l^*_{N,n}) - 2\gamma_0 \right)^{\frac{1}{2}} \exp\left\{ -\frac{\varepsilon - l^*_{N,n}}{\sqrt{2\gamma_0}} + \frac{1}{\alpha} \right\} \end{aligned} \tag{54}$$

holds true for

$$\varepsilon > \frac{\sqrt{2\gamma_0}}{\alpha} + l^*_{N,n}, \quad \alpha \in (0, 1].$$

Proof

The difference $L(\tau) - \hat{L}_{N,n}(\tau)$ can be presented as

$$L(\tau) - \hat{L}_{N,n}(\tau) = (L(\tau) - E\hat{L}_{N,n}(\tau)) - (\hat{L}_{N,n}(\tau) - E\hat{L}_{N,n}(\tau)) = l_{N,n}(\tau) - \hat{Z}_{N,n}(\tau).$$

Then

$$L(\tau) - \hat{L}_{N,n}(\tau) \geq \varepsilon \quad \Leftrightarrow \quad \hat{Z}_T(\tau) \leq -(\varepsilon - l_{N,n}(\tau)),$$

$$L(\tau) - \hat{L}_{N,n}(\tau) \leq -\varepsilon \quad \Leftrightarrow \quad \hat{Z}_T(\tau) \geq +\varepsilon + l_{N,n}(\tau).$$

Hence, for $\varepsilon > l^*_{N,n}$ we obtain

$$\{|L(\tau) - \hat{L}_{N,n}(\tau)| \geq \varepsilon\} \subset \{|\hat{Z}_T(\tau)| \geq \varepsilon - l^*_{N,n}\}$$

and

$$P\{\sup_{\tau\in[0,A]} |L(\tau) - \hat{L}_{N,n}(\tau)| \geq \varepsilon\} \leq P\{\sup_{\tau\in[0,A]} |\hat{Z}_T(\tau)| \geq \varepsilon - l^*_{N,n}\}. \tag{55}$$

In the frame of proving put

$$x = \varepsilon - l^*_{N,n}.$$

Since from Lemma 5.1 follows that $\hat{Z}_T(\tau)$ is a square Gaussian process then Theorem 4.1 can be used. It follows from (48) that as a function $\sigma(h)$ the function $\sigma(h) = C_g(N,n) \cdot h^\alpha$ could be considered, where $C_g(N,n)$ is defined in (49). In this case the inverse function equals $\sigma^{(-1)}(h) = \left(\frac{h}{C_g(N,n)}\right)^{1/\alpha}$. The metric massiveness of the segment $[0, A]$ with respect to the metric $\rho(t,s) = |t-s|$ can be majorized as

$$N(u) \leq \frac{A}{2u} + 1.$$

Therefore,

$$\begin{aligned} N(\sigma^{(-1)}(u)) &\leq \left(\frac{A}{2\sigma^{(-1)}(u)} + 1\right) \\ &= \left(\frac{A}{2}\Big(\frac{C_g(N,n)}{u}\Big)^{1/\alpha} + 1\right). \end{aligned}$$

Consider the function $r(u) = u^\beta - 1$, where $\beta \in (0,\alpha)$. It's clear that the conditions of Theorem 4.1 for the function $r(u)$ are satisfied. Since $0 < p < 1$ and $t_0 = C_g(N,n)\left(\frac{A}{2}\right)^\alpha$ then $\frac{A}{2}\left(\frac{C_g(N,n)}{pt_0}\right)^{1/\alpha} > 1$. Hence, it follows from the assumption $0 < u < t_0 p$ that the inequality

$$N(\sigma^{(-1)}(u)) \leq A\Big(\frac{C_g(N,n)}{u}\Big)^{1/\alpha}$$

holds. Since the inverse function of $r(u)$ is $r^{(-1)}(u) = (u+1)^{1/\beta}$, then

$$\begin{aligned} r^{(-1)}\left(\frac{1}{t_0 p}\int_0^{t_0 p} r\left(N\left(\sigma^{(-1)}(\nu)\right)\right) d\nu\right) &= \left(\frac{1}{t_0 p}\int_0^{t_0 p} \big[\big(\frac{A}{2}\Big(\frac{C_g(N,n)}{u}\Big)^{1/\alpha} + 1\big)^\beta\big] du\right)^{1/\beta} \\ &\leq \left(\frac{1}{t_0 p}\int_0^{t_0 p} \big[A\Big(\frac{C_g(N,n)}{u}\Big)^{1/\alpha}\big]^\beta du\right)^{1/\beta} \\ &= 2\Big(\frac{\alpha}{\alpha-\beta}\Big)^{1/\beta} p^{-1/\alpha}. \end{aligned}$$

Find the minimum with respect to β of the right-hand side of the relationship above

$$\inf_{\beta\in(0,\alpha)} \Big(\frac{\alpha}{\alpha-\beta}\Big)^{1/\beta} = \lim_{\beta\to 0}\Big(\frac{1}{1-\beta/\alpha}\Big)^{1/\beta} = e^{1/\alpha}.$$

From inequality (51) and equity above follows that for $x > 0$

$$\begin{aligned} &P\left\{\sup_{\tau\in[0,A]} |\hat{Z}_T(\tau)| > x\right\} \leq \\ &\leq 4e^{1/\alpha} C^{-\frac{1}{2}} \inf_{0<p<1}\left\{\frac{\sqrt{C + \sqrt{2}x(1-p)}}{p^{1/\alpha}} \exp\left\{-\frac{x(1-p)}{\sqrt{2}\gamma_0}\right\}\right\}. \end{aligned} \tag{56}$$

The minimum point of right-hand side of (56) is

$$p_{\min} = \frac{\sqrt{2}\gamma_0}{\alpha x}.$$

Since by the condition of Theorem $p \in (0,1)$, then for

$$x > \frac{\sqrt{2}\gamma_0}{\alpha}$$

the value $p_{\min}$ can be substituted in (56) and the following inequality is obtained

$$\begin{aligned} P & \left\{ \sup_{\tau\in[0,A]} |\hat{Z}_T(\tau)| > x \right\} \leq \\ \leq & \ M_\alpha x^{\frac{1}{\alpha}} \sqrt{C\alpha + \sqrt{2}\alpha x - 2\gamma_0} \exp\left\{ -\frac{x}{\sqrt{2}\gamma_0} + \frac{1}{\alpha} \right\}, \end{aligned} \tag{57}$$

where the constant M_α equals

$$M_\alpha = 2^{2-\frac{1}{2\alpha}} e^{1/\alpha} \gamma_0^{-\frac{1}{2}-\frac{1}{\alpha}} \alpha^{1/\alpha-1/2}.$$

Taking into account that $x = \varepsilon - l^*_{N,n}$, the assertion of Theorem follows from (57). □

Theorem 4.2 can be used to obtain the following estimate for the cross-correlogram in the space $L_p([0,A])$, where the constant $A > 0$ is fixed.

Consider a parametric set $\mathbf{T} = [0, A]$ and let $\{[0,A], \mathfrak{A}, \mu\}$ be a metric space with Euclidean measure μ.

Theorem 5.2

Assume that the conditions **A** and **B** are satisfied. Then for

$$\varepsilon \geq \left(\left(\frac{p}{\sqrt{2}} + \sqrt{\left(\frac{p}{2}+1\right)p}\right) A^{\frac{1}{p}} \gamma_0^{\frac{1}{2}} + A l^*_{N,n} \right)^p$$

the estimate

$$P\left\{ \int_0^A |L(\tau) - \hat{L}_{N,n}(\tau)|^p d\tau > \varepsilon \right\} \leq 2\sqrt{1 + \frac{(\varepsilon^{\frac{1}{p}} - A l^*_{N,n})\sqrt{2}}{A^{\frac{1}{p}}\gamma_0^{\frac{1}{2}}}} \exp\left\{ -\frac{\varepsilon^{\frac{1}{p}} - A l^*_{N,n}}{\sqrt{2}A^{\frac{1}{p}}\gamma_0^{\frac{1}{2}}} \right\} \tag{58}$$

holds true.

Proof

We show first that the result of Theorem 4.2 can be applied to the process $\hat{Z}_{N,n}(\tau)$, $\tau > 0$. Really, by Lemma 5.1 $\hat{Z}_{N,n}(\tau)$ is a square Gaussian stochastic process. Prove that the Lebesgue integral

$$\int_0^A (\mathbf{E}\hat{Z}^2_{N,n}(\tau))^{\frac{p}{2}} d\mu(\tau)$$

is correctly defined. From inequality (27) follows that

$$\begin{aligned} \int_0^A (\mathbf{E}\hat{Z}^2_{N,n}(\tau))^{\frac{p}{2}} d\mu(\tau) &= \int_0^A (Var\hat{Z}_{N,n}(\tau))^{\frac{p}{2}} d\tau \\ &< A\left(\frac{K^4 I_L^2}{n} \left(N\frac{N^2-1}{2N^2} + \frac{(2\sqrt{N}-2)^2}{N} \right) \right)^{\frac{p}{2}}. \end{aligned}$$

Therefore, inequality (52) for the process $\hat{Z}_{N,n}(\tau)$ as $x \geq (\frac{p}{\sqrt{2}} + \sqrt{(\frac{p}{2}+1)p})^p C_p$ can be rewritten as

$$P\left\{\int_0^A |\hat{Z}_{N,n}(\tau)|^p d\mu(\tau) > x\right\} \leq 2\sqrt{1 + \frac{x^{1/p}\sqrt{2}}{C_p^{\frac{1}{p}}}} \exp\left\{-\frac{x^{1/p}}{\sqrt{2}C_p^{\frac{1}{p}}}\right\}, \tag{59}$$

where $C_p = \int_0^A (\mathbf{E}\hat{Z}_{N,n}^2(\tau))^{\frac{p}{2}} d\tau$.

From the Minkowski inequality follows that

$$\begin{aligned}\left(\int_0^A |L(\tau) - \hat{L}_{N,n}(\tau)|^p d\mu(\tau)\right)^{\frac{1}{p}} &= \left(\int_0^A |L(\tau) \pm \mathbf{E}\hat{L}_{N,n}(\tau) - \hat{L}_{N,n}(\tau)|^p d\mu(\tau)\right)^{\frac{1}{p}} \\ &\leq \left(\int_0^A |L(\tau) - \mathbf{E}\hat{L}_{N,n}(\tau)|^p d\mu(\tau)\right)^{\frac{1}{p}} \\ &+ \left(\int_0^A |\mathbf{E}\hat{L}_{N,n}(\tau) - \hat{L}_{N,n}(\tau)|^p d\mu(\tau)\right)^{\frac{1}{p}} \\ &\leq l_{N,n}^* \cdot A + \left(\int_0^A |\hat{Z}_{N,n}(\tau)|^p d\mu(\tau)\right)^{\frac{1}{p}}.\end{aligned}$$

Since from the relationship above follows that

$$\begin{aligned}\left\{\int_0^A |L(\tau) - \hat{L}_{N,n}(\tau)|^p d\mu(\tau) > \varepsilon\right\} &= \left\{\left(\int_0^A |L(\tau) - \hat{L}_{N,n}(\tau)|^p d\mu(\tau)\right)^{\frac{1}{p}} > \varepsilon^{\frac{1}{p}}\right\} \\ &\subset \left\{Al_{N,n}^* + \left(\int_0^A |\hat{Z}_{N,n}(\tau)|^p d\mu(\tau)\right)^{\frac{1}{p}} > \varepsilon^{\frac{1}{p}}\right\} \\ &= \left\{\int_0^A |\hat{Z}_{N,n}(\tau)|^p d\mu(\tau) > \left(\varepsilon^{\frac{1}{p}} - Al_{N,n}^*\right)^p\right\},\end{aligned}$$

then substituting $x = \left(\varepsilon^{\frac{1}{p}} - Al_{N,n}^*\right)^p$ in (59), we obtain

$$P\left\{\int_0^A |L(\tau) - \hat{L}_{N,n}(\tau)|^p d\mu(\tau) > \varepsilon\right\} \leq 2\sqrt{1 + \frac{(\varepsilon^{\frac{1}{p}} - Al_{N,n}^*)\sqrt{2}}{C_p^{\frac{1}{p}}}} \exp\left\{-\frac{\varepsilon^{\frac{1}{p}} - Al_{N,n}^*}{\sqrt{2}C_p^{\frac{1}{p}}}\right\}. \tag{60}$$

Since (34) implies $\sup_{\tau\in[0,A]} Var\hat{Z}_{N,n}(\tau) \leq \gamma_0$, where γ_0 is from (53), then

$$C_p = \int_0^A (\mathbf{E}\hat{Z}_{N,n}^2(\tau))^{\frac{p}{2}} d\tau \leq A\gamma_0^{\frac{p}{2}}.$$

To complete the proof it's enough to substitute the above value C_p in (60). Hence,

$$P\left\{\int_0^A |L(\tau) - \hat{L}_{N,n}(\tau)|^p d\mu(\tau) > \varepsilon\right\} \leq 2\sqrt{1 + \frac{(\varepsilon^{\frac{1}{p}} - Al_{N,n}^*)\sqrt{2}}{A^{\frac{1}{p}}\gamma_0^{\frac{1}{2}}}} \exp\left\{-\frac{\varepsilon^{\frac{1}{p}} - Al_{N,n}^*}{\sqrt{2}A^{\frac{1}{p}}\gamma_0^{\frac{1}{2}}}\right\}.$$

□

6. Testing hypotheses on the impulse response function

Using Theorem 5.1 and Theorem 5.2 it is possible to test hypothesis on the shape of impulse response function.

Let the null hypothesis $\mathbf{H_0}$ state that an impulse response function is $L(\tau)$, $\tau \in [0, A]$, and the alternative $\mathbf{H_a}$ implies the opposite statement.

Denote

$$g_1(\varepsilon) = M_\alpha(\varepsilon - l^*_{N,n})^{\frac{1}{\alpha}} \left(C\alpha + \sqrt{2}\alpha(\varepsilon - l^*_{N,n}) - 2\gamma_0\right)^{\frac{1}{2}} \exp\left\{-\frac{\varepsilon - l^*_{N,n}}{\sqrt{2}\gamma_0} + \frac{1}{\alpha}\right\}.$$

From Theorem 5.1 follows that if $\varepsilon > z_{N,n} = \frac{\sqrt{2}\gamma_0}{\alpha} + l^*_{N,n}$, $\alpha \in (0, 1]$, then

$$P\left\{\sup_{\tau\in[0,A]} |L(\tau) - \hat{L}_{N,n}(\tau)| > \varepsilon\right\} \le g_1(\varepsilon).$$

Let $\varepsilon_{1,\delta}$ be a solution of the equation

$$g_1(\varepsilon_{1,\delta}) = \delta, \quad 0 < \delta < 1.$$

Put

$$\varepsilon^*_{1,\delta} = \max\{\varepsilon_{1,\delta}, z_{N,n}\}. \tag{61}$$

It is obvious that $g_1(\varepsilon^*_{1,\delta}) \le \delta$ and

$$P\left\{\sup_{\tau\in[0,A]} |L(\tau) - \hat{L}_{N,n}(\tau)| > \varepsilon^*_{1,\delta}\right\} \le \delta.$$

From Theorem 5.1 it follows that to test the hypothesis $\mathbf{H_0}$, we can use the following criterion.

Criterion 1. *For a given level of confidence* $1 - \delta$, $\delta \in (0, 1)$, *the hypothesis* $\mathbf{H_0}$ *is rejected if*

$$\sup_{\tau\in[0,A]} |L(\tau) - \hat{L}_{N,n}(\tau)| > \varepsilon^*_{1,\delta},$$

otherwise the hypothesis $\mathbf{H_0}$ *is accepted, where* $\varepsilon^*_{1,\delta}$ *is from* (61).

Remark 6.1. The equation $g_1(\varepsilon_{1,\delta}) = \delta$, has a solution for any $\delta > 0$ since the function $g_1(\varepsilon)$ decreases. We can find solution of the equation using numerical methods.

Denote

$$g_2(\varepsilon) = 2\sqrt{1 + \frac{(\varepsilon^{\frac{1}{p}} - Al^*_{N,n})\sqrt{2}}{A^{\frac{1}{p}}\gamma_0^{\frac{1}{2}}}} \exp\left\{-\frac{\varepsilon^{\frac{1}{p}} - Al^*_{N,n}}{\sqrt{2}A^{\frac{1}{p}}\gamma_0^{\frac{1}{2}}}\right\}.$$

From Theorem 5.2 follows that if

$$\varepsilon > t_{N,n}(p) = \left(\left(\frac{p}{\sqrt{2}} + \sqrt{\left(\frac{p}{2} + 1\right)p}\right)A^{\frac{1}{p}}\gamma_0^{\frac{1}{2}} + Al^*_{N,n}\right)^p,$$

then

$$P\left\{\int_0^A |L(\tau) - \hat{L}_{N,n}(\tau)|^p d\tau > \varepsilon\right\} \le g_2(\varepsilon).$$

Let $\varepsilon_{2,\delta}$ be a solution of the equation $g_2(\varepsilon_{2,\delta}) = \delta$, $0 < \delta < 1$. Put

$$\varepsilon^*_{2,\delta} = \max\{\varepsilon_{2,\delta}, t_{N,n}(p)\}. \tag{62}$$

It is clear that $g_2(\varepsilon^*_{2,\delta}) \leq \delta$ and

$$P\left\{ \sup_{\tau\in[0,A]} |L(\tau) - \hat{L}_{N,n}(\tau)| > \varepsilon^*_{2,\delta} \right\} \leq \delta.$$

From Theorem 5.2 it follows that to test the hypothesis $\mathbf{H_0}$, we can use the following criterion.
Criterion 2. *For a given level of confidence* $1-\delta$, $\delta \in (0,1)$, *the hypothesis* $\mathbf{H_0}$ *is rejected if*

$$\int_0^A |L(\tau) - \hat{L}_{N,n}(\tau)|^p d\tau > \varepsilon^*_{2,\delta},$$

otherwise the hypothesis $\mathbf{H_0}$ *is accepted, where* $\varepsilon^*_{2,\delta}$ *is from* (62).

7. Conclusions

In this paper we considere time-invariant continuous linear system in which the impulse response function was estimated applying a new proposed method. The input signal process is supposed to be a zero mean Gaussian stochastic process which is represented as a series with respect to an orthonormal basis in $L_2(\mathbf{R})$. A particular case where the orthonormal basis is given by the Hermite functions is studied in details. Some characteristics of the estimator of impulse function such as mathematical expectation, variance and variance of the increments are described. We also investigated the convergence rate for the estimator of unknown impulse response function in the space of continuous functions and in the space $L_p([0,A])$. For this reason the theory of square Gaussian random variables and processes is applied, namely we use inequalities for the $C(T)$ and $L_p(T)$ norms of square Gaussian stochastic process. It gives us an opportunity to construct two criteria of testing hypothesis on the shape of the impulse response function.

REFERENCES

1. H. Akaike, *On the statistical estimation of the frequence response function of a system having multiple input*, Ann. Inst. Statist. Math., vol. 17, pp. 185–210, 1965.
2. H. Akaike, *On the use of non-Gaussian process in the identification of a linear dynamic system*, Ann. Inst. Statist. Math., vol. 18, pp. 269–276, 1966.
3. J. S. Bendat, and A.G. Piersol, *Engineering Applications of Correlation and Spectral Analysis*, Wiley, New York, 1980.
4. I.P. Blazhievska, *Asymptotic unbiasedness and consystency of cross-correlogram estimator for impulse response functions of linear system*, Naukovi Visti NTUU KPI? vol. 4, pp. 7–12, 2014.(in Ukrainian)
5. V.V. Buldygin, and I.P. Blazhievska, *On correlation properties of the cross-correlogram estimators of impulse response functions,* Naukovi Visti NTUU KPI? vol. 5, pp. 120–128, 2009.(in Ukrainian)
6. V.V. Buldygin, and I.P. Blazhievska, *Asymptotic properties of cross-correlogram estimator for impulse response functions of linear system*, Naukovi Visti NTUU KPI? vol. 4, pp. 16–27, 2010.(in Ukrainian)
7. V.V. Buldygin, and Yu.V. Kozachenko, *Metric characterization of random variables and random processes*, Amer. Math. Soc., Providence, RI, 2000.
8. V.V. Buldygin, and V.G. Kurotschka, *On cross-coorrelogram estimators of the response function in continuous linear systems from discrete observations*, Random Oper. and Stoch. Equ., vol.7, no. 1, pp. 71–90, 1999.
9. V.V. Buldygin, and Fu Li, *On asymptotic normality of an estimation of unit impulse responses of linear system I* Theor. Probability and Math. Statist., vol. 54, pp. 3–17, 1997.
10. V.V. Buldygin, and Fu Li, *On asymptotic normality of an estimation of unit impulse responses of linear system II* Theor. Probability and Math. Statist., vol. 55, pp. 30–37, 1997.
11. V. Buldygin, L. Utzet, and V. Zaiats, *Asymptotic normality of cross-coorrelogram estimators of the response function*, Statistical Infernce for Stochastic Processes, vol. 7, pp. 1–34, 2004.
12. V. Buldygin, L. Utzet, and V. Zaiats, *A note on the application of intergals involving cyclic products of kernels*, QESTIIO, vol. 26, no. 1-2, pp. 3—14, 2002.
13. A. Erdelyi, *Higher transcendental functions*, Mc Graw-Hill Book Company, Inc, 1953.
14. I.I. Gikhman, and A.V. Skorokhod, *Introduction to the theory of random processes*, Dover Publications Inc., Mineola, New York,1996.
15. E. J. Hannan, and M. Deistler, *The Statistical Theory of Linear Systems*, Wiley, New York, 1988.
16. Yu. V. Kozachenko, and O.M. Moklyachuk, *Large deviation probabilities for square-Gaussian stochastic processes*, Extremes, vol. 2, no. 3, pp. 269–293, 1999.

17. Yu.V Kozachenko, and O.M. Moklyachuk, *Square-Gaussian stochastic processes*, Theory of Stoch. Processes, vol. 6(22), no.3-4, pp. 98–121, 2000.
18. Yu.V. Kozachenko, A.O. Pashko, and I.V. Rozora, *Simulation of stochastic processes and fields*, Zadruga, Kyiv, 2007. (in Ukrainian)
19. Yu. Kozacenko, and I. Rozora, *On cross-correlogram estimators of impulse response function*, Theor. Probability and Math. Statist., vol. 93, pp. 75–83, 2015.
20. Yu.V. Kozachenko, and O.V. Stus, *Square-Gaussian random processes and estimators of covariance functions,* Math. Communications., vol. 3, no. 1, pp. 83–94, 1998.
21. Yu. Kozachenko, and V. Troshki, *A criterion for testing hypotheses about the covariance function of a stationary Gaussian stochastic process,* Modern Stochastics: Theory and Applications, vol. 1, no. 1, pp. 139–149, 2014.
22. M. Schetzen, *The Volterra and Wiener Theories of Nonlinear Systems*, Wiley, New York, 1980.
23. T. Söderström, and P. Stoica, *System Identification, Prentice-Hall*, London, 1989.

6

Relations for moments of generalized record values from additive Weibull distribution and associated inference

R.U. Khan*, M.A. Khan, M.A.R. Khan

Department of Statistics and Operations Research, Aligarh Muslim University, India.

Abstract In this note we give some simple recurrence relations satisfied by single and product moments of k-th upper record values from the additive Weibull distribution. These relations are deduced for moments of upper record values. Further, conditional expectation and recurrence relation for single moments are used to characterize the additive Weibull distribution and some computational works are also carried out.

Keywords Order statistics, k-th upper record values, additive Weibull distribution, single moments, product moments, recurrence relations and characterization

1. Introduction

The statistical study of record values in a sequence of independent and identically distributed (iid) continuous random variables was first carried out by Chandler [5]. For an extensive study in this area one can refer to the works of Arnold *et. al* [3], Ahsanullah [1] and Ahsanullah and Nevzorov [2]. Dziubdziela and Kopociński [7] have generalized the concept of record values of Chandler [5] by random variables of a more generalized nature and we may call them as generalized record values or k-th record values. Setting $k = 1$, we obtain ordinary record statistics.

Record values and associated statistics are of great importance in several real life problems involving weather, economic studies, sports and so on. The prediction of a future record value is an interesting problem with many real life applications. For example the predicted value of the amount of next record level of water that a dam will capture from rain and hold or discharge is helpful for future planning purposes, predicted intensity of the next strongest earthquake is essential for disaster management planning, prediction of next level of new record in athletic events is helpful for subjecting the prospective athletes to rigorous training and practice and so on.

Several applications of k-th record values can be found in the literature, for instance, see the examples cited in Kamps [11] or Danielak and Raqab [6] in reliability theory. Suppose that a technical system or piece of equipment is subject to shocks, e.g. peaks of voltages. If the shocks are viewed as realizations of an iid sequence, then the model of ordinary records is adequate. If it is not the records themselves, but second or third values are of special interest, then the model of k-th record values is adequate. When record values themselves are viewed as outliers, then the second or third largest values are of special interest. Record statistics are applied in estimating strength of materials, predicting natural disasters, sport achievements, etc. For statistical inference based on ordinary records,

*Correspondence to: Department of Statistics and Operations Research, Aligarh Muslim University, Aligarh-202 002, India. E-mails: aruke@rediffmail.com, khanazam2808@gmail.com

serious difficulties arise if expected values of inter arrival time of records is infinite and occurrences of records are very rare in practice. This problem is avoided once we consider the model of k-th record statistics.

Let $\{X_n,\ n \geq 1\}$ be a sequence of *iid* random variables with distribution function (df) $F(x)$ and probability density function (pdf) $f(x)$. The $j-$th order statistic of a sample $X_1, X_2, \ldots, X_n$ is denoted by $X_{j:n}$. For a fixed positive integer k , Dziubdziela and Kopociński [7] define the sequences $\{U_n^{(k)},\ n \geq 1\}$ of k-th upper record times for the sequence $\{X_n,\ n \geq 1\}$ as follows:

$U_1^{(k)} = 1$

$$U_{n+1}^{(k)} = min\{j > U_n^{(k)} : X_{j:j+k-1} > X_{U_n^{(k)}:U_n^{(k)}+k-1}\}.$$

The sequence $\{Y_n^{(k)},\ n \geq 1\}$, where $Y_n^{(k)} = X_{U_n^{(k)}}$ is called the sequence of generalized upper record values or k-th upper record values of $\{X_n,\ n \geq 1\}$. Note that for $k = 1$, we have $Y_n^{(1)} = X_{U_n},\ n \geq 1$, which are the record values of $\{X_n,\ n \geq 1\}$ (Ahsanullah [1]). Moreover, we see that $Y_0^{(k)} = 0$ and $Y_1^{(k)} = min(X_1, X_2, ..., X_k) = X_{1:k}$. The *pdf* of $Y_n^{(k)}$ and the joint *pdf* of $Y_m^{(k)}$ and $Y_n^{(k)}$ are given by (Dziubdziela and Kopociński [7], Grudzień [8])

$$f_{Y_n^{(k)}}(x) = \frac{k^n}{(n-1)!}\,[-ln\bar{F}(x)]^{n-1}\,[\bar{F}(x)]^{k-1}\,f(x),\ \ n \geq 1, \tag{1}$$

and

$$f_{Y_m^{(k)},Y_n^{(k)}}(x,y) = \frac{k^n}{(m-1)!(n-m-1)!}\,[-ln\bar{F}(x)]^{m-1}\frac{f(x)}{\bar{F}(x)}$$
$$\times[ln\bar{F}(x) - ln\bar{F}(y)]^{n-m-1}[\bar{F}(y)]^{k-1}\,f(y),\ \ x < y,\ 1 \leq m < n,\ n \geq 2, \tag{2}$$

where

$\bar{F}(x) = 1 - F(x)$.

For some recent developments on generalized record values with special reference to those arising from Pareto, generalized Pareto, Weibull, exponential-Weibull and modified Weibull distributions, see Pawlas and Szynal [16, 17], Khan *et. al* [13] and Khan and Khan [12] respectively. In this work we mainly focus on the study of generalized record values arising from the additive Weibull distribution.

A random variable X is said to have a additive Weibull distribution (Lemonte *et. al* [14]) if its *pdf* is of the form

$$f(x) = (\alpha\beta x^{\beta-1} + \theta\delta x^{\delta-1})e^{-(\alpha x^{\beta}+\theta x^{\delta})},\ \ x > 0,\ \ \alpha > 0,\ \ \theta > 0\ \text{ and }\ \delta,\ \beta > 0 \tag{3}$$

and the corresponding *df* is

$$F(x) = 1 - e^{-(\alpha x^{\beta}+\theta x^{\delta})},\ \ x > 0,\ \ \alpha > 0,\ \ \theta > 0\ \text{ and }\ \delta,\ \beta > 0. \tag{4}$$

It is easy to see that

$$f(x) = (\alpha\beta x^{\beta-1} + \theta\delta x^{\delta-1})\bar{F}(x). \tag{5}$$

The exponential-Weibull and Weibull distributions are the special cases for $\delta = 1$ or $\beta = 1$ and $\theta = 0$ or $\alpha = 0$, respectively. The exponential distribution arises when $\beta = 1,\ \theta = 0$ or $\alpha = 0,\ \delta = 1$. The Rayleigh and two-parameter linear failure rate distributions are obtained when $\alpha = 0,\ \delta = 2$ or $\theta = 0,\ \beta = 2$ and $\beta = 2,\ \delta = 1$ or $\beta = 1,\ \ \delta = 2$, respectively.

The relation in (5) will be exploited in this paper to derive some recurrence relations for the moments of k-th upper record values from the additive Weibull distribution and to give a characterization of this distribution.

2. Relations for Single Moments

Before coming to the main result we shall prove the following Lemma.

Lemma 2.1. Fix a positive integer $k \geq 1$, for $n \geq 1$ and $j = 0, 1, ...$

$$E(Y_n^{(k)})^j - E(Y_{n-1}^{(k)})^j = \frac{jk^n}{(n-1)!}\int_{\alpha}^{\beta} x^{j-1}[-ln\bar{F}(x)]^{n-1}[\bar{F}(x)]^k dx. \tag{6}$$

Proof. From (1), we have

$$E(Y_n^{(k)})^j - E(Y_{n-1}^{(k)})^j = \frac{k^n}{(n-1)!}\int_\alpha^\beta x^j[-ln\bar{F}(x)]^{n-1}[\bar{F}(x)]^{k-1}f(x)dx$$

$$-\frac{k^{n-1}}{(n-2)!}\int_\alpha^\beta x^j[-ln\bar{F}(x)]^{n-2}[\bar{F}(x)]^{k-1}f(x)dx$$

$$= \frac{k^n}{(n-1)!}\int_\alpha^\beta x^j[-ln\bar{F}(x)]^{n-2}[\bar{F}(x)]^{k-1}f(x)\Big\{-ln\bar{F}(x) - \frac{n-1}{k}\Big\}dx. \tag{7}$$

Let

$$h(x) = -\frac{1}{k}[-ln\bar{F}(x)]^{n-1}[\bar{F}(x)]^k. \tag{8}$$

Differentiating both sides of (8), we get

$$h'(x) = [-ln\bar{F}(x)]^{n-2}[\bar{F}(x)]^{k-1}f(x)\Big\{-ln\bar{F}(x) - \frac{n-1}{k}\Big\}dx.$$

Thus,

$$E(Y_n^{(k)})^j - E(Y_{n-1}^{(k)})^j = \frac{k^n}{(n-1)!}\int_\alpha^\beta x^j h'(x)dx. \tag{9}$$

Now integrating in (9) by parts and using the value of $h(x)$ from (8), we have the result given in (6).
Theorem 2.1. For distribution given in (4). Fix a positive integer $k \geq 1$, for $n \geq 1$ and $j = 0, 1, ...$

$$E(Y_n^{(k)})^j = \frac{\alpha\beta k}{(j+\beta)}\{E(Y_n^{(k)})^{j+\beta} - E(Y_{n-1}^{(k)})^{j+\beta}\} + \frac{\theta\delta k}{(j+\delta)}\{E(Y_n^{(k)})^{j+\delta} - E(Y_{n-1}^{(k)})^{j+\delta}\}. \tag{10}$$

Proof. From (1) and (5), we have

$$E(Y_n^{(k)})^j = \frac{k^n}{(n-1)!}\Big\{\alpha\beta\int_0^\infty x^{j+\beta-1}[-ln\bar{F}(x)]^{n-1}[\bar{F}(x)]^k dx + \theta\delta\int_0^\infty x^{j+\delta-1}[-ln\bar{F}(x)]^{n-1}[\bar{F}(x)]^k dx\Big\}.$$

Making use of Lemma 2.1, we establish the result given in (10).

Remarks

i) Setting $\delta = 1$ or $\beta = 1$ in (10), we get the recurrence relation for single moments of k-th upper record values from the exponential-Weibull lifetime distribution as obtained by Khan *et. al* [13].

ii) Setting $\theta = 0$ or $\alpha = 0$ in (10), we get the recurrence relation for single moments of k-th upper record values from the Weibull distribution, which verify the results obtained by Pawlas and Szynal [17].

iii) Putting $\beta = 1$, $\theta = 0$ or $\alpha = 0$, $\delta = 1$ in (10), we deduce the recurrence relation for single moments of k-th upper record values from the exponential distribution, established by Pawlas and Szynal [15].

iv) Putting $\alpha = 0$, $\delta = 2$ or $\theta = 0$, $\beta = 2$ in (10), we get the recurrence relation for single moments of k-th upper record values from the Rayleigh distribution as given by Khan *et. al* [13].

v) Setting $\beta = 2$, $\delta = 1$ or $\beta = 1$, $\delta = 2$ in (10), the result for single moments of k-th upper record values is deduced for linear failure rate distribution as established by Khan *et. al* [13].

Corollary 2.1. The recurrence relation for single moments of upper record values from the additive Weibull lifetime distribution has the form

$$EX_{U_n}^{j}=\frac{\alpha\beta}{j+\beta}(EX_{U_n}^{j+\beta}-EX_{U_{n-1}}^{j+\beta})+\frac{\theta\delta}{j+\delta}(EX_{U_n}^{j+\delta}-EX_{U_{n-1}}^{j+\delta}). \tag{11}$$

Remarks

i) If $\delta=1$ or $\beta=1$ in (11), we get the recurrence relation for single moments of record values from the exponential-Weibull lifetime distribution, obtained by Khan *et. al* [13].

ii) If $\theta=0$ or $\alpha=0$ in (11), the relation for single moments of record values obtained by Pawlas and Szynal [17] for Weibull distribution is deduced.

iii) If $\beta=1$, $\theta=0$ or $\alpha=0,\ \delta=1$ in (11), we get the recurrence relation for single moments of record values from exponential distribution as obtained by Balakrishnan and Ahsanullah [4] .

iv) If $\alpha=0,\ \delta=2$ or $\theta=0,\ \beta=2$ in (11), the recurrence relation for single moments of record values is deduced for Rayleigh distribution as given by Khan *et. al* [13].

v) If $\beta=2,\ \delta=1$ or $\beta=1,\ \delta=2$ in (11), the recurrence relation for single moments of record values is deduced for linear failure rate distribution, established by Khan *et. al* [13].

Numerical computations for the first four moments of upper record values from additive Weibull distribution for arbitrary chosen values of $\alpha,\ \beta,\ \theta,\ \delta$ and various sample size $n=1,2,...,5$ are given in Table 2.1.

Table 2.1. First four moments of upper record values

n	$\alpha=1,\beta=1,\theta=1$				$\alpha=2,\beta=1,\theta=1$			
	$\delta=2$				$\delta=2$			
	$E(X)$	$E(X^2)$	$E(X^3)$	$E(X^4)$	$E(X)$	$E(X^2)$	$E(X^3)$	$E(X^4)$
1	0.54564	0.45435	0.47769	0.59025	0.37893	0.24212	0.20521	0.21063
2	0.93205	1.06794	1.41557	2.10089	0.68946	0.62106	0.67101	0.83110
3	1.24176	1.75823	2.76050	4.72074	0.95656	1.08686	1.40716	2.02388
4	1.50624	2.49375	4.46824	8.56976	1.19363	1.61274	2.40879	3.91384
5	1.74049	3.25950	6.50467	13.7311	1.40856	2.18286	3.66646	6.60269
n	$\alpha=1,\beta=2,\theta=2$				$\alpha=2,\beta=2,\theta=2$			
	$\delta=3$				$\delta=3$			
	$E(X)$	$E(X^2)$	$E(X^3)$	$E(X^4)$	$E(X)$	$E(X^2)$	$E(X^3)$	$E(X^4)$
1	0.57912	0.39544	0.30227	0.25173	0.49670	0.29764	0.20235	0.15126
2	0.80575	0.70103	0.64948	0.63459	0.70724	0.54649	0.45350	0.39951
3	0.95933	0.96662	1.01668	1.11094	0.85302	0.77003	0.72996	0.72245
4	1.07952	1.20789	1.39605	1.66202	0.96828	0.97703	1.02296	1.10753
5	1.18002	1.43216	1.78391	2.27632	1.06525	1.17196	1.32804	1.54653

3. Relations for Product Moments

Lemma 3.1. Fix a positive integer $k\geq 1$, for $1\leq m\leq n-2,\ \ i,j=0,1,...$

$$E[(Y_m^{(k)})^i(Y_n^{(k)})^j] - E[(Y_m^{(k)})^i(Y_{n-1}^{(k)})^j] = \frac{jk^{n-1}}{(m-1)!(n-m-1)!}\int_\alpha^\beta\int_x^\beta x^i y^{j-1}$$

$$\times [-ln\bar{F}(x)]^{m-1}\frac{f(x)}{\bar{F}(x)}[ln\bar{F}(x) - ln\bar{F}(y)]^{n-m-1}[\bar{F}(y)]^k dydx. \quad (12)$$

Proof. From (2), we have

$$E[(Y_m^{(k)})^i(Y_n^{(k)})^j] - E[(Y_m^{(k)})^i(Y_{n-1}^{(k)})^j] = \frac{k^n}{(m-1)!(n-m-1)!}$$

$$\times \int_\alpha^\beta\int_x^\beta x^i y^j[-ln\bar{F}(x)]^{m-1}\frac{f(x)}{\bar{F}(x)}[ln\bar{F}(x) - ln\bar{F}(y)]^{n-m-2}$$

$$\times [\bar{F}(y)]^{k-1}f(y)\left\{[ln\bar{F}(x) - ln\bar{F}(y)] - \frac{n-m-1}{k}\right\}dydx. \quad (13)$$

Let

$$h(x,y) = -\frac{1}{k}[ln\bar{F}(x) - ln\bar{F}(y)]^{n-m-1}[\bar{F}(y)]^k \quad (14)$$

$$\frac{\partial}{\partial y}h(x,y) = [ln\bar{F}(x) - ln\bar{F}(y)]^{n-m-2}[\bar{F}(y)]^{k-1}f(y)$$

$$\times\left\{[ln\bar{F}(x) - ln\bar{F}(y)] - \frac{n-m-1}{k}\right\}. \quad (15)$$

Taking into account the value of (15) in (13), we get

$$E[(Y_m^{(k)})^i(Y_n^{(k)})^j] - E[(Y_m^{(k)})^i(Y_{n-1}^{(k)})^j] = \frac{k^n}{(m-1)!(n-m-1)!}$$

$$\times \int_\alpha^\beta x^i[-ln\bar{F}(x)]^{m-1}\frac{f(x)}{\bar{F}(x)}\left\{\int_x^\beta y^j\frac{\partial}{\partial y}h(x,y)dy\right\}dx. \quad (16)$$

Now in view of (14)

$$\int_x^\beta y^j\frac{\partial}{\partial y}h(x,y)dy = \frac{j}{k}\int_x^\beta y^{j-1}[ln\bar{F}(x) - ln\bar{F}(y)]^{n-m-1}[\bar{F}(y)]^k dy. \quad (17)$$

After substituting (17) in (16), the required expression is obtained.

Theorem 3.1. For distribution given in (4) and $m \geq 1$ and $i,j = 0,1,...$

$$E[(Y_m^{(k)})^i(Y_{m+1}^{(k)})^j] = \frac{\alpha\beta k}{j+\beta}\left\{E[(Y_m^{(k)})^i(Y_{m+1}^{(k)})^{j+\beta}] - E(Y_m^{(k)})^{i+j+\beta}\right\}$$

$$+ \frac{\theta\delta k}{j+\delta}\left\{E[(Y_m^{(k)})^i(Y_{m+1}^{(k)})^{j+\delta}] - E(Y_m^{(k)})^{i+j+\delta}\right\} \quad (18)$$

and for $1 \leq m \leq n-2,\ i,j = 0,1,...$

$$E[(Y_m^{(k)})^i(Y_n^{(k)})^j] = \frac{\alpha\beta k}{j+\beta}\left\{E[(Y_m^{(k)})^i(Y_n^{(k)})^{j+\beta}] - E[(Y_m^{(k)})^i(Y_{n-1}^{(k)})^{j+\beta}]\right\}$$

$$+ \frac{\theta\delta k}{j+\delta}\left\{E[(Y_m^{(k)})^i(Y_n^{(k)})^{j+\delta}] - E[(Y_m^{(k)})^i(Y_{n-1}^{(k)})^{j+\delta}]\right\} \quad (19)$$

Proof. From (2) and (5), we have

$$E[(Y_m^{(k)})^i(Y_n^{(k)})^j] = \frac{\alpha\beta k^n}{(m-1)!(n-m-1)!}\int_0^\infty\int_x^\infty x^i y^{j+\beta-1}[-ln\bar{F}(x)]^{m-1}\frac{f(x)}{\bar{F}(x)}$$
$$\times[ln\bar{F}(x)-ln\bar{F}(y)]^{n-m-1}[\bar{F}(y)]^k dydx + \frac{\theta\delta k^n}{(m-1)!(n-m-1)!}\int_0^\infty\int_x^\infty x^i y^{j+\delta-1}$$
$$\times[-ln\bar{F}(x)]^{m-1}\frac{f(x)}{\bar{F}(x)}[ln\bar{F}(x)-ln\bar{F}(y)]^{n-m-1}[\bar{F}(y)]^k dydx.$$

Making use of Lemma 3.1, we derive the relation given in (19). Proceeding in a similar manner for the case $n = m+1$, the recurrence relation given in (18) can easily be established.

One can also note that Theorem 2.1 can be deduced from Theorem 3.1 by putting $i = 0$.

Remarks

i) Setting $\delta = 1$ or $\beta = 1$ in (19), we get the recurrence relation for product moments of k-th upper record values from the exponential-Weibull lifetime distribution as obtained by Khan *et. al* [13].

ii) Setting $\theta = 0$ or $\alpha = 0$ in (19), we get the recurrence relation for single moments of k-th upper record values obtained by Pawlas and Szynal [17] for Weibull distribution is deduced.

iii) Putting $\beta = 1$, $\theta = 0$ or $\alpha = 0$, $\delta = 1$ in (19), result for product moments of k-th upper record values is deduced for the exponential distribution as established by Pawlas and Szynal [15].

iv) Putting $\alpha = 0$, $\delta = 2$ or $\theta = 0$, $\beta = 2$ in (19), the result for product moments of k-th upper record values is deduced for the Rayleigh distribution as given by Khan *et. al* [13].

v) Setting $\beta = 2$, $\delta = 1$ or $\beta = 1$, $\delta = 2$ in (19), the result for product moments of k-th upper record values is deduced for linear failure rate distribution as established by Khan *et. al* [13].

Corollary 3.1. The recurrence relation for product moments of upper record values from the additive Weibull distribution has the form

$$E(X_{U_m}^i X_{U_n}^j) = \frac{\alpha\beta}{j+\beta}\{E(X_{U_m}^i X_{U_n}^{j+\beta}) - E(X_{U_m}^i X_{U_{n-1}}^{j+\beta})\} + \frac{\theta\delta}{j+\delta}\{E(X_{U_m}^i X_{U_n}^{j+\delta}) - E(X_{U_m}^i X_{U_{n-1}}^{j+\delta})\} \quad (20)$$

Remarks

i) If $\delta = 1$ or $\beta = 1$ in (20), we obtain the recurrence relation for product moments of record values from the exponential-Weibull lifetime distribution as established by Khan *et. al* [13].

ii) If $\theta = 0$ or $\alpha = 0$ in (20), the result for product moments of record values obtained by Pawlas and Szynal [17] for Weibull distribution is deduced.

iii) If $\beta = 1$, $\theta = 0$ or $\alpha = 0$, $\delta = 1$ in (20), the result for product moments of record values obtained by Balakrishnan and Ahsanullah [4] for the exponential distribution is deduced.

iv) If $\alpha = 0$, $\delta = 2$ or $\theta = 0$, $\beta = 2$ in (20),the recurrence relation for product moments of record values is deduced for Rayleigh distribution as given by Khan *et. al* [13].

v) If $\beta = 2$, $\delta = 1$ or $\beta = 1$, $\delta = 2$ in (20), the result for product moments of record values is deduced for linear failure rate distribution as established by Khan *et. al* [13].

4. Characterizations

Theorem 4.1. Fix a positive integer $k \geq 1$ and let j be a non-negative integer. A necessary and sufficient condition for a random variable X to be distributed with *pdf* given by (3) is that

$$E(Y_n^{(k)})^j = \frac{\alpha\beta k}{j+\beta}\{E(Y_n^{(k)})^{j+\beta} - E(Y_{n-1}^{(k)})^{j+\beta}\} + \frac{\theta\delta k}{j+\delta}\{E(Y_n^{(k)})^{j+\delta} - E(Y_{n-1}^{(k)})^{j+\delta}\}. \tag{21}$$

for $n = 1, 2, ...$

Proof. The necessary part follows immediately from (10). On the other hand if the recurrence relation (21) is satisfied, then on using (1), we have

$$\begin{aligned}
&\frac{k^n}{(n-1)!}\int_0^\infty x^j[-ln\bar{F}(x)]^{n-1}[\bar{F}(x)]^{k-1}f(x)dx \\
&= \frac{\alpha\beta k^{n+1}}{(n-1)!(j+\beta)}\int_0^\infty x^{j+\beta}[-ln\bar{F}(x)]^{n-1}[\bar{F}(x)]^{k-1}f(x)dx \\
&-\frac{\alpha\beta k^n}{(n-2)!(j+\beta)}\int_0^\infty x^{j+\beta}[-ln\bar{F}(x)]^{n-2}[\bar{F}(x)]^{k-1}f(x)dx \\
&+\frac{\theta\delta k^{n+1}}{(n-1)!(j+\delta)}\int_0^\infty x^{j+\delta}[-ln\bar{F}(x)]^{n-1}[\bar{F}(x)]^{k-1}f(x)dx \\
&-\frac{\theta\delta k^n}{(n-2)!(j+\delta)}\int_0^\infty x^{j+\delta}[-ln\bar{F}(x)]^{n-2}[\bar{F}(x)]^{k-1}f(x)dx.
\end{aligned} \tag{22}$$

Integrating the first and third integral on right side of (22) by parts and simplifying the resulting expression, we find that

$$\int_0^\infty x^j[-ln\bar{F}(x)]^{n-1}[\bar{F}(x)]^{k-1}\{f(x) - (\alpha\beta x^{\beta-1} + \theta\delta x^{\delta-1})\bar{F}(x)\}dx = 0. \tag{23}$$

Now applying a generalization of the Müntz-Szász Theorem (see for example Hwang and Lin [10]) to (23), we obtain

$$f(x) = (\alpha\beta x^{\beta-1} + \theta\delta x^{\delta-1})\bar{F}(x),$$

which proves that $f(x)$ has the form as in (5).

Remark 4.1. Theorem 4.1 can be used to characterize the exponential-Weibull lifetime and Weibull distributions by setting $\delta = 1$ or $\beta = 1$ and $\theta = 0$ or $\alpha = 0$, respectively. The exponential distribution when $\beta = 1$, $\theta = 0$ or $\alpha = 0$, $\delta = 1$. The Rayleigh and two-parameter linear failure rate distributions when $\alpha = 0$, $\delta = 2$ or $\theta = 0$, $\beta = 2$ and $\beta = 2$, $\delta = 1$ or $\beta = 1$, $\delta = 2$, respectively.

Corollary 4.1. Under the assumptions of Theorem 4.1 with $j = 0$ the following equations

$$E(Y_n^{(k)})^\delta = E(Y_{n-1}^{(k)})^\delta - \frac{\alpha}{\theta}\{E(Y_n^{(k)})^\beta - E(Y_{n-1}^{(k)})^\beta\} + \frac{1}{\theta k}, \quad n = 1, 2, ...$$

characterize the additive Weibull distribution.

Remark 4.2. If $k = 1$ we obtain the following characterization of the additive Weibull distribution

$$EX_{U_n}^\delta = EX_{U_{n-1}}^\delta - \frac{\alpha}{\theta}(EX_{U_n}^\beta - EX_{U_{n-1}}^\beta) + \frac{1}{\theta}, \quad n = 1, 2, ...$$

Now we shall show how the Theorem 4.1 can be used in a characterization of the additive-Weibull distribution in terms of moments of minimal order statistics. Putting $n = 1$ in (21), we get

$$EX^j_{1:k} = \frac{\alpha\beta k}{j+\beta}EX^{j+\beta}_{1:k} + \frac{\theta\delta k}{j+\delta}EX^{j+\delta}_{1:k},$$

for any fixed integer $k \geq 1$. This result leads to the following theorem.

Theoram 4.2. Let j be a non-negative integer. A necessary and sufficient condition for a random variable X to be distributed with *pdf* given by (3) is that

$$EX^j_{1:k} = \frac{\alpha\beta k}{j+\beta}EX^{j+\beta}_{1:k} + \frac{\theta\delta k}{j+\delta}EX^{j+\delta}_{1:k}, \tag{24}$$

for $k = 1, 2, \dots$.

Proof. The necessary part follows immediately from (10).On the other hand if the recurrence relation (24) is satisfied, then

$$\int_0^\infty x^j[\bar{F}(x)]^{k-1}f(x)dx = \frac{\alpha\beta k}{j+\beta}\int_0^\infty x^{j+\beta}[\bar{F}(x)]^{k-1}f(x)dx + \frac{\theta\delta k}{j+\delta}\int_0^\infty x^{j+\delta}[\bar{F}(x)]^{k-1}f(x)dx \tag{25}$$

Integrating the integrals on the right-hand side in (25) by parts, we get

$$\int_0^\infty x^j[\bar{F}(x)]^{k-1}f(x)dx = \alpha\beta\int_0^\infty x^{j+\beta-1}[\bar{F}(x)]^k dx + \theta\delta\int_0^\infty x^{j+\delta-1}[\bar{F}(x)]^k dx$$

which further reduces to

$$\int_0^\infty x^j[\bar{F}(x)]^{k-1}\{f(x) - (\alpha\beta x^{\beta-1} + \theta\delta x^{\delta-1})\bar{F}(x)\}dx = 0, \ \ k = 1, 2, \dots . \tag{26}$$

Now applying a generalization of the Müntz-Szász Theorem, (see for example Hwang and Lin [10]) to (26), we obtain

$$f(x) = (\alpha\beta x^{\beta-1} + \theta\delta x^{\delta-1})\bar{F}(x)$$

which proves that

$$F(x) = 1 - e^{-(\alpha x^\beta + \theta x^\delta)}, \quad x > 0, \alpha > 0, \theta > 0 \text{ and } \delta, \beta > 0.$$

Theorem 4.3. Let X be a non-negative random variable having an absolutely continuous *df* $F(x)$ with $F(0) = 0$ and $0 \leq F(x) \leq 1$ for all $x > 0$, then

$$E[\xi(Y_n^{(k)}) \mid (Y_l^{(k)}) = x] = e^{-(\alpha x^\beta + \theta x^\delta)}\left(\frac{k}{k+1}\right)^{n-l}, \ \ l = m, \ \ m+1 \tag{27}$$

if and only if

$F(x) = 1 - e^{-(\alpha x^\beta + \theta x^\delta)}$,

where

$\xi(y) = e^{-(\alpha y^\beta + \theta y^\delta)}$,

Proof. From (2) and (1), we have

$$E[\xi(Y_n^{(k)}) \mid (Y_m^{(k)}) = x] = \frac{k^{n-m}}{(n-m-1)!}\int_x^\infty e^{-(\alpha y^\beta + \theta y^\delta)}[ln\bar{F}(x) - ln\bar{F}(y)]^{n-m-1}\left(\frac{\bar{F}(y)}{\bar{F}(x)}\right)^{k-1}\frac{f(y)}{\bar{F}(x)}dy. \tag{28}$$

By setting $u = \frac{\bar{F}(y)}{\bar{F}(x)} = \frac{e^{-(\alpha y^\beta + \theta y^\delta)}}{e^{-(\alpha x^\beta + \theta x^\delta)}}$ from (4) in (28), we have

$$E[\xi(Y_n^{(k)}) \mid (Y_m^{(k)}) = x] = \frac{k^{n-m}}{(n-m-1)!}e^{-(\alpha x^\beta + \theta x^\delta)}\int_0^1 u^k(-lnu)^{n-m-1}du. \tag{29}$$

We have Gradshteyn and Ryzhik ([6], p-551)

$$\int_0^1 (-lnx)^{\mu-1} x^{\nu-1} dx = \frac{\Gamma\mu}{\nu^{\mu}}, \quad \mu > 0, \ \nu > 0. \tag{30}$$

On using (30) in (29), we have the result given in (27).
To prove sufficient part, we have

$$\frac{k^{n-m}}{(n-m-1)!} \int_x^{\infty} e^{-(\alpha y^{\beta}+\theta y^{\delta})} [ln\bar{F}(x) - ln\bar{F}(y)]^{n-m-1} [\bar{F}(y)]^{k-1} f(y) dy = [\bar{F}(x)]^k g_{n|m}(x), \tag{31}$$

where
$g_{n|m}(x) = e^{-(\alpha x^{\beta}+\theta x^{\delta})} \left(\frac{k}{k+1}\right)^{n-m}$.
Differentiating (31) both the sides with respect to x , we get

$$-\frac{k^{n-m} f(x)}{\bar{F}(x)(n-m-2)!} \int_x^{\infty} e^{-(\alpha y^{\beta}+\theta y^{\delta})} [ln\bar{F}(x) - ln\bar{F}(y)]^{n-m-2}$$

$$\times [\bar{F}(y)]^{k-1} f(y) dy = g'_{n|m}(x) [\bar{F}(x)]^k - k \, g_{n|m}(x) [\bar{F}(x)]^{k-1} f(x)$$

or

$$-k \, g_{n|m+1}(x) [\bar{F}(x)]^{k-1} f(x) = g'_{n|m}(x) [\bar{F}(x)]^k - k \, g_{n|m}(x) [\bar{F}(x)]^{k-1} f(x).$$

Therefore,

$$\frac{f(x)}{\bar{F}(x)} = -\frac{g'_{n|m}(x)}{k[g_{n|m+1}(x) - g_{n|m}(x)]} = (\alpha\beta x^{\beta-1} + \theta\delta x^{\delta-1}),$$

where

$$g'_{n|m}(x) = -(\alpha\beta x^{\beta-1} + \theta\delta x^{\delta-1}) e^{-(\alpha x^{\beta}+\theta x^{\delta})} \left(\frac{k}{k+1}\right)^{n-m},$$

$$g_{n|m+1}(x) - g_{n|m}(x) = \frac{1}{k} e^{-(\alpha x^{\beta}+\theta x^{\delta})} \left(\frac{k}{k+1}\right)^{n-m},$$

which proves that
$F(x) = 1 - e^{-(\alpha x^{\beta}+\theta x^{\delta})}, \quad x > 0, \ \alpha > 0, \ \theta > 0 \text{ and } \delta, \ \beta > 0.$

Acknowledgement

The authors acknowledge their gratefulness to the Chief-Editor and learned referee for their valuable comments and suggestions.

REFERENCES

1. M. Ahsanullah, *Record Statistics*, Nova Science Publishers, New York, 1995.
2. M. Ahsanullah and V.B. Nevzorov, *Record via Probability Theory*, Atlantis Press, Paris, 2015.
3. B.C. Arnold, N. Balakrishnan and H.N. Nagaraja, *Records*, John Wiley, New York, 1998.
4. N. Balakrishnan and M. Ahsanullah, *Relations for single and product moments of record values from exponential distribution*, Journal of Applied Statistical Science, vol. 2, pp. 73-87, 1995.
5. K.N. Chandler, *The distribution and frequency of record values*, Journal of the Royal Statistical Society. Series B, vol. 14, pp. 220-228, 1952.
6. K. Danielak and M.Z. Raqab, *Sharp bounds for expectations of k-th record increments*, Australian and New Zealand Journal of Statistics, vol. 46, pp. 665-673, 2004.
7. W. Dziubdziela and B. Kopociński, *Limiting properties of the k-th record value*, Applicationes Mathematicae, vol. 15, pp. 187-190,1976.
8. I.S. Gradshteyn and I.M. Ryzhik, *Table of Integrals, Series and Products*, Academic Press, New York, 2007.

9. Z. Grudzień, *Characterization of distribution of time limits in record statistics as well as distributions and moments of linear record statistics from the samples of random numbers,* Praca Doktorska, UMCS, Lublin, 1982.
10. J.S. Hwang and G.D. Lin, *On a generalized moments problem II,* Proceedings of the American Mathematical Society, vol. 91, pp. 577-580, 1984.
11. U. Kamps, *A Concept of Generalized Order Statistics,* B.G. Teubner Stuttgart, Germany, 1995.
12. M.A. Khan and R.U. Khan, *k-th upper record values from modified Weibull distribution and characterization,* International Journal of Computational and Theoretical Statistics, vol. 3, pp. 75-80, 2016.
13. R.U. Khan, A. Kulshrestha and M.A. Khan, *Relations for moments of k-th record values from exponential-Weibull lifetime distribution and a characterization,* Journal of the Egyptian Mathematical Society, vol. 23, pp. 558-562, 2015.
14. A.J. Lemonte, G.M. Cordeiro and E.M.M. Ortega, *On the additive Weibull distribution,* Journal of Statistical Computation and Simulation, vol. 43, pp. 2066-2080, 2014.
15. P. Pawlas and D. Szynal, *Relations for single and product moments of k-th record values from exponential and Gumble distributions,* Journal of Applied Statistical Science, vol. 7, pp. 53-62, 1998.
16. P. Pawlas and D. Szynal, *Recurrence relations for single and product moments of k-th record values from Pareto, generalized Pareto and Burr distributions,* Communications in Statistics - Theory and Methods, vol. 28, pp. 1699-1709, 1999.
17. P. Pawlas and D. Szynal, *Recurrence relations for single and product moments of k-th record values from Weibull distribution and a characterization,* Journal of Applied Statistical Science, vol. 10, pp. 17-26, 2000.

7

System of nonlinear variational inclusion problems with (A, η)-maximal monotonicity in Banach spaces

N. K. Sahu [1,*], N. K. Mahato [2], Ram N. Mohapatra [3]

[1]Dhirubhai Ambani Institute of Information and Communication Technology, Gandhinagar, India
[2] PDPM IIITDM, Jabalpur, India
[3] University of Central Florida, Orlando, FL. 32816, USA

Abstract This paper deals with a new system of nonlinear variational inclusion problems involving (A, η)-maximal relaxed monotone and relative (A, η)-maximal monotone mappings in 2-uniformly smooth Banach spaces. Using the generalized resolvent operator technique, the approximation solvability of the proposed problem is investigated. An iterative algorithm is constructed to approximate the solution of the problem. Convergence analysis of the proposed algorithm is investigated. Similar results are also proved for other system of variational inclusion problems involving relative (A, η)-maximal monotone mappings and (H, η)-maximal monotone mappings.

Keywords Variational inclusion, Generalized resolvent operator, 2-uniformly smooth Banach space, Semi-inner product space

1. Introduction

Variational inequalities have been well studied and generalized to different directions due to its large association with partial differential equations and optimization problems. Variational inclusion problem is a natural generalization of variational inequality problem and it is of recent interest. Monotonicity plays a prominent role in the solvability of variational inclusion problem. Researchers have been studied different kind of monotonicity such as maximal monotonicity, relaxed monotonicity, H-monotonicity, A-monotonicity etc., and solved different variational inclusion problems taking the help of such monotonicity of the underlying operator. The method based on the resolvent operator technique is a generalization of projection method and has been widely used to solve variational inclusion problems (see [1], [3][5], [6], [7], [9], [11] and [12]).

Peng and Zhu [14] introduced and studied a new system of variational inclusion problems involving (H, η)-monotone operators. Verma [16] solved a new class of set valued variational inclusions involving (A, η)-monotone operators in Hilbert space. He also studied the notion of (A, η)-maximal relaxed monotonicity in Verma [17]. He examined the sensitivity analysis for quasi-variational inclusion problems using (A, η)-maximal relaxed monotonicity. Recently, Agarwal and Verma [2] solved a new system of variational inclusion problems involving (A, η)-maximal relaxed monotone mappings and relative (A, η)-maximal monotone mappings in Hilbert space.

*Correspondence to: N. K. Sahu (Email: nabindaiict@gmail.com). Dhirubhai Ambani Institute of Information and Communication Technology, Gandhinagar, India.

More recently, Sahu et al. [15] have studied a class of A-monotone implicit variational inclusion problems in semi-inner product spaces.

In this paper, we generalize the work of Agarwal and Verma [2] from Hilbert space to 2-uniformly smooth Banach space. Although the results are similar, here we use different notion and techniques to solve those system of variational inclusion problems in 2-uniformly smooth Banach space. People generally take the help of bounded linear functionals to study those type of problems in Banach space. Instead of using arbitrary bounded linear functionals, we take the help of semi-inner product structure introduced by Lumer [13]. Semi-inner product helps us in a number of aspects for the smooth study of the problem. So first we afford a view of semi-inner product and its important features that we use in our work.

Definition 1.1

(Lumer [13]) Let X be a vector space over the field F of real or complex numbers. A functional $[.,.]: X \times X \to F$ is called a semi-inner product if it satisfies the following:

1. $[x+y,z] = [x,z] + [y,z], \ \forall x,y,z \in X$;
2. $[\lambda x, y] = \lambda[x,y], \ \forall \lambda \in F$ and $x, y \in X$;
3. $[x,x] > 0$, for $x \neq 0$;
4. $|[x,y]|^2 \leqslant [x,x][y,y]$.

The pair $(X, [.,.])$ is called a semi-inner product space.

A semi-inner product space is a normed linear space with the norm $\|x\| = [x,x]^{\frac{1}{2}}$. Every normed linear space can be made into a semi-inner product space in infinitely many different ways. Giles [8] had shown that if the underlying space X is a uniformly convex smooth Banach space then it is possible to define a semi-inner product uniquely. For a detailed study and fundamental results on semi-inner product spaces, one may refer to Lumer [13], Giles [8] and Koehler [10].

Example 1.1

The real Banach space $L^p(\mathbb{R})$ for $1 < p < \infty$ is a uniformly convex smooth Banach space with a unique semi-inner product defined by

$$[f,g] = \frac{1}{\|g\|_p^{p-2}} \int_{\mathbb{R}} f(t)|g(t)|^{p-1} sgn(g(t))dt, \quad f,g \in L^p(\mathbb{R}).$$

Example 1.2

The real sequence space l^p for $1 < p < \infty$ is a uniformly convex smooth space with a unique semi-inner product defined by

$$[x,y] = \frac{1}{\|y\|_p^{p-2}} \sum_i x_i y_i |y_i|^{p-2}, \quad x,y \in l^p.$$

Definition 1.2

(Xu [18]) Let X be a real Banach space. The modulus of smoothness of X is defined as

$$\rho_X(t) = \sup\left\{ \frac{\|x+y\| + \|x-y\|}{2} - 1 : \ \|x\| = 1, \|y\| = t, t > 0 \right\}.$$

X is said to be uniformly smooth if $\lim\limits_{t \to 0} \dfrac{\rho_X(t)}{t} = 0$.

X is said to be p-uniformly smooth if there exists a positive real constant c such that $\rho_X(t) \leq ct^p, \ p > 1$.

X is said to be 2-uniformly smooth if there exists a positive real constant c such that $\rho_X(t) \leq ct^2$.

Lemma 1.1

(Xu [18]) Let X be a smooth Banach space. Then the following statements are equivalent:

(i) X is 2-uniformly smooth.

(ii) There is a constant $c > 0$ such that for every $x, y \in X$, the following inequality holds

$$\|x+y\|^2 \leq \|x\|^2 + 2\langle y, f_x \rangle + c\|y\|^2, \tag{1.1}$$

where $f_x \in J(x)$ and $J(x) = \{x^* \in X^* : \langle x, x^* \rangle = \|x\|^2 \text{ and } \|x^*\| = \|x\|\}$ is the normalized duality mapping, where X^* denotes the dual space of X and $\langle x, x^* \rangle$ denotes the value of the functional x^* at x, that is $x^*(x)$.

Remark 1.1
Every normed linear space is a semi-inner product space (see Lumer [13]). In fact by Hahn Banach theorem, for each $x \in X$, there exists at least one functional $f_x \in X^*$ such that $\langle x, f_x \rangle = \|x\|^2$. Given any such mapping f from X into X^*, it has been verified that $[y, x] = \langle y, f_x \rangle$ defines a semi-inner product. Hence we can write the inequality (1.1) as

$$\|x + y\|^2 \leq \|x\|^2 + 2[y, x] + c\|y\|^2, \forall x, y \in X. \tag{1.2}$$

The constant c is chosen with best possible minimum value. We call c, as the constant of smoothness of X.

Example 1.3
The functions space L^p is 2-uniformly smooth for $p \geq 2$ and it is p-uniformly smooth for $1 < p < 2$.
If $2 \leq p < \infty$, then we have for all $x, y \in L^p$,

$$\|x + y\|^2 \leq \|x\|^2 + 2[y, x] + (p-1)\|y\|^2.$$

Here the constant of smoothness is $p - 1$.

2. Preliminaries

In the vicinity of the paper, we assume that X, is a real 2-uniformly smooth Banach space endowed with norm $\|.\|$ and semi-inner product $[.,.]$, d is the metric induced by the norm, 2^X is the power set of X, $CB(X)$ is the set of all closed and bounded subsets of X and $D(.,.)$ is the Housdorff metric on $CB(X)$ defined by

$$D(A, B) = \max\big\{\sup_{x\in A} d(x, B), \sup_{y\in B} d(A, y)\big\},$$

where $d(x, B) = \inf_{y\in B} d(x, y)$ and $d(A, y) = \inf_{x\in A} d(x, y)$.

Definition 2.1
Let $T, A : X \to X$ be single valued mappings. The operator T is said to be
(i) monotone if $[Tx - Ty, x - y] \geq 0, \forall x, y \in X$;
(ii) monotone with respect to A if $[Tx - Ty, Ax - Ay] \geq 0, \forall x, y \in X$;
(iii) strictly monotone if T is monotone and $[Tx - Ty, x - y] = 0$ if and only if $x = y$;
(iv) r-strongly monotone if there exists a constant $r > 0$ such that

$$[Tx - Ty, x - y] \geq r\|x - y\|^2, \forall x, y \in X;$$

(v) r-strongly monotone with respect to A if there exists a constant $r > 0$ such that

$$[Tx - Ty, Ax - Ay] \geq r\|x - y\|^2, \forall x, y \in X;$$

(vi) m-relaxed monotone if there exists a constant $m > 0$ such that

$$[Tx - Ty, x - y] \geq (-m)\|x - y\|^2, \forall x, y \in X;$$

(vii) m-relaxed monotone with respect to A if there exists a constant $m > 0$ such that

$$[Tx - Ty, Ax - Ay] \geq (-m)\|x - y\|^2, \forall x, y \in X;$$

(viii) (c, μ)-relaxed cocoercive if there exist constant $c, \mu > 0$ such that

$$[Tx - Ty, x - y] \geq (-c)\|Tx - Ty\|^2 + \mu\|x - y\|^2, \forall x, y \in X;$$

(ix) (c, μ)-relaxed cocoercive with respect to A if there exist constant $c, \mu > 0$ such that

$$[Tx - Ty, Ax - Ay] \geq (-c)\|Tx - Ty\|^2 + \mu\|x - y\|^2, \forall x, y \in X;$$

(x) s-Lipschitz continuous if there exists a constant $s > 0$ such that

$$\|Tx - Ty\| \leq s\|x - y\|, \forall x, y \in X.$$

In the following example we see that a map is not monotone but it is monotone with respect to another operator.

Example 2.1
Let $T : \mathbb{R} \to \mathbb{R}$ be a map defined by $T(x) = -x$ and $A : \mathbb{R} \to \mathbb{R}$ be defined by $A(x) = -\frac{1}{2}x$. One can easily see that T is not monotone, but it is monotone with respect to A.

Definition 2.2
A single valued map $\eta : X \times X \to X$ is said to be τ-Lipschitz continuous if there exists a constant $\tau > 0$ such that $\|\eta(x, y)\| \leq \tau\|x - y\|$ for all $x, y \in X$.

Let $M : X \to 2^X$ be a set valued map. We denote both the mapping and its graph by M, that is $M = \{(x, y) : y \in M(x)\}$. The domain of M is defined by

$$D(M) = \{x \in X : \exists y \in X : (x, y) \in M\}.$$

The range of M is defined by

$$R(M) = \{y \in X : \exists x \in X : (x, y) \in M\}.$$

The inverse M^{-1} of M is $\{(y, x) : (x, y) \in M\}$.
For any two set valued mappings $N, M : X \to 2^X$ and any real number ρ, we define

$$N + M = \{(x, y + z) : (x, y) \in N, (x, z) \in M\},$$
$$\rho M = \{(x, \rho y) : (x, y) \in M\}.$$

For a map $A : X \to X$ and a set valued map $M : X \to 2^X$, we define

$$A + M = \{(x, y + z) : Ax = y \text{ and } (x, z) \in M\}.$$

Definition 2.3
Let $\eta : X \times X \to X$ and $A : X \to X$ be single valued mappings. Let $M : X \to 2^X$ be a set valued map. Then the map M is said to be:
(i) monotone if $[u - v, x - y] \geq 0$ for all $(x, u) \in M$, $(y, v) \in M$;
(ii) monotone with respect to A if $[u - v, Ax - Ay] \geq 0$ for all $(x, u) \in M$, $(y, v) \in M$;
(iii) η-monotone if $[u - v, \eta(x, y)] \geq 0$ for all $(x, u) \in M$, $(y, v) \in M$;
(iv) η-monotone with respect to A if $[u - v, \eta(Ax, Ay)] \geq 0$ for all $(x, u) \in M$, $(y, v) \in M$;
(v) strictly η-monotone if M is η-monotone and equality holds if and only if $x = y$;
(vi) (r, η)-strongly monotone if there exists a constant $r > 0$ such that $[u - v, \eta(x, y)] \geq r\|x - y\|^2$ for all $(x, u) \in M$, $(y, v) \in M$;
(vii) (r, η)-strongly monotone with respect to A if there exists a constant $r > 0$ such that $[u - v, \eta(Ax, Ay)] \geq r\|x - y\|^2$ for all $(x, u) \in M$, $(y, v) \in M$;
(viii) (m, η)-relaxed monotone if there exists a constant $m > 0$ such that $[u - v, \eta(x, y)] \geq (-m)\|x - y\|^2$ for all $(x, u) \in M$, $(y, v) \in M$;
(ix) (m, η)-relaxed monotone with respect to A if there exists a constant $m > 0$ such that $[u - v, \eta(Ax, Ay)] \geq (-m)\|x - y\|^2$ for all $(x, u) \in M$, $(y, v) \in M$;
(x) maximal monotone if M is monotone and $(I + \rho M)(X) = X$ for all $\rho > 0$, where I is the identity mapping on X;
(xi) maximal η-monotone if M is η-monotone and $(I + \rho M)(X) = X$ for all $\rho > 0$.

Definition 2.4
Let $A : X \to X$, $H : X \to X$ and $\eta : X \times X \to X$ be single valued mappings. Let $M : X \to 2^X$ be a set valued map. Then M is said to be:
(i) A-maximal m-relaxed monotone if M is m-relaxed monotone and $(A + \rho M)(X) = X$ for all $\rho > 0$;
(ii) Relative A-maximal m-relaxed monotone if M is m-relaxed monotone with respect to A and $(A + \rho M)(X) = X$ for all $\rho > 0$;
(iii) (A, η)-maximal m-relaxed monotone if M is (m, η)-relaxed monotone and $(A + \rho M)(X) = X$ for all $\rho > 0$;
(iv) Relative (A, η)-maximal m-relaxed monotone if M is (m, η)-relaxed monotone with respect to A and $(A + \rho M)(X) = X$ for all $\rho > 0$;
(v) H-maximal monotone if M is monotone and $(H + \rho M)(X) = X$ for all $\rho > 0$;
(vi) (H, η)-maximal monotone if M is η-monotone and $(H + \rho M)(X) = X$ for all $\rho > 0$.

Lemma 2.1
Let $\eta : X \times X \to X$ be a single valued mapping, $A : X \to X$ be a strictly η-monotone mapping, and $M : X \to 2^X$ be an (A, η)-maximal m-relaxed monotone mapping. Then the mapping $(A + \rho M)^{-1}$ is single valued.

Proof
For a given $u \in X$, assume that $x, y \in (A + \rho M)^{-1}(u)$. Then we have

$$\frac{1}{\rho}(u - A(x)) \in M(x); \quad \frac{1}{\rho}(u - A(y)) \in M(y).$$

Since M is (m, η)-relaxed monotone, it follows that

$$\frac{1}{\rho}[u - A(x) - (u - A(y)), \eta(x, y)] \geq (-m)\|x - y\|^2$$
$$\Rightarrow [Ax - Ay, \eta(x, y)] \leq (\rho m)\|x - y\|^2.$$

Again A is strictly η-monotone, that is $[Ax - Ay, \eta(x, y)] \geq 0, \ \forall x, y \in X$ and and equality hold if and only if $x = y$. Hence we have

$$0 \leq [Ax - Ay, \eta(x, y)] \leq (\rho m)\|x - y\|^2.$$

This is possible only when $x = y$, since ρ and m are positive. Hence one can conclude that the operator $(A + \rho M)^{-1}$ is single valued. □

Now we can define the generalized resolvent operator associated with (A, η)-maximal m-relaxed monotone mapping.

Definition 2.5
Let $\eta : X \times X \to X$ be a single valued mapping, $A : X \to X$ be a strictly η-monotone mapping, and $M : X \to 2^X$ be an (A, η)-maximal m-relaxed monotone mapping. Then the generalized resolvent operator $J_{\rho,M}^{A,\eta} : X \to X$ is defined by

$$J_{\rho,M}^{A,\eta}(x) = (A + \rho M)^{-1}(x) \ \text{ for all } x \in X,$$

where ρ is a positive constant.

Lemma 2.2
Let $\eta : X \times X \to X$ be a τ-Lipschitz continuous map, $A : X \to X$ be an (r, η)-strongly monotone mapping and $M : X \to 2^X$ be an (A, η)-maximal m-relaxed monotone mapping. Then the generalized resolvent operator $J_{\rho,M}^{A,\eta} : X \to X$ is $\frac{\tau}{r - \rho m}$-Lipschitz continuous for $0 < \rho < \frac{r}{m}$.

Proof
For any $x, y \in X$, we have

$$J_{\rho,M}^{A,\eta}(x) = (A+\rho M)^{-1}(x)$$
$$J_{\rho,M}^{A,\eta}(y) = (A+\rho M)^{-1}(y).$$

This implies that

$$\frac{1}{\rho}\{x - AJ_{\rho,M}^{A,\eta}(x)\} \in MJ_{\rho,M}^{A,\eta}(x)$$
$$\frac{1}{\rho}\{y - AJ_{\rho,M}^{A,\eta}(y)\} \in MJ_{\rho,M}^{A,\eta}(y).$$

Since M is (A, η)-maximal m-relaxed monotone, we have

$$\begin{aligned} & \frac{1}{\rho}[x - AJ_{\rho,M}^{A,\eta}(x) - \{y - AJ_{\rho,M}^{A,\eta}(y)\}, \eta(J_{\rho,M}^{A,\eta}(x), J_{\rho,M}^{A,\eta}(y))] \\ = & \frac{1}{\rho}[x - y - \{AJ_{\rho,M}^{A,\eta}(x) - AJ_{\rho,M}^{A,\eta}(y)\}, \eta(J_{\rho,M}^{A,\eta}(x), J_{\rho,M}^{A,\eta}(y))] \\ \geq & (-m)\|J_{\rho,M}^{A,\eta}(x) - J_{\rho,M}^{A,\eta}(y)\|^2. \end{aligned}$$

Again using the above inequality and (r, η)-strong monotonicity of A, we get

$$\begin{aligned} & \|x-y\|\|\eta(J_{\rho,M}^{A,\eta}(x), J_{\rho,M}^{A,\eta}(y))\| \\ \geq & [x-y, \eta(J_{\rho,M}^{A,\eta}(x), J_{\rho,M}^{A,\eta}(y))] \\ = & [x - y - \{AJ_{\rho,M}^{A,\eta}(x) - AJ_{\rho,M}^{A,\eta}(y)\}, \eta(J_{\rho,M}^{A,\eta}(x), J_{\rho,M}^{A,\eta}(y))] \\ & +[AJ_{\rho,M}^{A,\eta}(x) - AJ_{\rho,M}^{A,\eta}(y), \eta(J_{\rho,M}^{A,\eta}(x), J_{\rho,M}^{A,\eta}(y))] \\ \geq & -\rho m\|J_{\rho,M}^{A,\eta}(x) - J_{\rho,M}^{A,\eta}(y)\|^2 + r\|J_{\rho,M}^{A,\eta}(x) - J_{\rho,M}^{A,\eta}(y)\|^2 \\ = & (r-\rho m)\|J_{\rho,M}^{A,\eta}(x) - J_{\rho,M}^{A,\eta}(y)\|^2. \end{aligned}$$

Applying the τ-Lipschitz continuity of η in the above inequality, we get

$$\begin{aligned} \|x-y\|\tau\|J_{\rho,M}^{A,\eta}(x) - J_{\rho,M}^{A,\eta}(y)\| & \geq \|x-y\|\|\eta(J_{\rho,M}^{A,\eta}(x), J_{\rho,M}^{A,\eta}(y))\| \\ & \geq (r-\rho m)\|J_{\rho,M}^{A,\eta}(x) - J_{\rho,M}^{A,\eta}(y)\|^2. \end{aligned}$$

This implies that

$$\Rightarrow \|J_{\rho,M}^{A,\eta}(x) - J_{\rho,M}^{A,\eta}(y)\| \leq \frac{\tau}{r-\rho m}\|x-y\|, \text{ where } 0 < \rho < \frac{r}{m}.$$

□

Lemma 2.3
Let $\eta : X \times X \to X$ be t-strongly monotone and τ-Lipschitz continuous, $A : X \to X$ be r-strongly monotone, and let $M : X \to 2^X$ be a relative (A, η)-maximal monotone mapping. Then the resolvent operator $J_{\rho,M}^{A,\eta} : X \to X$ is $\frac{\tau}{rt}$-Lipschitz continuous.

Proof
For any $x, y \in X$, we have

$$J_{\rho,M}^{A,\eta}(x) = (A+\rho M)^{-1}(x)$$
$$J_{\rho,M}^{A,\eta}(y) = (A+\rho M)^{-1}(y).$$

This implies that

$$\frac{1}{\rho}\{x - AJ^{A,\eta}_{\rho,M}(x)\} \in MJ^{A,\eta}_{\rho,M}(x)$$
$$\frac{1}{\rho}\{y - AJ^{A,\eta}_{\rho,M}(y)\} \in MJ^{A,\eta}_{\rho,M}(y).$$

Since M is η-monotone with respect to A, we have

$$\frac{1}{\rho}[x - y - \{AJ^{A,\eta}_{\rho,M}(x) - AJ^{A,\eta}_{\rho,M}(y)\}, \eta\big(AJ^{A,\eta}_{\rho,M}(x), AJ^{A,\eta}_{\rho,M}(y)\big)] \geq 0.$$

Using the t-strongly monotonicity of η in the above inequality, we get

$$\begin{aligned}[x - y, \eta\big(AJ^{A,\eta}_{\rho,M}(x), AJ^{A,\eta}_{\rho,M}(y)\big)] &\geq [AJ^{A,\eta}_{\rho,M}(x) - AJ^{A,\eta}_{\rho,M}(y), \eta\big(AJ^{A,\eta}_{\rho,M}(x), AJ^{A,\eta}_{\rho,M}(y)\big)] \\ &\geq t\|AJ^{A,\eta}_{\rho,M}(x) - AJ^{A,\eta}_{\rho,M}(y)\|^2.\end{aligned}$$

Again using the τ-Lipschitz continuity of η in the above inequality, we get

$$\begin{aligned}\|x - y\|\tau\|AJ^{A,\eta}_{\rho,M}(x) - AJ^{A,\eta}_{\rho,M}(y)\| &\geq \|x - y\|\|\eta\big(AJ^{A,\eta}_{\rho,M}(x), AJ^{A,\eta}_{\rho,M}(y)\big)\| \\ &\geq [x - y, \eta\big(AJ^{A,\eta}_{\rho,M}(x), AJ^{A,\eta}_{\rho,M}(y)\big)] \\ &\geq t\|AJ^{A,\eta}_{\rho,M}(x) - AJ^{A,\eta}_{\rho,M}(y)\|^2.\end{aligned}$$

$$\Rightarrow \|AJ^{A,\eta}_{\rho,M}(x) - AJ^{A,\eta}_{\rho,M}(y)\| \leq \frac{\tau}{t}\|x - y\|.$$

Since A is r-strongly monotone, we have

$$\begin{aligned}r\|J^{A,\eta}_{\rho,M}(x) - J^{A,\eta}_{\rho,M}(y)\|^2 &\leq [AJ^{A,\eta}_{\rho,M}(x) - AJ^{A,\eta}_{\rho,M}(y), J^{A,\eta}_{\rho,M}(x) - J^{A,\eta}_{\rho,M}(y)] \\ &\leq \|AJ^{A,\eta}_{\rho,M}(x) - AJ^{A,\eta}_{\rho,M}(y)\|\|J^{A,\eta}_{\rho,M}(x) - J^{A,\eta}_{\rho,M}(y)\| \\ &\leq \frac{\tau}{t}\|x - y\|\|J^{A,\eta}_{\rho,M}(x) - J^{A,\eta}_{\rho,M}(y)\|.\end{aligned}$$

$$\Rightarrow \|J^{A,\eta}_{\rho,M}(x) - J^{A,\eta}_{\rho,M}(y)\| \leq \frac{\tau}{rt}\|x - y\|.$$

□

As the Lemma 2.1 and Lemma 2.2, we can have the following result for (H, η)-maximal monotone mapping.

Lemma 2.4
Let $\eta : X \times X \to X$ be a single valued mapping, $H : X \to X$ be a strictly η-monotone mapping, and $M : X \to 2^X$ be an (H, η)-maximal monotone mapping. Then the mapping $(H + \rho M)^{-1}$ is single valued.

Definition 2.6
Let $\eta : X \times X \to X$ be a single valued mapping, $H : X \to X$ be a strictly η-monotone mapping, and $M : X \to 2^X$ be an (H, η)-maximal monotone mapping. Then the generalized resolvent operator $J^{H,\eta}_{\rho,M} : X \to X$ is defined by

$$J^{H,\eta}_{\rho,M}(x) = (H + \rho M)^{-1}(x) \text{ for all } x \in X,$$

where ρ is a positive constant.

Lemma 2.5
Let $\eta : X \times X \to X$ be a τ-Lipschitz continuous map, $H : X \to X$ be an (r, η)-strongly monotone mapping and $M : X \to 2^X$ be an (H, η)-maximal monotone mapping. Then the generalized resolvent operator $J^{H,\eta}_{\rho,M} : X \to X$ is $\frac{\tau}{r}$-Lipschitz continuous.

Housdorff pseudo-metric $\mathcal{D} : 2^X \times 2^X \to \mathbb{R} \cup \{+\infty\}$ is defined as

$$\mathcal{D}(A, B) = \max\left\{ \sup_{x\in A} \inf_{y\in B} \|x - y\|, \sup_{x\in B} \inf_{y\in A} \|x - y\| \right\}$$

for that any two subsets A and B of X. When the domain 2^X is restricted to the closed bounded subsets of X, then $\mathcal{D}$ is the Housdorff metric.

Definition 2.7
A set valued map $M : X \to 2^X$ is said to be $\mathcal{D}$-Lipschitz continuous if there exists a constant $\nu > 0$ such that

$$\mathcal{D}(M(x), M(y)) \leq \nu\|x - y\| \text{ for all } x, y \in X.$$

3. System of nonlinear variational inclusion problems

In this section we state the problem that we are intended to solve and discuss its various forms. We consider the following general system of nonlinear variational inclusion problems involving (A, η)-maximal m-relaxed monotone mappings.

Let X_1 and X_2 be 2-uniformly smooth real Banach spaces, and K_1 and K_2 be two nonempty, closed and convex subsets X_1 and X_2, respectively. Let $F : X_1 \times X_2 \to X_1$, $G : X_1 \times X_2 \to X_2$, $A_i : X_i \to X_i$ and $\eta_i : X_i \times X_i \to X_i$ $(i = 1, 2)$ be nonlinear single valued mappings. Let $U : X_1 \to 2^{X_1}$, $V : X_2 \to 2^{X_2}$ be set valued mappings, and let $M_i : X_i \to 2^{X_i}$ be (A_i, η_i)-maximal m_i-relaxed monotone mappings $(i = 1, 2)$. We consider the problem of finding an element $(a, b) \in X_1 \times X_2$, $u \in U(a)$, $v \in V(b)$ such that

$$\begin{cases} 0 \in F(a, v) + M_1(a) \\ 0 \in G(u, b) + M_2(b). \end{cases} \tag{3.3}$$

Special Cases:
Case (I): When $M_1(x) = \partial\varphi(x)$ and $M_2(y) = \partial\phi(y)$ for all $x \in X_1$ and $y \in X_2$, where $\varphi : X_1 \to \mathbb{R} \cup \{+\infty\}$ and $\phi : X_2 \to \mathbb{R} \cup \{+\infty\}$ are two proper, convex and lower semi-continuous functionals, $\partial\varphi$ and $\partial\phi$ are subdifferentials of φ and ϕ, respectively. Then the problem (3.3) reduces to the following problem:
Find $(a, b) \in X_1 \times X_2$, $u \in U(a)$, $v \in V(b)$ such that

$$\begin{cases} [F(a, v), x - a] + \varphi(x) - \varphi(a) \geq 0, \forall x \in X_1 \\ [G(u, b), y - b] + \phi(y) - \phi(b) \geq 0, \forall y \in X_2. \end{cases} \tag{3.4}$$

The above system is called a system of set valued mixed variational inequality problems.
Case (II): If U and V are both identity mappings, then problem (3.4) reduces to the problem:
Find $(a, b) \in X_1 \times X_2$ such that

$$\begin{cases} [F(a, b), x - a] + \varphi(x) - \varphi(a) \geq 0, \forall x \in X_1 \\ [G(a, b), y - b] + \phi(y) - \phi(b) \geq 0, \forall y \in X_2. \end{cases} \tag{3.5}$$

The above nonlinear variational inequality problem was studied by Cho et al. [4].
Case (III): If $M_1(x) = \partial\delta_{K_1}(x)$ and $M_2(y) = \partial\delta_{K_2}(y)$ for all $x \in K_1$ and $y \in K_2$, where K_1 and K_2 are nonempty closed convex subsets of X_1 and X_2, respectively, and δ_{K_1} and δ_{K_2} are indicator functions of K_1 and K_2, respectively. Then the problem (3.4) reduces to the following problem:
Find $(a, b) \in\in K_1 \times K_2$ such that

$$\begin{cases} [F(a, b), x - a] \geq 0, \forall x \in K_1 \\ [G(a, b), y - b] \geq 0, \forall y \in K_2. \end{cases} \tag{3.6}$$

4. Existence of solution and approximation solvability

This section delineates about the existence of solution of the proposed variational inclusion problem (3.3). We start with converting the system of variational inclusion problems into fixed point type of problems by taking the help of generalized resolvent operator. Then we prove the existence of solutions for the system. We construct iterative algorithm to approximate the solution of the proposed problem and discuss its convergency.

In the following theorem, an equivalent formulation of the problem (3.3) is given. Through out this section we assume that X_1 and X_2 are 2- uniformly smooth real Banach spaces, and K_1 and K_2 be two nonempty, closed and convex subsets X_1 and X_2, respectively. Let $F : X_1 \times X_2 \to X_1$, $G : X_1 \times X_2 \to X_2$, $A_i : X_i \to X_i$ and $\eta_i : X_i \times X_i \to X_i$ $(i = 1, 2)$ be nonlinear single valued mappings. Let $U : X_1 \to 2^{X_1}$, $V : X_2 \to 2^{X_2}$ be set valued mappings, and let $M_i : X_i \to 2^{X_i}$ be (A_i, η_i)-maximal m_i-relaxed monotone mappings $(i = 1, 2)$.

Theorem 4.1
Let $(a, b) \in X_1 \times X_2$, $u \in U(a)$, $v \in V(b)$. Then (a, b, u, v) is a solution of the problem (3.3) if and only if (a, b, u, v) satisfies

$$\begin{cases} a = J^{A_1,\eta_1}_{\rho_1,M_1}\big(A_1(a) - \rho_1 F(a, v)\big), \\ b = J^{A_2,\eta_2}_{\rho_2,M_2}\big(A_2(a) - \rho_2 G(u, b)\big), \end{cases} \tag{4.7}$$

where ρ_1 and ρ_2 are two positive constants.

Proof
Let (a, b, u, v) be a solution of (3.3). That is

$$0 \in F(a, v) + M_1(a)$$
$$0 \in G(u, b) + M_2(b).$$

This implies that

$$\begin{aligned} & -\rho_1 F(a, v) \in \rho_1 M_1(a). \\ \Rightarrow\ & A_1(a) - \rho_1 F(a, v) \in A_1(a) + \rho_1 M_1(a) = (A_1 + \rho_1 M_1)(a). \\ \Rightarrow\ & J^{A_1,\eta_1}_{\rho_1,M_1}\big(A_1(a) - \rho_1 F(a, v)\big) = a. \end{aligned}$$

Similarly, we can show that $J^{A_2,\eta_2}_{\rho_2,M_2}\big(A_2(a) - \rho_2 G(u, b)\big) = b$.

Conversely, assume that

$$a = J^{A_1,\eta_1}_{\rho_1,M_1}\big(A_1(a) - \rho_1 F(a, v)\big),$$
$$b = J^{A_2,\eta_2}_{\rho_2,M_2}\big(A_2(a) - \rho_2 G(u, b)\big).$$

This implies that

$$\begin{aligned} & (A_1 + \rho_1 M_1)(a) = A_1(a) - \rho_1 F(a, v). \\ \Rightarrow\ & 0 \in F(a, v) + M_1(a). \end{aligned}$$

Similarly, one can show that $0 \in G(u, b) + M_2(b)$. □

Using the above theorem, we can able to construct the following algorithm:

Algorithm 4.1
Step 1: Choose initial approximation $(a_0, b_0) \in X_1 \times X_2$ and $u_0 \in U(a_0)$, $v_0 \in V(b_0)$.

Step 2: Construct the iteration

$$\begin{cases} a_{k+1} = (1-\lambda_k-\delta_k)a_k + \lambda_k J^{A_1,\eta_1}_{\rho_1,M_1}(A_1(a_k)-\rho_1 F(a_k,v_k)), \\ b_{k+1} = (1-\lambda_k-\delta_k)b_k + \lambda_k J^{A_2,\eta_2}_{\rho_2,M_2}(A_2(b_k)-\rho_2 G(u_k,b_k)), \end{cases} \tag{4.8}$$

where λ_k and δ_k are positive real constant such that $0<\lambda_k+\delta_k\le 1$, and $\limsup_{k\ge 0}\lambda_k<1$.
Step 3: Choose $u_{k+1}\in U(a_{k+1})$ and $v_{k+1}\in V(b_{k+1})$ such that

$$\begin{cases} \|u_{k+1}-u_k\| \le (1+(1+k)^{-1})\mathcal{D}_1(U(a_{k+1}),U(a_k)), \\ \|v_{k+1}-v_k\| \le (1+(1+k)^{-1})\mathcal{D}_2(V(b_{k+1}),V(b_k)), \end{cases} \tag{4.9}$$

where $\mathcal{D}_i(.,.)$ is the Hausdorff pseudo-metric on 2^{X_i} for $i=1,2$.
Step 4: If $a_{k+1}, b_{k+1}, u_{k+1}, v_{k+1}$ satisfy (4.8) to a sufficient degree of accuracy, then stop; else set $k=k+1$ and return to Step 2.

Theorem 4.2
Let $\eta_i: X_i\times X_i\to X_i$ be τ_i-Lipschitz continuous, $A_i: X_i\to X_i$ be (r_i,η_i)-strongly monotone and β_i-Lipschitz continuous, and let $M_i: X_i\to 2^{X_i}$ be (A_i,η_i)-maximal m_i-relaxed monotone mappings for $i=1,2$. Let $U: X_1\to C(X_1)$ be $\mathcal{D}_1-\gamma_1$-Lipschitz continuous and $V: X_2\to C(X_2)$ be $\mathcal{D}_2-\gamma_2$-Lipschitz continuous. Let $F: X_1\times X_2\to X_1$ be a nonlinear mapping such that for any given $(a,b)\in X_1\times X_2$, $F(.,b)$ is (c_1,μ_1)-relaxed cocoercive with respect to A_1 and α_1-Lipschitz continuous, and $F(a,.)$ is ξ_1-Lipschitz continuous. Let $G: X_1\times X_2\to X_2$ be another nonlinear mapping such that for any given $(x,y)\in X_1\times X_2$, $G(x,.)$ is (c_2,μ_2)-relaxed cocoercive with respect to A_2 and α_2-Lipschitz continuous, and $G(.,y)$ is ξ_2-Lipschitz continuous. If there exist positive real constants ρ_1,ρ_2 such that

$$\begin{cases} \tau_1(r_2-\rho_2 m_2)\sqrt{\beta_1^2+2c_1\rho_1\alpha_1^2-2\rho_1\mu_1+C\rho_1^2\alpha_1^2}+\tau_2(r_1-\rho_1 m_1)\rho_2\xi_2\gamma_1 < e, \\ \tau_2(r_1-\rho_1 m_1)\sqrt{\beta_2^2+2c_2\rho_2\alpha_2^2-2\rho_2\mu_2+C\rho_2^2\alpha_2^2}+\tau_1(r_2-\rho_2 m_2)\rho_1\xi_1\gamma_2 < e, \end{cases} \tag{4.10}$$

where $e=(r_1-\rho_1 m_1)(r_2-\rho_2 m_2)$. Then the variational inclusion problem (3.3) has a solution (a,b,u,v). The iterative sequences $\{a_k\},\{b_k\},\{u_k\},\{v_k\}$ generated by Algorithm 4.1 converge strongly to the solution (a,b,u,v).

Proof
Using the Algorithm 4.1 and Lemma 2.2, we have

$$\begin{aligned} & \|a_{k+1}-a_k\| \\ = {} & \|(1-\lambda_k-\delta_k)a_k+\lambda_k J^{A_1,\eta_1}_{\rho_1,M_1}(A_1(a_k)-\rho_1F(a_k,v_k)) \\ & -(1-\lambda_k-\delta_k)a_{k-1}+\lambda_k J^{A_1,\eta_1}_{\rho_1,M_1}(A_1(a_{k-1})-\rho_1F(a_{k-1},v_{k-1}))\| \\ \le {} & (1-\lambda_k-\delta_k)\|a_k-a_{k-1}\| \\ & +\lambda_k\|J^{A_1,\eta_1}_{\rho_1,M_1}(A_1(a_k)-\rho_1F(a_k,v_k))-J^{A_1,\eta_1}_{\rho_1,M_1}(A_1(a_{k-1})-\rho_1F(a_{k-1},v_{k-1}))\| \\ \le {} & (1-\lambda_k-\delta_k)\|a_k-a_{k-1}\| \\ & +\lambda_k\frac{\tau_1}{r_1-\rho_1m_1}\|A_1(a_k)-A_1(a_{k-1})-\rho_1(F(a_k,v_k)-F(a_{k-1},v_{k-1}))\| \\ \le {} & (1-\lambda_k)\|a_k-a_{k-1}\| \\ & +\lambda_k\frac{\tau_1}{r_1-\rho_1m_1}\|A_1(a_k)-A_1(a_{k-1})-\rho_1(F(a_k,v_k)-F(a_{k-1},v_k))\| \\ & +\lambda_k\frac{\tau_1}{r_1-\rho_1m_1}\rho_1\|F(a_{k-1},v_k)-F(a_{k-1},v_{k-1})\|. \end{aligned} \tag{4.11}$$

Similarly, we can estimate

$$\begin{aligned}
&\|b_{k+1}-b_k\| \\
\le\ & (1-\lambda_k)\|b_k-b_{k-1}\| \\
&+\lambda_k\frac{\tau_2}{r_2-\rho_2 m_2}\|A_2(b_k)-A_2(b_{k-1})-\rho_2\big(G(u_k,b_k)-G(u_k,b_{k-1})\big)\| \\
&+\lambda_k\frac{\tau_2}{r_2-\rho_2 m_2}\rho_2\|G(u_k,b_{k-1})-G(u_{k-1},b_{k-1})\|. \qquad (4.12)
\end{aligned}$$

It is given that A_i are β_i-Lipschitz continuous for $i=1,2$, $F(.,b)$ is (c_1,μ_1)-relaxed cocoercive with respect to A_1 and α_1-Lipschitz continuous, $G(x,.)$ is (c_2,μ_2)-relaxed cocoercive with respect to A_2 and α_2-Lipschitz continuous. Hence we estimate

$$\begin{aligned}
&\|A_1(a_k)-A_1(a_{k-1})-\rho_1\big(F(a_k,v_k)-F(a_{k-1},v_k)\big)\|^2 \\
\le\ & \|A_1(a_k)-A_1(a_{k-1})\|^2-2\rho_1[F(a_k,v_k)-F(a_{k-1},v_k),A_1(a_k)-A_1(a_{k-1})] \\
&+C\rho_1^2\|F(a_k,v_k)-F(a_{k-1},v_k)\|^2 \\
\le\ & \beta_1^2\|a_k-a_{k-1}\|^2-2\rho_1\{-c_1\|F(a_k,v_k)-F(a_{k-1},v_k)\|^2+\mu_1\|a_k-a_{k-1}\|^2\} \\
&+C\rho_1^2\alpha_1^2\|a_k-a_{k-1}\|^2 \\
\le\ & \beta_1^2\|a_k-a_{k-1}\|^2+2c_1\rho_1\alpha_1^2\|a_k-a_{k-1}\|^2-2\rho_1\mu_1\|a_k-a_{k-1}\|^2+C\rho_1^2\alpha_1^2\|a_k-a_{k-1}\|^2 \\
=\ & \big(\beta_1^2+2c_1\rho_1\alpha_1^2-2\rho_1\mu_1+C\rho_1^2\alpha_1^2\big)\|a_k-a_{k-1}\|^2
\end{aligned}$$

This implies that

$$\begin{aligned}
&\|A_1(a_k)-A_1(a_{k-1})-\rho_1\big(F(a_k,v_k)-F(a_{k-1},v_k)\big)\| \\
\le\ & \sqrt{\beta_1^2+2c_1\rho_1\alpha_1^2-2\rho_1\mu_1+C\rho_1^2\alpha_1^2}\,\|a_k-a_{k-1}\|. \qquad (4.13)
\end{aligned}$$

Similarly, we have

$$\begin{aligned}
&\|A_2(b_k)-A_2(b_{k-1})-\rho_2\big(G(u_k,b_k)-G(u_k,b_{k-1})\big)\| \\
\le\ & \sqrt{\beta_2^2+2c_2\rho_2\alpha_2^2-2\rho_2\mu_2+C\rho_2^2\alpha_2^2}\,\|b_k-b_{k-1}\|. \qquad (4.14)
\end{aligned}$$

Again, since $F(a,.)$ is ξ_1-Lipschitz continuous and V is $\mathcal{D}_2-\gamma_2$-Lipschitz continuous, we have

$$\begin{aligned}
\|F(a_{k-1},v_k)-F(a_{k-1},v_{k-1})\| &\le \xi_1\|v_k-v_{k-1}\| \\
&\le \xi_1(1+n^{-1})\mathcal{D}_2(V(b_k,V(b_{k-1})) \\
&\le \xi_1(1+n^{-1})\gamma_2\|b_k-b_{k-1}\|. \qquad (4.15)
\end{aligned}$$

Similarly, since $G(.,y)$ is ξ_2-Lipschitz continuous and U is $\mathcal{D}_1-\gamma_1$-Lipschitz continuous, we have

$$\|G(u_k,b_{k-1})-G(u_{k-1},b_{k-1})\|\le\xi_2(1+n^{-1})\gamma_1\|a_k-a_{k-1}\|. \qquad (4.16)$$

Putting (4.13) and (4.15) in (4.11), we get

$$\begin{aligned}
&\|a_{k+1}-a_k\| \\
\le\ & (1-\lambda_k)\|a_k-a_{k-1}\|+\lambda_k\frac{\tau_1}{r_1-\rho_1 m_1}\sqrt{\beta_1^2+2c_1\rho_1\alpha_1^2-2\rho_1\mu_1+C\rho_1^2\alpha_1^2}\,\|a_k-a_{k-1}\| \\
&+\lambda_k\frac{\tau_1}{r_1-\rho_1 m_1}\rho_1\xi_1(1+n^{-1})\gamma_2\|b_k-b_{k-1}\|. \qquad (4.17)
\end{aligned}$$

Similarly, using (4.14) and (4.16) in (4.12), we get

$$\begin{aligned}
&\|b_{k+1}-b_k\| \\
\leq\ &(1-\lambda_k)\|b_k-b_{k-1}\|+\lambda_k\frac{\tau_2}{r_2-\rho_2 m_2}\sqrt{\beta_2^2+2c_2\rho_2\alpha_2^2-2\rho_2\mu_2+C\rho_2^2\alpha_2^2}\,\|b_k-b_{k-1}\| \\
&+\lambda_k\frac{\tau_2}{r_2-\rho_2 m_2}\rho_2\xi_2(1+n^{-1})\gamma_1\|a_k-a_{k-1}\|. \qquad (4.18)
\end{aligned}$$

Hence

$$\begin{aligned}
&\|a_{k+1}-a_k\|+\|b_{k+1}-b_k\| \\
\leq\ &\Big(1-\lambda_k+\lambda_k\frac{\tau_1}{r_1-\rho_1 m_1}\sqrt{\beta_1^2+2c_1\rho_1\alpha_1^2-2\rho_1\mu_1+C\rho_1^2\alpha_1^2} \\
&+\lambda_k\frac{\tau_2}{r_2-\rho_2 m_2}\rho_2\xi_2(1+n^{-1})\gamma_1\Big)\|a_k-a_{k-1}\| \\
&+\Big(1-\lambda_k+\lambda_k\frac{\tau_2}{r_2-\rho_2 m_2}\sqrt{\beta_2^2+2c_2\rho_2\alpha_2^2-2\rho_2\mu_2+C\rho_2^2\alpha_2^2} \\
&+\lambda_k\frac{\tau_1}{r_1-\rho_1 m_1}\rho_1\xi_1(1+n^{-1})\gamma_2\Big)\|b_k-b_{k-1}\| \\
\leq\ &(1-\lambda_k(1-\theta_k))(\|a_k-a_{k-1}\|+\|b_k-b_{k-1}\|) \\
\leq\ &(1-\triangle(1-\theta_k))(\|a_k-a_{k-1}\|+\|b_k-b_{k-1}\|), \qquad (4.19)
\end{aligned}$$

where

$$\begin{aligned}
\theta_k=\max\Big\{&\frac{\tau_1}{r_1-\rho_1 m_1}\sqrt{\beta_1^2+2c_1\rho_1\alpha_1^2-2\rho_1\mu_1+C\rho_1^2\alpha_1^2}+\frac{\tau_2}{r_2-\rho_2 m_2}\rho_2\xi_2(1+n^{-1})\gamma_1, \\
&\frac{\tau_2}{r_2-\rho_2 m_2}\sqrt{\beta_2^2+2c_2\rho_2\alpha_2^2-2\rho_2\mu_2+C\rho_2^2\alpha_2^2}+\frac{\tau_1}{r_1-\rho_1 m_1}\rho_1\xi_1(1+n^{-1})\gamma_2\Big\},
\end{aligned}$$

and $\triangle=\limsup_{k\geq 0}\lambda_k<1$.
We see that $\theta_k\to\theta$ as $k\to\infty$, where

$$\begin{aligned}
\theta=\max\Big\{&\frac{\tau_1}{r_1-\rho_1 m_1}\sqrt{\beta_1^2+2c_1\rho_1\alpha_1^2-2\rho_1\mu_1+C\rho_1^2\alpha_1^2}+\frac{\tau_2}{r_2-\rho_2 m_2}\rho_2\xi_2\gamma_1, \\
&\frac{\tau_2}{r_2-\rho_2 m_2}\sqrt{\beta_2^2+2c_2\rho_2\alpha_2^2-2\rho_2\mu_2+C\rho_2^2\alpha_2^2}+\frac{\tau_1}{r_1-\rho_1 m_1}\rho_1\xi_1\gamma_2\Big\}.
\end{aligned}$$

The assumption (4.10) assures that $0<\theta<1$. Therefore, by (4.19) and $0<\lambda_k+\delta_k\leq 1$ implies that $\{a_k\}$ and $\{b_k\}$ are both Cauchy sequences. Since the spaces X_1 and X_2 are complete, there exist $a\in X_1$ and $b\in X_2$ such that $a_k\to a$ and $b_k\to b$ as $k\to\infty$.
Next we prove that $u_k\to u\in U(a)$ and $v_k\to v\in V(b)$ as $k\to\infty$. The inequalities (4.15) and (4.16) guarantee that $\{u_k\}$ and $\{v_k\}$ are also Cauchy sequences. Therefore there exist $u\in X_1$ and $v\in X_2$ such that $u_k\to u$ and $v_k\to v$ as $k\to\infty$. Moreover,

$$\begin{aligned}
d(u,U(a))=\inf\{\|u-t\|:t\in U(a)\}\ &\leq\ \|u-u_k\|+d(u_k,U(a)) \\
&\leq\ \|u-u_k\|+\mathcal{D}_1(U(a_k),U(a)) \\
&\leq\ \|u-u_k\|+\xi_1\|a_k-a\|\to 0.
\end{aligned}$$

Since $U(a)$ is closed, we have $u\in U(a)$. Similarly, we can show that $v\in V(b)$.
Finally, applying the continuity property of A_1,F,A_2,G and the continuity of the generalized resolvent operator, we get

$$\begin{cases} a=J_{\rho_1,M_1}^{A_1,\eta_1}\big(A_1(a)-\rho_1F(a,v)\big), \\ b=J_{\rho_2,M_2}^{A_2,\eta_2}\big(A_2(a)-\rho_2G(u,b)\big). \end{cases}$$

Because of Theorem 4.1, we conclude that (a,b,u,v) is a solution of the variational inclusion problem (3.3). □

If we replace the relaxed cocoercivity condition by strong monotonicity condition on the maps $F(.,b)$ and $G(x,.)$, then we get the following result. We write it in the form of a corollary.

Corollary 4.1
Let $\eta_i : X_i \times X_i \to X_i$ be τ_i-Lipschitz continuous, $A_i : X_i \to X_i$ be (r_i, η_i)-strongly monotone and β_i-Lipschitz continuous, and let $M_i : X_i \to 2^{X_i}$ be (A_i, η_i)-maximal m_i-relaxed monotone mappings for $i = 1, 2$. Let $U : X_1 \to C(X_1)$ be $\mathcal{D}_1 - \gamma_1$-Lipschitz continuous and $V : X_2 \to C(X_2)$ be $\mathcal{D}_2 - \gamma_2$-Lipschitz continuous. Let $F : X_1 \times X_2 \to X_1$ be a nonlinear mapping such that for any given $(a, b) \in X_1 \times X_2$, $F(.,b)$ is μ_1-strongly monotone with respect to A_1 and α_1-Lipschitz continuous, and $F(a,.)$ is ξ_1-Lipschitz continuous. Let $G : X_1 \times X_2 \to X_2$ be another nonlinear mapping such that for any given $(x, y) \in X_1 \times X_2$, $G(x,.)$ is μ_2-strongly monotone with respect to A_2 and α_2-Lipschitz continuous, and $G(.,y)$ is ξ_2-Lipschitz continuous. If there exist positive real constants ρ_1, ρ_2 such that

$$\begin{cases} \tau_1(r_2 - \rho_2 m_2)\sqrt{\beta_1^2 - 2\rho_1\mu_1 + C\rho_1^2\alpha_1^2} + \tau_2(r_1 - \rho_1 m_1)\rho_2\xi_2\gamma_1 < e, \\ \tau_2(r_1 - \rho_1 m_1)\sqrt{\beta_2^2 - 2\rho_2\mu_2 + C\rho_2^2\alpha_2^2} + \tau_1(r_2 - \rho_2 m_2)\rho_1\xi_1\gamma_2 < e, \end{cases} \tag{4.20}$$

where $e = (r_1 - \rho_1 m_1)(r_2 - \rho_2 m_2)$. Then the variational inclusion problem (3.3) has a solution (a, b, u, v). The iterative sequences $\{a_k\}, \{b_k\}, \{u_k\}, \{v_k\}$ generated by Algorithm 4.1 converge strongly to the solution (a, b, u, v).

In the following theorem we impose the relative (A_i, η_i)-maximal monotonicity on the set valued map M. The proof is similar as that of Theorem 4.5, but as the relax condition on M is removed and different conditions on η and A are imposed, so we include the proof.

Theorem 4.3
Let $\eta_i : X_i \times X_i \to X_i$ be t_i-strongly monotone and τ_i-Lipschitz continuous, $A_i : X_i \to X_i$ be r_i-strongly monotone and β_i-Lipschitz continuous, and let $M_i : X_i \to 2^{X_i}$ be relative (A_i, η_i)-maximal monotone mappings for $i = 1, 2$. Let $U : X_1 \to C(X_1)$ be $\mathcal{D}_1 - \gamma_1$-Lipschitz continuous and $V : X_2 \to C(X_2)$ be $\mathcal{D}_2 - \gamma_2$-Lipschitz continuous. Let $F : X_1 \times X_2 \to X_1$ be a nonlinear mapping such that for any given $(a, b) \in X_1 \times X_2$, $F(.,b)$ is (c_1, μ_1)-relaxed cocoercive with respect to A_1 and α_1-Lipschitz continuous, and $F(a,.)$ is ξ_1-Lipschitz continuous. Let $G : X_1 \times X_2 \to X_2$ be another nonlinear mapping such that for any given $(x, y) \in X_1 \times X_2$, $G(x,.)$ is (c_2, μ_2)-relaxed cocoercive with respect to A_2 and α_2-Lipschitz continuous, and $G(.,y)$ is ξ_2-Lipschitz continuous. If there exist positive real constants ρ_1, ρ_2 such that

$$\begin{cases} \tau_1 r_2 t_2\sqrt{\beta_1^2 + 2c_1\rho_1\alpha_1^2 - 2\rho_1\mu_1 + C\rho_1^2\alpha_1^2} + \tau_2 r_1 t_1\rho_2\xi_2\gamma_1 < r_1 r_2 t_1 t_2, \\ \tau_2 r_1 t_1\sqrt{\beta_2^2 + 2c_2\rho_2\alpha_2^2 - 2\rho_2\mu_2 + C\rho_2^2\alpha_2^2} + \tau_1 r_2 t_2\rho_1\xi_1\gamma_2 < r_1 r_2 t_1 t_2. \end{cases} \tag{4.21}$$

Then the variational inclusion problem (3.3) has a solution (a, b, u, v). The iterative sequences $\{a_k\}, \{b_k\}, \{u_k\}, \{v_k\}$ generated by Algorithm 4.1 converge strongly to the solution (a, b, u, v).

Proof
Using the Algorithm 4.1 and Lemma 2.3, we have

$$\begin{aligned}
& \|a_{k+1} - a_k\| \\
= & \|(1-\lambda_k-\delta_k)a_k + \lambda_k J^{A_1,\eta_1}_{\rho_1,M_1}\big(A_1(a_k) - \rho_1 F(a_k,v_k)\big) \\
& -(1-\lambda_k-\delta_k)a_{k-1} + \lambda_k J^{A_1,\eta_1}_{\rho_1,M_1}\big(A_1(a_{k-1}) - \rho_1 F(a_{k-1},v_{k-1})\big)\| \\
\le & (1-\lambda_k-\delta_k)\|a_k - a_{k-1}\| \\
& +\lambda_k \|J^{A_1,\eta_1}_{\rho_1,M_1}\big(A_1(a_k) - \rho_1 F(a_k,v_k)\big) - J^{A_1,\eta_1}_{\rho_1,M_1}\big(A_1(a_{k-1}) - \rho_1 F(a_{k-1},v_{k-1})\big)\| \\
\le & (1-\lambda_k-\delta_k)\|a_k - a_{k-1}\| \\
& +\lambda_k \frac{\tau_1}{r_1 t_1}\|A_1(a_k) - A_1(a_{k-1}) - \rho_1\big(F(a_k,v_k) - F(a_{k-1},v_{k-1})\big)\| \\
\le & (1-\lambda_k)\|a_k - a_{k-1}\| \\
& +\lambda_k \frac{\tau_1}{r_1 t_1}\|A_1(a_k) - A_1(a_{k-1}) - \rho_1\big(F(a_k,v_k) - F(a_{k-1},v_k)\big)\| \\
& +\lambda_k \frac{\tau_1}{r_1 t_1}\rho_1\|F(a_{k-1},v_k) - F(a_{k-1},v_{k-1})\|. \qquad (4.22)
\end{aligned}$$

Similarly, we can estimate

$$\begin{aligned}
& \|b_{k+1} - b_k\| \\
\le & (1-\lambda_k)\|b_k - b_{k-1}\| \\
& +\lambda_k \frac{\tau_2}{r_2 t_2}\|A_2(b_k) - A_2(b_{k-1}) - \rho_2\big(G(u_k,b_k) - G(u_k,b_{k-1})\big)\| \\
& +\lambda_k \frac{\tau_2}{r_2 t_2}\rho_2\|G(u_k,b_{k-1}) - G(u_{k-1},b_{k-1})\|. \qquad (4.23)
\end{aligned}$$

It is given that A_i are β_i-Lipschitz continuous for $i = 1, 2$, $F(.,b)$ is (c_1, μ_1)-relaxed cocoercive with respect to A_1 and α_1-Lipschitz continuous, $G(x,.)$ is (c_2, μ_2)-relaxed cocoercive with respect to A_2 and α_2-Lipschitz continuous. Hence we estimate

$$\begin{aligned}
& \|A_1(a_k) - A_1(a_{k-1}) - \rho_1\big(F(a_k,v_k) - F(a_{k-1},v_k)\big)\|^2 \\
\le & \|A_1(a_k) - A_1(a_{k-1})\|^2 - 2\rho_1[F(a_k,v_k) - F(a_{k-1},v_k), A_1(a_k) - A_1(a_{k-1})] \\
& +C\rho_1^2\|F(a_k,v_k) - F(a_{k-1},v_k)\|^2 \\
\le & \beta_1^2\|a_k - a_{k-1}\|^2 - 2\rho_1\{-c_1\|F(a_k,v_k) - F(a_{k-1},v_k)\|^2 + \mu_1\|a_k - a_{k-1}\|^2\} \\
& +C\rho_1^2\alpha_1^2\|a_k - a_{k-1}\|^2 \\
\le & \beta_1^2\|a_k - a_{k-1}\|^2 + 2c_1\rho_1\alpha_1^2\|a_k - a_{k-1}\|^2 - 2\rho_1\mu_1\|a_k - a_{k-1}\|^2 + C\rho_1^2\alpha_1^2\|a_k - a_{k-1}\|^2 \\
= & \big(\beta_1^2 + 2c_1\rho_1\alpha_1^2 - 2\rho_1\mu_1 + C\rho_1^2\alpha_1^2\big)\|a_k - a_{k-1}\|^2
\end{aligned}$$

This implies that

$$\begin{aligned}
& \|A_1(a_k) - A_1(a_{k-1}) - \rho_1\big(F(a_k,v_k) - F(a_{k-1},v_k)\big)\| \\
\le & \sqrt{\beta_1^2 + 2c_1\rho_1\alpha_1^2 - 2\rho_1\mu_1 + C\rho_1^2\alpha_1^2}\,\|a_k - a_{k-1}\|. \qquad (4.24)
\end{aligned}$$

Similarly, we have

$$\begin{aligned}
& \|A_2(b_k) - A_2(b_{k-1}) - \rho_2\big(G(u_k,b_k) - G(u_k,b_{k-1})\big)\| \\
\le & \sqrt{\beta_2^2 + 2c_2\rho_2\alpha_2^2 - 2\rho_2\mu_2 + C\rho_2^2\alpha_2^2}\,\|b_k - b_{k-1}\|. \qquad (4.25)
\end{aligned}$$

Again, since $F(a,.)$ is ξ_1-Lipschitz continuous and V is $\mathcal{D}_2-\gamma_2$-Lipschitz continuous, we have

$$\begin{aligned}\|F(a_{k-1},v_k)-F(a_{k-1},v_{k-1})\| &\leq \xi_1\|v_k-v_{k-1}\| \\ &\leq \xi_1(1+n^{-1})\mathcal{D}_2(V(b_k,V(b_{k-1})) \\ &\leq \xi_1(1+n^{-1})\gamma_2\|b_k-b_{k-1}\|. \end{aligned} \tag{4.26}$$

Similarly, since $G(.,y)$ is ξ_2-Lipschitz continuous and U is $\mathcal{D}_1-\gamma_1$-Lipschitz continuous, we have

$$\|G(u_k,b_{k-1})-G(u_{k-1},b_{k-1})\| \leq \xi_2(1+n^{-1})\gamma_1\|a_k-a_{k-1}\|. \tag{4.27}$$

Putting (4.24) and (4.26) in (4.22), we get

$$\begin{aligned}&\|a_{k+1}-a_k\| \\ \leq\ & (1-\lambda_k)\|a_k-a_{k-1}\|+\lambda_k\frac{\tau_1}{r_1t_1}\sqrt{\beta_1^2+2c_1\rho_1\alpha_1^2-2\rho_1\mu_1+C\rho_1^2\alpha_1^2}\,\|a_k-a_{k-1}\| \\ &+\lambda_k\frac{\tau_1}{r_1t_1}\rho_1\xi_1(1+n^{-1})\gamma_2\|b_k-b_{k-1}\|. \end{aligned} \tag{4.28}$$

Similarly, using (4.25) and (4.27) in (4.23), we get

$$\begin{aligned}&\|b_{k+1}-b_k\| \\ \leq\ & (1-\lambda_k)\|b_k-b_{k-1}\|+\lambda_k\frac{\tau_2}{r_2t_2}\sqrt{\beta_2^2+2c_2\rho_2\alpha_2^2-2\rho_2\mu_2+C\rho_2^2\alpha_2^2}\,\|b_k-b_{k-1}\| \\ &+\lambda_k\frac{\tau_2}{r_2t_2}\rho_2\xi_2(1+n^{-1})\gamma_1\|a_k-a_{k-1}\|. \end{aligned} \tag{4.29}$$

Hence

$$\begin{aligned}&\|a_{k+1}-a_k\|+\|b_{k+1}-b_k\| \\ \leq\ & \big(1-\lambda_k+\lambda_k\frac{\tau_1}{r_1t_1}\sqrt{\beta_1^2+2c_1\rho_1\alpha_1^2-2\rho_1\mu_1+C\rho_1^2\alpha_1^2} \\ &+\lambda_k\frac{\tau_2}{r_2t_2}\rho_2\xi_2(1+n^{-1})\gamma_1\big)\|a_k-a_{k-1}\| \\ &+\big(1-\lambda_k+\lambda_k\frac{\tau_2}{r_2t_2}\sqrt{\beta_2^2+2c_2\rho_2\alpha_2^2-2\rho_2\mu_2+C\rho_2^2\alpha_2^2} \\ &+\lambda_k\frac{\tau_1}{r_1t_1}\rho_1\xi_1(1+n^{-1})\gamma_2\big)\|b_k-b_{k-1}\| \\ \leq\ & (1-\lambda_k(1-\theta_k))(\|a_k-a_{k-1}\|+\|b_k-b_{k-1}\|) \\ \leq\ & (1-\triangle(1-\theta_k))(\|a_k-a_{k-1}\|+\|b_k-b_{k-1}\|), \end{aligned} \tag{4.30}$$

where

$$\begin{aligned}\theta_k &= \max\big\{\frac{\tau_1}{r_1t_1}\sqrt{\beta_1^2+2c_1\rho_1\alpha_1^2-2\rho_1\mu_1+C\rho_1^2\alpha_1^2}+\frac{\tau_2}{r_2t_2}\rho_2\xi_2(1+n^{-1})\gamma_1, \\ &\frac{\tau_2}{r_2t_2}\sqrt{\beta_2^2+2c_2\rho_2\alpha_2^2-2\rho_2\mu_2+C\rho_2^2\alpha_2^2}+\frac{\tau_1}{r_1t_1}\rho_1\xi_1(1+n^{-1})\gamma_2\big\},\end{aligned}$$

and $\triangle=\limsup_{k\geq 0}\lambda_k<1$.
We see that $\theta_k\to\theta$ as $k\to\infty$, where

$$\begin{aligned}\theta=\max\big\{&\frac{\tau_1}{r_1t_1}\sqrt{\beta_1^2+2c_1\rho_1\alpha_1^2-2\rho_1\mu_1+C\rho_1^2\alpha_1^2}+\frac{\tau_2}{r_2t_2}\rho_2\xi_2\gamma_1, \\ &\frac{\tau_2}{r_2t_2}\sqrt{\beta_2^2+2c_2\rho_2\alpha_2^2-2\rho_2\mu_2+C\rho_2^2\alpha_2^2}+\frac{\tau_1}{r_1t_1}\rho_1\xi_1\gamma_2\big\}.\end{aligned}$$

The assumption (4.21) assures that $0 < \theta < 1$. Therefore, by (4.30) and $0 < \lambda_k + \delta_k \leq 1$ implies that $\{a_k\}$ and $\{b_k\}$ are both Cauchy sequences. Since the spaces X_1 and X_2 are complete, there exist $a \in X_1$ and $b \in X_2$ such that $a_k \to a$ and $b_k \to b$ as $k \to \infty$.
Next we prove that $u_k \to u \in U(a)$ and $v_k \to v \in V(b)$ as $k \to \infty$. The inequalities (4.26) and (4.27) guarantee that $\{u_k\}$ and $\{v_k\}$ are also Cauchy sequences. Therefore there exist $u \in X_1$ and $v \in X_2$ such that $u_k \to u$ and $v_k \to v$ as $k \to \infty$. Moreover,

$$\begin{aligned} d(u, U(a)) = \inf\{\|u - t\| : t \in U(a)\} &\leq \|u - u_k\| + d(u_k, U(a)) \\ &\leq \|u - u_k\| + \mathcal{D}_1(U(a_k), U(a)) \\ &\leq \|u - u_k\| + \xi_1 \|a_k - a\| \to 0. \end{aligned}$$

Since $U(a)$ is closed, we have $u \in U(a)$. Similarly, we can show that $v \in V(b)$.
Finally, applying the continuity property of A_1, F, A_2, G and the continuity of the generalized resolvent operator, we get

$$\begin{cases} a = J_{\rho_1, M_1}^{A_1, \eta_1}\big(A_1(a) - \rho_1 F(a, v)\big), \\ b = J_{\rho_2, M_2}^{A_2, \eta_2}\big(A_2(a) - \rho_2 G(u, b)\big). \end{cases}$$

Because of Theorem 4.1, we conclude that (a, b, u, v) is a solution of the variational inclusion problem (3.3). □

If we replace the relaxed cocoercivity condition by strong monotonicity condition on the maps $F(.,b)$ and $G(x,.)$, then we get the following result. We write it in the form of a corollary.

Corollary 4.2
Let $\eta_i : X_i \times X_i \to X_i$ be t_i-strongly monotone and τ_i-Lipschitz continuous, $A_i : X_i \to X_i$ be r_i-strongly monotone and β_i-Lipschitz continuous, and let $M_i : X_i \to 2^{X_i}$ be relative (A_i, η_i)-maximal monotone mappings for $i = 1, 2$. Let $U : X_1 \to C(X_1)$ be $\mathcal{D}_1 - \gamma_1$-Lipschitz continuous and $V : X_2 \to C(X_2)$ be $\mathcal{D}_2 - \gamma_2$-Lipschitz continuous. Let $F : X_1 \times X_2 \to X_1$ be a nonlinear mapping such that for any given $(a, b) \in X_1 \times X_2$, $F(.,b)$ is μ_1-strongly monotone with respect to A_1 and α_1-Lipschitz continuous, and $F(a,.)$ is ξ_1-Lipschitz continuous. Let $G : X_1 \times X_2 \to X_2$ be another nonlinear mapping such that for any given $(x, y) \in X_1 \times X_2$, $G(x,.)$ is μ_2-strongly monotone with respect to A_2 and α_2-Lipschitz continuous, and $G(.,y)$ is ξ_2-Lipschitz continuous. If there exist positive real constants ρ_1, ρ_2 such that

$$\begin{cases} \tau_1 r_2 t_2 \sqrt{\beta_1^2 - 2\rho_1\mu_1 + C\rho_1^2\alpha_1^2} + \tau_2 r_1 t_1 \rho_2 \xi_2 \gamma_1 < r_1 r_2 t_1 t_2, \\ \tau_2 r_1 t_1 \sqrt{\beta_2^2 - 2\rho_2\mu_2 + C\rho_2^2\alpha_2^2} + \tau_1 r_2 t_2 \rho_1 \xi_1 \gamma_2 < r_1 r_2 t_1 t_2. \end{cases} \tag{4.31}$$

Then the variational inclusion problem (3.3) has a solution (a, b, u, v). The iterative sequences $\{a_k\}, \{b_k\}, \{u_k\}, \{v_k\}$ generated by Algorithm 4.1 converge strongly to the solution (a, b, u, v).

Next we study the solvability of the system of variational inclusion problems (3.3) involving (H, η)-maximal monotonicity instead of (A, η)-maximal relaxed monotonicity. As in the earlier case, we convert the problem to another equivalent form. Then construct an iterative algorithm using the equivalent form. Finally, we study the convergency of the iterative method to the solution of the problem (3.3). The idea of the proof are similar. So we state the results without proof.

Let X_1 and X_2 are 2-uniformly smooth Banach spaces. Let $F : X_1 \times X_2 \to X_1, G : X_1 \times X_2 \to X_2, H_i : X_i \to X_i$ and $\eta_i : X_i \times X_i \to X_i$ $(i = 1, 2)$ be nonlinear single valued mappings. Let $U : X_1 \to 2^{X_1}$, $V : X_2 \to 2^{X_2}$ be set valued mappings, and let $M_i : X_i \to 2^{X_i}$ be (H_i, η_i)-maximal monotone mappings $(i = 1, 2)$.

Theorem 4.4
Let $(a, b) \in X_1 \times X_2$, $u \in U(a)$, $v \in V(b)$. Then (a, b, u, v) is a solution of the problem (3.3) if and only if (a, b, u, v) satisfies

$$\begin{cases} a = J_{\rho_1, M_1}^{H_1, \eta_1}\big(H_1(a) - \rho_1 F(a, v)\big), \\ b = J_{\rho_2, M_2}^{H_2, \eta_2}\big(H_2(a) - \rho_2 G(u, b)\big), \end{cases}$$

where ρ_1 and ρ_2 are two positive constants.

Using the above theorem, we can able to construct the following algorithm:

Algorithm 4.2
Step 1: Choose initial approximation $(a_0, b_0) \in X_1 \times X_2$ and $u_0 \in U(a_0)$, $v_0 \in V(b_0)$.
Step 2: Construct the iteration

$$\begin{cases} a_{k+1} = (1 - \lambda_k - \delta_k)a_k + \lambda_k J^{H_1,\eta_1}_{\rho_1,M_1}(H_1(a_k) - \rho_1 F(a_k, v_k)), \\ b_{k+1} = (1 - \lambda_k - \delta_k)b_k + \lambda_k J^{H_2,\eta_2}_{\rho_2,M_2}(H_2(b_k) - \rho_2 G(u_k, b_k)), \end{cases} \tag{4.32}$$

where λ_k and δ_k are positive real constant such that $0 < \lambda_k + \delta_k \leq 1$, and $\limsup_{k\geq 0} \lambda_k < 1$.
Step 3: Choose $u_{k+1} \in U(a_{k+1})$ and $v_{k+1} \in V(b_{k+1})$ such that

$$\begin{cases} \|u_{k+1} - u_k\| \leq (1 + (1 + k)^{-1})\mathcal{D}_1(U(a_{k+1}), U(a_k)), \\ \|v_{k+1} - v_k\| \leq (1 + (1 + k)^{-1})\mathcal{D}_2(V(b_{k+1}), V(b_k)), \end{cases}$$

where $\mathcal{D}_i(.,.)$ is the Hausdorff pseudo-metric on 2^{X_i} for $i = 1, 2$.
Step 4: If $a_{k+1}, b_{k+1}, u_{k+1}, v_{k+1}$ satisfy (4.32) to a sufficient degree of accuracy, then stop; else set $k = k + 1$ and return to Step 2.

Theorem 4.5
Let $\eta_i : X_i \times X_i \to X_i$ be τ_i-Lipschitz continuous, $H_i : X_i \to X_i$ be (r_i, η_i)-strongly monotone and β_i-Lipschitz continuous, and let $M_i : X_i \to 2^{X_i}$ be (H_i, η_i)-maximal monotone mappings for $i = 1, 2$. Let $U : X_1 \to C(X_1)$ be $\mathcal{D}_1 - \gamma_1$-Lipschitz continuous and $V : X_2 \to C(X_2)$ be $\mathcal{D}_2 - \gamma_2$-Lipschitz continuous. Let $F : X_1 \times X_2 \to X_1$ be a nonlinear mapping such that for any given $(a, b) \in X_1 \times X_2$, $F(., b)$ is (c_1, μ_1)-relaxed cocoercive with respect to H_1 and α_1-Lipschitz continuous, and $F(a, .)$ is ξ_1-Lipschitz continuous. Let $G : X_1 \times X_2 \to X_2$ be another nonlinear mapping such that for any given $(x, y) \in X_1 \times X_2$, $G(x, .)$ is (c_2, μ_2)-relaxed cocoercive with respect to H_2 and α_2-Lipschitz continuous, and $G(., y)$ is ξ_2-Lipschitz continuous. If there exist positive real constants ρ_1, ρ_2 such that

$$\begin{cases} \tau_1 r_2 \sqrt{\beta_1^2 + 2c_1\rho_1\alpha_1^2 - 2\rho_1\mu_1 + C\rho_1^2\alpha_1^2} + \tau_2 r_1 \rho_2 \xi_2 \gamma_1 < r_1 r_2, \\ \tau_2 r_1 \sqrt{\beta_2^2 + 2c_2\rho_2\alpha_2^2 - 2\rho_2\mu_2 + C\rho_2^2\alpha_2^2} + \tau_1 r_2 \rho_1 \xi_1 \gamma_2 < r_1 r_2. \end{cases}$$

Then the variational inclusion problem (3.3) has a solution (a, b, u, v). The iterative sequences $\{a_k\}, \{b_k\}, \{u_k\}, \{v_k\}$ generated by Algorithm 4.2 converge strongly to the solution (a, b, u, v).

5. Conclusion

A new system of nonlinear variational inclusion problems involving (A, η)-maximal relaxed monotone mappings has been introduced in 2-uniformly smooth Banach spaces. Using the generalized resolvent operator technique, an iterative algorithm has been constructed to solve the proposed system, and the convergence analysis of the algorithm has been investigated. Moreover the obtained results are generalized to solve the system of variational inclusions involving relative (A, η)-maximal relaxed monotone mappings. The obtained results generalize most of the results investigated in the literature, and offer a wide range of applications to future research on the sensitivity analysis, variational inclusion problems, variational inequality problems in Banach spaces.

Acknowledgement

The authors are thankful to the referees for their valuable suggestions that improved the presentation of the paper.

REFERENCES

1. S. Adly, *Perturbed algorithm and sensitivity analysis for a general class of variational inclusions,* J. Math. Anal. Appl., vol. 201, pp. 609–630, 1996.
2. R. P. Agarwal, and R. U. Verma, *General system of (A, η)-maximal relaxed monotone variational inclusion problems based on generalized hybrid algorithms,* Commun. Nonlinear Sci. Numer. Simulat., vol. 15, pp. 238–251, 2010.
3. I. Ahmad, M. Rahaman, and R. Ahmad, *Relaxed resolvent operator for solving variational inclusion problem,* Stat. Optim. Inf. Comput., vol. 4, pp. 183–193, 2016.
4. Y. J. Cho, Y. P. Fang, N. J. Huang, and H. J. Hwang, *Algorithms for systems of nonlinear variational inequalities,* J. Korean Math. Soc., vol. 41, pp. 489–499, 2004.
5. X. P. Ding, *Perturbed proximal point for generalized quasi-variational inclusions,* J. Math. Anal. Appl., vol. 210, pp. 88–101, 1997.
6. X. P. Ding, and C. L. Luo, *Perturbed proximal point algorithms for generalized quasi-variational like inclusions,* J. Comput. Appl. Math., vol. 210, pp. 153–165, 2000.
7. Y. P. Fang, and N. J. Huang, *H-monotone operator and resolvent operator technique for variational inclusions,* Appl. Math. Comput., vol. 145, pp. 795–803, 2003.
8. J. R. Giles, *Classes of semi-inner product spaces,* Trans. Amer. Math. Soc., vol. 129, pp. 436–446, 1967.
9. N. J. Huang, *A new class of generalized set valued implicit variational inclusions in Banach spaces with an application,* Comput. Math. Appl., vol. 41, pp. 937–943, 2001.
10. D. O. Koehler, *A note on some operator theory in certain semi-inner product spaces,* Proc. Amer. Math. Soc., vol. 30, pp. 363–366, 1971.
11. H. Y. Lan, J. H. Kim, and Y. J. Cho, *On a new system of nonlinear A-monotone multivalued variational inclusions,* J. Math. Anal. Appl., vol. 327, pp. 481–493, 2007.
12. C. H. Lee, Q. H. Ansari, and J. C. Yao, *A perturbed algorithm for strongly nonlinear variational like inclusions,* Bull. Aust. Math. Soc., vol.62, pp. 417–426, 2000.
13. G. Lumer, *Semi-inner product spaces,* Trans. Amer. Math. Soc., vol. 100, pp. 29–43, 1961.
14. J. Peng, and D. Zhu, *A new system of generalized mixed quasi-variational inclusions with (H, η)-monotone operators,* J. Math. Anal. Appl., vol. 327, pp. 175–187, 2007.
15. N. K. Sahu, R. N. Mohapatra, C. Nahak, and S. Nanda, *Approximation solvability of a class of A-monotone implicit variational inclusion problems in semi-inner product spaces,* Appl. Math. Comput., vol. 236, pp. 109–117, 2014.
16. R. U. Verma, *Approximation solvability of a class of nonlinear set-valued variational inclusions involving (A, η)-monotone mappings,* J. Math. Anal. Appl., vol. 337, pp. 969–975, 2008.
17. R. U. Verma, *Sensitivity analysis for generalized strongly monotone variational inclusions based on (A, η)-resolvent operator technique,* Appl. Math. Lett., vol. 19, pp. 1409–1413, 2006.
18. H. K. Xu, *Inequalities in Banach spaces with applications,* Nonlinear Analysis: Theory, Methods and Applications, vol. 16, pp. 1127–1138, 1991.

8

Modified random errors S-iterative process for stochastic fixed point theorems in a generalized convex metric space

Plern Saipara[1], Wiyada Kumam[2,*], Parin Chaipunya[1]

[1]*KMUTT Fixed Point Research Laboratory, Department of Mathematics, King Mongkut's University of Technology Thonburi, Thailand*
[2]*Program in Applied Statistics, Department of Mathematics and Computer Science, Faculty of Science and Technology, Rajamangala University of Technology Thanyaburi, Thailand*

Abstract In this paper, we suggest the modified random S-iterative process and prove some stochastic fixed point theorems of a finite family of random uniformly quasi-Lipschitzian operators in a generalized convex metric space. Our results improves and extends various results in the literature.

Keywords Modified random S-iteration, Common random fixed point theorems, Generalized convex metric spaces, Convex structure

1. Introduction

Fixed point theorems are a very performance tool of the present mathematical applications. Also, random fixed point theorems are stochastic generalizations of Banach's fixed point theorem and Banach's type fixed point theorems in complete metric spaces. Random nonlinear analysis is an important mathematical methodology which is mainly concerned with the study of random nonlinear operators and their properties and its development is required for study of wide classes of random operator equations. Random Techniques have been crucial in various areas from pure mathematics to applied sciences. The study of random fixed point theorem was first introduced by Prague school of probability in the 1950s. Later, Spacek [1] and Hans [2, 3] first proved random fixed point theorems for random contraction mappings in separable complete metric spaces. Moreover, there were many authors who have studied about random fixed point theorems and its application, for instance, in [4, 5, 6, 7, 8, 9, 10, 11, 12, 13, 14, 15, 16, 17, 18, 19, 20, 21]. In the sense of another fixed point theorems, there are many mathematicians who studied common fixed point theorems in various mappings and spaces (see in [22, 23, 24, 25, 26])

On the other hand, in 1970, Takahashi [27] first suggested the knowledge of convex metric spaces and first studied the fixed point theorems for non-expansive mappings in this spaces. Later, the iterative processes for non-expansive mappings in the hyperbolic type space was studied by Kirk and Goebel, see in [28, 29]. Later, many paper of Liu [30, 31, 32] showed some sufficient and necessary conditions for two schemes of Ishikawa iterative process of asymptotically quasi-non-expansive mappings to converge to fixed point in a convex Banach space.

*Correspondence to: Wiyada Kumam (Email: wiyada.kum@mail.rmutt.ac.th), Department of Mathematics and Computer Science, Faculty of Science and Technology, Rajamangala University of Technology Thanyaburi (RMUTT) Rungsit-Nakorn Nayok Rd., Klong 6, Thanyaburi, Pathumthani 12110, Thailand.

Next, Tian [33] give some sufficient and necessary conditions for Ishikawa iterative process of two asymptotically quasi-non-expansive mappings which introduced by Das and Debats [34] to converge to fixed points in convex metric spaces.

Recently, in a convex metric spaces, Khan [35] suggested the iterative process for common fixed points of asymptotically quasi-non-expansive mappings $\{T_i : i \in J\}$ where $J = \{1, 2, 3, ..., k\}$ as follows:

$$\begin{cases} x_{n+1} = W(T_k^n y_{(k-1)n}, x_n; \alpha_{kn}), \\ y_{(k-1)n} = W(T_{k-1}^n y_{(k-2)n}, x_n; \alpha_{(k-1)n}), \\ y_{(k-2)n} = W(T_{k-2}^n y_{(k-3)n}, x_n; \alpha_{(k-2)n}), \\ \quad \ldots \\ y_{2n} = W(T_1^n y_{1n}, x_n; \alpha_{2n}), \\ y_{1n} = W(T_1^n y_{0n}, x_n; \alpha_{1n}), \end{cases} \tag{1}$$

where $y_{0n} = x_n$ and $\{\alpha_{in}\}$ are real sequences in $[0, 1]$.

Also, Wang and Liu [36] introduced an Ishikawa type iterative process with errors to estimate the fixed point mappings of T and S which were uniformly quasi-Lipschitzian mappings in generalized convex metric spaces as follows:

$$\begin{cases} x_{n+1} = W(x_n, Sy_n, u_n, a_n, b_n, c_n), \\ y_n = W(x_n, Tx_n, v_n, a'_n, b'_n, c'_n), \end{cases} \tag{2}$$

where $\{a_n\}$, $\{b_n\}$, $\{c_n\}$, $\{a'_n\}$, $\{b'_n\}$ and $\{c'_n\}$ are real sequences in [0, 1] such that $a_n + b_n + c_n = a'_n + b'_n + c'_n = 1$ and $\{u_n\}$, $\{v_n\}$ are two bounded sequence.

Recently, the S-iterative process was suggested by Agarwal, O'Regan and Sahu [37] in a Banach space. They proved that this iterative process converges faster than Mann iterative process and Ishikawa iterative process.

$$\begin{cases} x_1 \in K, \\ x_{n+1} = (1 - \alpha_n)Tx_n + \alpha_n T(y_n), \\ y_n = (1 - \beta_n)x_n + \beta_n T(x_n), n \in \mathbb{N}, \end{cases} \tag{3}$$

where $\{\alpha_n\}$ and $\{\beta_n\}$ are the sequences in (0, 1). Also, there were many authors who have studied about this iterative process for estimate a fixed point in various spaces and them results showed that the rate of convergence for this iterative process is much quicker than another iterative process (see e.g.,[38, 39, 40, 41, 42]).

Motivated and inspired by (1), (2) and (3), we modified the following random S-iterative process in a generalized convex metric space. Let $\{T_i : i \in J\}$ where $J = \{1, 2, 3, ..., k\}$ be a finite family of random uniformly quasi-Lipschitzian operators such that $T_i : \Omega \times K \to K$, where K is a non-empty closed convex subset of a separable generalized convex metric space (X, d) with a convex structure W (see in definition 3 and 4). Let $\zeta_1 : \Omega \to K$ be a measurable mapping, the sequence $\{\zeta_n(\omega)\}$ is generated by

$$\begin{cases} \zeta_{n+1} = W(T_k^n(\omega, \zeta_n(\omega)), T_k^n(\omega, \chi_{(k-1)n}(\omega)), v_{kn}(\omega); \alpha_{kn}, \beta_{kn}, \gamma_{kn}), \\ \chi_{(k-1)n} = W(\zeta_n(\omega), T_{k-1}^n(\omega, \chi_{(k-2)n}(\omega)), v_{(k-1)n}(\omega); \alpha_{(k-1)n}, \beta_{(k-1)n}, \gamma_{(k-1)n}), \\ \chi_{(k-2)n} = W(\zeta_n(\omega), T_{k-2}^n(\omega, \chi_{(k-3)n}(\omega)), v_{(k-2)n}(\omega); \alpha_{(k-2)n}, \beta_{(k-2)n}, \gamma_{(k-2)n}), \\ \quad \ldots \\ \chi_{2n} = W(\zeta_n(\omega), T_2^n(\omega, \chi_{1n}(\omega)), v_{2n}(\omega); \alpha_{2n}, \beta_{2n}, \gamma_{2n}), \\ \chi_{1n} = W(\zeta_n(\omega), T_1^n(\omega, \chi_{0n}(\omega)), v_{1n}(\omega); \alpha_{1n}, \beta_{1n}, \gamma_{1n}), \end{cases} \tag{4}$$

where $\chi_{0n}(\omega) = \zeta_n(\omega)$, for any given $i \in J$, $\{\alpha_{in}\}$, $\{\beta_{in}\}$, $\{\gamma_{in}\}$ are real sequences in $[0,1]$ with $\alpha_{in} + \beta_{in} + \gamma_{in} = 1$ and the bounded sequence $\{v_{in}(\omega)\} : \Omega \to K$ is a sequence of measurable mappings which $\{v_{in}(\omega)\} \in K$, $\forall \omega \in \Omega$ and $\forall n \in \mathbb{N}$.

The purposes of this paper are to suggest the modified random S-iterative process and prove some stochastic fixed point theorems of a finite family of random uniformly quasi-Lipschitzian operators in a generalized convex metric space. In this paper was organized as follows. In section 2 and 3, we present preliminaries and main results, respectively.

2. Preliminaries

Overview on this paper, we denote the notation that $T^n(\omega, x)$ is the n th iteration $T(\omega, T(\omega, ...T(\omega, x)...))$ of T and the letter I denotes the random mapping $T : \Omega \times K \to K$ defined by $I(\omega, x) = x$ and $T^0 = I$. First, we present about the following definition of a random fixed point operator T.

Definition 1
[43] Let (Ω, Σ) be a measurable space with Σ be a σ-algebra of subsets of Ω, and let K be a non-empty subset of a metric space (X, d).

i) A mapping $\xi : \Omega \to X$ is measurable if $\xi^{-1}(U) \in \Sigma$ for any open subset U of X;
ii) the operator $T : \Omega \times K \to K$ is a random mapping iff for any fixed $x \in K$, $T(\cdot, x) : \Omega \to K$ is measurable and continuous if $\forall \omega \in \Omega$, $T(\omega, x) : K \to X$ is continuous;
iii) a measurable mapping $\xi : \Omega \to X$ is a random fixed point of the random operator $T : \Omega \times X \to X$ iff $T(\omega, \xi(\omega)) = \xi(\omega)$, $\forall \omega \in \Omega$.

By above definition 1, we denote the set of all random fixed points of a random operator T by $RF(T)$. Next, we present about the following definitions of operator T that using in our main results.

Definition 2
Let K be a non-empty subset of a separable metric space (X, d) and $T : \Omega \times K \to K$ be a random operator. The operator T is called

i) an asymptotically nonexpansive random operator if there exists a sequence of measurable mappings $\{k_n(\omega)\} : \Omega \to [1, \infty)$ with $\lim_{n\to\infty} k_n(\omega) = 0$ such that
$$d(T^n(\omega, x), T^n(\omega, y)) \leq (1 + k_n(\omega))d(x, y)$$
for all $\omega \in \Omega$ and $x, y \in K$;
ii) a uniformly L-Lipschitzian random operator if
$$d(T^n(\omega, x), T^n(\omega, y)) \leq Ld(x, y)$$
for all $\omega \in \Omega$, $x, y \in K$ and L is positive constant;
iii) an asymptotically nonexpansive random operator if there exists a sequence of measurable mappings $\{k_n(\omega)\} : \Omega \to [1, \infty)$ with $\lim_{n\to\infty} k_n(\omega) = 0$ such that
$$d(T^n(\omega, \eta(\omega)), \xi(\omega)) \leq (1 + k_n(\omega))d(\eta(\omega), \xi(\omega))$$
for all $\omega \in \Omega$, where $\xi : \Omega \to K$ is a random fixed point of operator T and $\eta : \Omega \to K$ is any measurable mapping;
iv) a uniformly quasi-Lipschitzian random operator if
$$d(T^n(\omega, \eta(\omega)), \xi(\omega)) \leq Ld(\eta(\omega), \xi(\omega))$$
for all $\omega \in \Omega$, where $\xi : \Omega \to K$ is a random fixed point of operator T, $\eta : \Omega \to K$ is any measurable mapping and L is positive constant;

v) an semi-compact random mapping if for any sequence of measurable mappings $\{\xi_n(\omega)\} : \Omega \to K$, with $\lim_{n\to\infty} d(T(\omega, \xi_n(\omega)), \xi_n(\omega)) = 0$, for all $\omega \in \Omega$ there exists a subsequence $\{\xi_{n_j}\}$ of $\{\xi_n\}$ which converges pointwise to ξ, where $\xi : \Omega \to K$ is a measurable mapping.

Now, we present about the following definition of a convex structure in a metric space and a generalized convex metric space.

Definition 3
[27] A convex structure in a metric space (X, d) is a mapping $W : X \times X \times [0, 1] \to X$ satisfying, for any $x, y, u \in X$ and any $\lambda \in [0, 1]$

$$d(u, W(x, y, \lambda)) \leq \lambda d(u, x) + (1 - \lambda)d(u, y).$$

A metric space together with a convex structure is said to be a convex metric space. A non-empty subset K of X is called convex if $W(x, y, \lambda) \in K$ for any $(x, y, \lambda) \in K \times K \times [0; 1]$.

Definition 4
[33, 36] Let X be a metric space, $I = [0, 1]$, $\{\alpha_n\}, \{\beta_n\}, \{\gamma_n\}$ be real sequence in $[0, 1]$ with $\alpha_n + \beta_n + \gamma_n = 1$. A mapping $W : X^3 \times I^3 \times \to X$ is called a convex structure on X, if it satisfies the following conditions: for each $(x, y, z; \alpha_n, \beta_n, \gamma_n) \in X^3 \times I^3$ and $u \in X$,

$$d(u, W(x, y, z; \alpha_n, \beta_n, \gamma_n)) \leq \alpha_n d(u, x) + \beta_n d(u, y) + \gamma_n d(u, z).$$

A metric space together with a convex structure is called a generalized convex metric space.

In the sense of random fixed point, a non-empty subset K of X is called convex if $W(x, y, z; \alpha_n, \beta_n, \gamma_n) \in K$ for any $(x, y, z; \alpha_n, \beta_n, \gamma_n) \in X^3 \times I^3$. The mapping $W : K^3 \times I^3 \to X$ is called a random convex structure if for any measurable mappings $\xi, \eta, \zeta : \Omega \to K$ and each fixed $\alpha_n, \beta_n, \gamma_n \in [0, 1]$ with $\alpha_n + \beta_n + \gamma_n = 1$, the mapping $W(\xi(\cdot), \eta(\cdot), \zeta(\cdot)) : \Omega \to K$ is measurable.

The last, we present the following lemmas for proving the main results as follow.

Lemma 1
[44] Let X be a separable metric space and Y a metric space. If $f : \Omega \times X \to Y$ is measurable in $\omega \in \Omega$ and continuous in $x \in X$, and if $x : \Omega \to X$ is measurable, then $f(\cdot, x(\cdot)) : \Omega \to Y$ is measurable.

Lemma 2
[31] Let $\{p_n\}, \{q_n\}, \{r_n\}$ be sequences of nonnegative real numbers satisfying the following conditions:

$$p_{n+1} \leq (1 + q_n)p_n + r_n, \sum_{n=0}^{\infty} q_n < \infty, \sum_{n=0}^{\infty} r_n < \infty$$

we have

i) $\lim_{n\to\infty} p_n$ exists;
ii) if $\liminf_{n\to\infty} p_n = 0$, then $\lim_{n\to\infty} p_n = 0$.

3. The main results

In this section, we state and prove some stochastic fixed point theorems of a finite family of random uniformly quasi-Lipschitzian operators in a generalized convex metric space as follow.

Lemma 3
Let K be a nonempty closed convex subset of a separable generalized convex metric space (X, d). Let $\{T_i : i \in J\}$ where $J = \{1, 2, 3, ..., k\}$ be a finite family of uniformly quasi- Lipschitzian random mappings $L_i > 0$. Suppose that the sequence $\{\zeta_n(\omega)\}$ is as in (4) and $\sum_{n=0}^{\infty}(\beta_{kn} + \gamma_{kn}) < \infty$. If $\mathcal{F} = \cap_{i=1}^{k} RF(T_i) \neq \emptyset$, then

1. there exist two positive constants $\mathcal{M}_0, \mathcal{M}_1$, such that;
$$d(\zeta_{n+1}(\omega, \zeta(\omega)) \leq (1+\Theta_n\mathcal{M}_0)d(\zeta_n(\omega), \zeta(\omega)) + \Theta_n\mathcal{M}_1 \tag{5}$$
where $\Theta_n = \beta_{kn} + \gamma_{kn}$ and $\alpha_{kn} \leq \gamma_{nk}$, for all $\zeta(\omega) \in \mathcal{F}$ and $n \in \mathbb{N}$;
2. there exist a positive constants $\mathcal{M}_2$, such that;
$$d(\zeta_{n+m}(\omega), \zeta(\omega)) \leq \mathcal{M}_2 d(\zeta_n(\omega), \zeta(\omega)) + \mathcal{M}_1\mathcal{M}_2\Sigma_{j=n}^{n+m-1}\Theta_j, \tag{6}$$
for all $\zeta(\omega) \in F$ and $n \in \mathbb{N}$.

Proof

1. Let $\zeta(\omega) \in \mathcal{F}$. Since $\{v_{in}(\omega)\}$ are bounded sequences in K for any $j \in J$, there exists $\mathcal{M} > 0$ such that
$$\mathcal{M} = \sup_{\omega\in\Omega} \max_{1\leq i\leq k} d(v_{in}(\omega), \zeta(\omega)).$$
Let $L = \max_{1\leq i\leq k}\{L_i\} > 0$. From the sequence in (4), we get

$$\begin{aligned}
& d(\chi_{1n}(\omega), \zeta(\omega)) \\
= \; & d(W(\zeta_n(\omega), T_1^n(\omega, \chi_{0n}(\omega)), v_{1n}(\omega); \alpha_{1n}, \beta_{1n}, \gamma_{1n}), \zeta(\omega)) \\
\leq \; & \alpha_{1n}d(\zeta_n(\omega), \zeta(\omega)) + \beta_{1n}d(T_1^n(\omega, \chi_{0n}(\omega)), \zeta(\omega)) + \gamma_{1n}d(v_{1n}(\omega), \zeta(\omega)) \\
\leq \; & \alpha_{1n}d(\zeta_n(\omega), \zeta(\omega)) + \beta_{1n}Ld(\chi_{0n}(\omega), \zeta(\omega)) + \gamma_{1n}\mathcal{M} \\
= \; & \alpha_{1n}d(\zeta_n(\omega), \zeta(\omega)) + \beta_{1n}Ld(\zeta_{0n}(\omega), \zeta(\omega)) + \gamma_{1n}\mathcal{M} \\
\leq \; & d(\zeta_n(\omega), \zeta(\omega)) + Ld(\zeta_{0n}(\omega), \zeta(\omega)) + \mathcal{M} \\
= \; & (1+L)d(\zeta_n(\omega), \zeta(\omega)) + \mathcal{M}.
\end{aligned}$$

For $1 \leq i \leq k-1$, suppose that

$$d(\chi_{in}(\omega), \zeta(\omega)) = (1+L)^i d(\zeta_n(\omega), \zeta(\omega)) + \Sigma_{j=0}^{i-1}L^j\mathcal{M}$$

holds. Then

$$\begin{aligned}
& d(\chi_{(i+1)n}(\omega), \zeta(\omega)) \\
= \; & d(W(\zeta_n(\omega), T_{i+1}^n(\omega, \chi_{in}(\omega)), v_{(i+1)n}(\omega); \alpha_{(i+1)n}, \beta_{(i+1)n}, \gamma_{(i+1)n}), \zeta(\omega)) \\
\leq \; & \alpha_{(i+1)n}d(\zeta_n(\omega), \zeta(\omega)) + \beta_{(i+1)n}d(T_{i+1}^n(\omega, \chi_{in}(\omega)), \zeta(\omega)) \\
& +\gamma_{(i+1)n}d(v_{(i+1)n}(\omega), \zeta(\omega)) \\
\leq \; & \alpha_{(i+1)n}d(\zeta_n(\omega), \zeta(\omega)) + \beta_{(i+1)n}Ld(\chi_{in}(\omega), \zeta(\omega)) \\
& +\gamma_{(i+1)n}d(v_{(i+1)n}(\omega), \zeta(\omega)) \\
\leq \; & \alpha_{(i+1)n}d(\zeta_n(\omega), \zeta(\omega)) + \beta_{(i+1)n}L\{(1+L)^i d(\zeta_n(\omega), \zeta(\omega)) + \Sigma_{j=0}^{i-1}L^j\mathcal{M}\} \\
& +\gamma_{(i+1)n}d(v_{1n}(\omega), \zeta(\omega)) \\
\leq \; & \{\alpha_{(i+1)n} + \beta_{(i+1)n}L(1+L)^i\}d(\zeta_n(\omega), \zeta(\omega)) + \beta_{(i+1)n}L\Sigma_{j=0}^{i-1}L^j\mathcal{M} + \gamma_{(i+1)n}\mathcal{M} \\
\leq \; & \{1 + L(1+L)^i\}d(\zeta_n(\omega), \zeta(\omega)) + \Sigma_{j=1}^{i}L^j\mathcal{M} + \mathcal{M} \\
= \; & \{1 + L(1+L)^i\}d(\zeta_n(\omega), \zeta(\omega)) + \Sigma_{j=1}^{i}L^j\mathcal{M} + L^0\mathcal{M} \\
\leq \; & \{(1+L)^i\}d(\zeta_n(\omega), \zeta(\omega)) + \Sigma_{j=1}^{i}L^j\mathcal{M}.
\end{aligned}$$

Therefore, for any $1 \le i \le k$, by mathematical induction, we get

$$d(\chi_{in}(\omega), \zeta(\omega)) \le (1+L)^i d(\zeta_n(\omega), \zeta(\omega)) + \Sigma_{j=0}^{i-1} L^j \mathcal{M}.$$

Then, it follows from (4) that

$$\begin{aligned}
& d(\zeta_{n+1}(\omega), \zeta(\omega)) \\
= \;& d(W(T_k^n(\omega, \zeta_n(\omega)), T_k^n(\omega, \chi_{(k-1)n}(\omega)), v_{kn}(\omega); \alpha_{kn}, \beta_{kn}, \gamma_{kn}), \zeta(\omega)) \\
\le \;& \alpha_{kn} d(T_k^n(\omega, \zeta_n(\omega)), \zeta(\omega)) + \beta_{kn} d(T_k^n(\omega, \chi_{(k-1)n}(\omega)), \zeta(\omega)) \\
& + \gamma_{kn} d(v_{kn}(\omega), \zeta(\omega)) \\
\le \;& \alpha_{kn} L d(\zeta_n(\omega), \zeta(\omega)) + \beta_{kn} L d(\chi_{(k-1)n}(\omega), \zeta(\omega)) \\
& + \gamma_{kn} d(v_{kn}(\omega), \zeta(\omega)) \\
\le \;& \alpha_{kn} L d(\zeta_n(\omega), \zeta(\omega)) + \beta_{kn} L \{(1+L)^{k-1} d(\zeta_n(\omega), \zeta(\omega)) \\
& + \Sigma_{j=0}^{k-2} L^j \mathcal{M}\} + \gamma_{kn} \mathcal{M} \\
\le \;& \{\alpha_{kn} L + \beta_{kn} L (1+L)^{k-1}\} d(\zeta_n(\omega), \zeta(\omega)) + \beta_{kn} L \Sigma_{j=0}^{k-2} L^j \mathcal{M} + \gamma_{kn} \mathcal{M} \\
\le \;& \{\alpha_{kn} L + \frac{\beta_{kn} L (1+L)^k}{1+L}\} d(\zeta_n(\omega), \zeta(\omega)) + (\beta_{kn} + \gamma_{kn})(\Sigma_{j=0}^{k-2} L^j \mathcal{M} + \mathcal{M}) \\
\le \;& \{\frac{\alpha_{kn} L (1+L)}{1+L} + \frac{\beta_{kn} L (1+L)^k}{1+L}\} d(\zeta_n(\omega), \zeta(\omega)) + \Theta_n \Sigma_{j=0}^{k-1} L^j \mathcal{M} \\
\le \;& \{\alpha_{kn}(1+L) + \beta_{kn}(1+L)^k\} d(\zeta_n(\omega), \zeta(\omega)) + + \Theta_n \Sigma_{j=0}^{k-1} L^j \mathcal{M} \\
\le \;& \{1 + \alpha_{kn}(1+L)^k + \beta_{kn}(1+L)^k\} d(\zeta_n(\omega), \zeta(\omega)) + \Theta_n \Sigma_{j=0}^{k-1} L^j \mathcal{M} \\
\le \;& \{1 + (\beta_{kn} + \gamma_{kn})(1+L)^k\} d(\zeta_n(\omega), \zeta(\omega)) + \Theta_n \Sigma_{j=0}^{k-1} L^j \mathcal{M} \\
= \;& (1 + \Theta_n \mathcal{M}_0) d(\zeta_n(\omega), \zeta(\omega)) + \Theta_n \mathcal{M}_1
\end{aligned}$$

where $\Theta_n = \beta_{kn} + \gamma_{kn}$, $\mathcal{M}_0 = (1+L)^k$ and $\mathcal{M}_1 = \Sigma_{j=0}^{k-1} L^j \mathcal{M}$.

2. By $1 + x \le e^x$ for any $x \ge 0$, we get

$$\begin{aligned}
& d(\zeta_{n+m}(\omega), \zeta(\omega)) \\
\le \;& (1 + \Theta_{n+m-1} \mathcal{M}_0) d(\zeta_{n+m-1}(\omega), \zeta(\omega)) + \Theta_{n+m-1} \mathcal{M}_1 \\
\le \;& e^{\Theta_{n+m-1} \mathcal{M}_0} \{(1 + \Theta_{n+m-2} \mathcal{M}_0) d(\zeta_{n+m-2}(\omega), \zeta(\omega)) + \Theta_{n+m-2} \mathcal{M}_1\} \\
& + \Theta_{n+m-1} \mathcal{M}_1 \\
\le \;& e^{\Theta_{n+m-1} \mathcal{M}_0} e^{\Theta_{n+m-2} \mathcal{M}_0} d(\zeta_{n+m-2}(\omega), \zeta(\omega)) + \Theta_{n+m-2} \mathcal{M}_1 + \Theta_{n+m-1} \mathcal{M}_1 \\
\le \;& e^{(\Theta_{n+m-1} + \Theta_{n+m-2}) \mathcal{M}_0} d(\zeta_{n+m-2}(\omega), \zeta(\omega)) + e^{(\Theta_{n+m-1} \mathcal{M}_0}(\Theta_{n+m-2} + \Theta_{n+m-1}) \mathcal{M}_1 \\
& \cdots \\
\le \;& e^{\mathcal{M}_0 \Sigma_{j=1}^{n+m-1} \Theta_j} d(\zeta_{n+m-2}(\omega), \zeta(\omega)) + e^{\mathcal{M}_0 \Sigma_{j=1}^{n+m-1} \Theta_j} \mathcal{M}_1 \Sigma_{j=1}^{n+m-1} \Theta_j \\
\le \;& e^{\mathcal{M}_0 \Sigma_{j=1}^{\infty} \Theta_j} d(\zeta_{n+m-2}(\omega), \zeta(\omega)) + e^{\mathcal{M}_0 \Sigma_{j=1}^{\infty} \Theta_j} \mathcal{M}_1 \Sigma_{j=1}^{n+m-1} \Theta_j \\
\le \;& \mathcal{M}_2 d(\zeta_{n+m-2}(\omega), \zeta(\omega)) + \mathcal{M}_1 \mathcal{M}_2 \Sigma_{j=1}^{n+m-1} \Theta_j,
\end{aligned}$$

where $\mathcal{M}_2 = e^{\mathcal{M}_0 \Sigma_{j=1}^{\infty} \Theta_j}$.

□

Theorem 1
Let K be a non-empty closed convex subset of a separable complete generalized convex metric space (X, d) with a random convex structure W. Let $\{T_i : i \in J\} : \Omega \times K \to K$ be a finite family of continuous uniformly quasi-Lipchitzian random operators with $L_i > 0$. Suppose that the sequence $\{\zeta_n(\omega)\}$ is generated by (4) and $\Sigma_{n=1}^{\infty}(\beta_{kn} + \gamma_{kn}) < \infty$. If $\mathcal{F} = \bigcap_{i=1}^{k} RF(T_i) \neq \emptyset$, then $\{\zeta_n(\omega)\}$ converges to a common fixed point of $\{T_i : i \in J\}$ if and only if $\liminf_{n\to\infty} d(\zeta_n(\omega), \mathcal{F}) = 0$, where $d(x, \mathcal{F}) = \inf\{d(x, y) : \forall y \in \mathcal{F}\}$.

Proof
By lemma 3, we have

$$d(\zeta_{n+1}(\omega, \mathcal{F}) \leq (1 + \Theta_n \mathcal{M}_0) d(\zeta_n(\omega), \mathcal{F}) + \Theta_n \mathcal{M}_1.$$

Because $\Sigma_{n=1}^{\infty}(\beta_{kn} + \gamma_{kn}) < \infty$, by Lemma 2, thus $\lim_{n\to\infty} d(\zeta_n(\omega), \mathcal{F})$ exists. From hypothesis, $\liminf_{n\to\infty} d(\zeta_n(\omega), \mathcal{F}) = 0$, we get

$$\lim_{n\to\infty} d(\zeta_n(\omega), \mathcal{F}) = 0.$$

Now, we prove that $\{\zeta_n(\omega)\}$ is a Cauchy sequence. Actually, for each $\varepsilon > 0$, there exists a constant N_0 such that for all $n \geq N_0$, we get

$$d(\zeta_n(\omega), \mathcal{F}) \leq \tfrac{\varepsilon}{4\mathcal{M}_2} \text{ and } \Sigma_{n=N_0}^{\infty} \Theta_n \leq \tfrac{\varepsilon}{4\mathcal{M}_1 \mathcal{M}_2}.$$

Especially, there exists $\varrho_1(\omega) \in \mathcal{F}$ and $N_1 > N_0$, where N_1 is constant such that

$$d(\zeta_{N_1}(\omega), \varrho_1(\omega)) \leq \tfrac{\varepsilon}{4\mathcal{M}_2}.$$

By Lemma 3, we get

$$\begin{aligned}
& d(\zeta_{n+m}(\omega), \zeta_n(\omega)) \\
= \; & d(\zeta_{n+m}(\omega), \varrho(\omega)) + d(\varrho(\omega), \zeta_n(\omega)) \\
\leq \; & \mathcal{M}_2 d(\zeta_{N_1}(\omega), \zeta(\omega)) + \mathcal{M}_1 \mathcal{M}_2 \Sigma_{j=N_1}^{n+m-1} \Theta_j + \mathcal{M}_2 d(\zeta_{N_1}(\omega), \zeta(\omega)) \\
& + \mathcal{M}_1 \mathcal{M}_2 \Sigma_{j=N_1}^{n-1} \Theta_j \\
\leq \; & 2\mathcal{M}_2 d(\zeta_{N_1}(\omega), \zeta(\omega)) + \mathcal{M}_1 \mathcal{M}_2 (\Sigma_{j=N_1}^{n+m-1} \Theta_j + \Sigma_{j=N_1}^{n-1} \Theta_j) \\
\leq \; & 2\mathcal{M}_2 d(\zeta_{N_1}(\omega), \zeta(\omega)) + 2\mathcal{M}_1 \mathcal{M}_2 \Sigma_{j=N_1}^{\infty} \Theta_j \\
\leq \; & 2\mathcal{M}_2 \left(\frac{\varepsilon}{4\mathcal{M}_2}\right) + 2\mathcal{M}_1 \mathcal{M}_2 \left(\frac{\varepsilon}{4\mathcal{M}_1 \mathcal{M}_2}\right) = \varepsilon.
\end{aligned}$$

That is $\{\zeta_n(\omega)\}$ is a Cauchy sequence in closed convex subset of complete generalized convex metric spaces. Therefore, $\{\zeta_n(\omega)\}$ converges to a point of K. Suppose $\lim_{n\to\infty} \zeta_n(\omega) = \varrho(\omega)$, $\forall \omega \in \Omega$. Since T_i are continuous by Lemma 1, we know that for any measurable mapping $f : \Omega \to K, T_i^n(\omega, f(\omega)) : \Omega \to K$ are measurable mappings. So, $\{\zeta_n(\omega)\}$ is a sequence of measurable mappings. Hence, $\varrho : \Omega \to K$ is also measurable. Now, we show that $\varrho(\omega) \in \mathcal{F}$. From

$$d(\varrho(\omega), \mathcal{F}) \leq d(\zeta_n(\omega), \varrho(\omega)) + d(\zeta_n(\omega), \mathcal{F}).$$

Since $d(\zeta_n(\omega), \varrho(\omega)) = 0$ and $d(\zeta_n(\omega), \mathcal{F}) = 0$, so, we get $d(\varrho(\omega), \mathcal{F}) = 0$. Hence, $\varrho(\omega) \in \mathcal{F}$. □

From definition 2, if T is an asymptotically quasi-nonexpansive random operator, then T is a uniformly quasi-Lipschitzian random operator $((L = \sup_{n\geq 1}\{k_n\}))$. And if $RF(T) \neq \emptyset$, then every uniformly L-Lipschitzian random operator is a uniformly quasi-Lipschitzian random operator, we get the following corollary.

Corollary 1

Let K be a non-empty closed convex subset of a separable complete generalized convex metric space (X, d) with a random convex structure W. Let $\{T_i : i \in J\} : \Omega \times K \to K$ be a finite family of continuous asymptotically quasi-nonexpansive random operator with $L_i > 0$. Suppose that the sequence $\{\zeta_n(\omega)\}$ is generated by (4) and $\Sigma_{n=1}^{\infty}(\beta_{kn} + \gamma_{kn}) < \infty$. If $\mathcal{F} = \bigcap_{i=1}^{k} RF(T_i) \neq \emptyset$, then $\{\zeta_n(\omega)\}$ converges to a common fixed point of $\{T_i : i \in J\}$ if and only if $\liminf_{n\to\infty} d(\zeta_n(\omega), \mathcal{F}) = 0$, where $d(x, \mathcal{F}) = \inf\{d(x, y) : \forall y \in \mathcal{F}\}$.

Theorem 2

Let K be a non-empty closed convex subset of a separable complete generalized convex metric space (X, d) with a random convex structure W. Let $\{T_i : i \in J\} : \Omega \times K \to K$ be a finite family of continuous uniformly quasi-Lipchitzian random mappings with $L_i > 0$. Suppose that the sequence $\{\zeta_n(\omega)\}$ generated by (4), $\Sigma_{n=1}^{\infty}(\beta_{kn} + \gamma_{kn}) < \infty$ and $\mathcal{F} = \bigcap_{i=1}^{k} RF(T_i) \neq \emptyset$. If for some given $1 \leq l \leq k$,

(i) $\lim_{n\to\infty} d(T_l(\omega, \zeta_n(\omega)), \zeta_n(\omega)) = 0$;
(ii) there exists a positive constant $\mathcal{M}_3$ such that

$$d(T_l(\omega, \zeta_n(\omega)), \zeta_n(\omega)) \geq \mathcal{M}_3 d(\zeta_n(\omega), \mathcal{F}).$$

Then $\{\zeta_n(\omega)\}$ converges to a common fixed point of $\{T_i : i \in J\}$.

Proof

By condition (i), we get

$$\lim_{n\to\infty} d(T_l(\omega, \zeta_n(\omega)), \zeta_n(\omega)) = 0,$$

that is,

$$d(T_l(\omega, \zeta_n(\omega)), \zeta_n(\omega)) = 0.$$

Also, by condition (ii), we get

$$\mathcal{M}_3 d(\zeta_n(\omega), \mathcal{F}) \leq d(T_l(\omega, \zeta_n(\omega)), \zeta_n(\omega)) = 0,$$

that is,

$$\mathcal{M}_3 d(\zeta_n(\omega), \mathcal{F}) = 0.$$

Thus, follow proof from Theorem 3, we get, $\{\zeta_n(\omega)\}$ converges to a common fixed point of $\{T_i : i \in J\}$. This completes the proof. □

Theorem 3

Let K be a non-empty closed convex subset of a separable complete generalized convex metric space (X, d) with a random convex structure W. Let $\{T_i : i \in J\} : \Omega \times K \to K$ be a finite family of continuous uniformly quasi-Lipchitzian random mappings with $L_i > 0$. Suppose that the sequence $\{\zeta_n(\omega)\}$ generated by (4), $\Sigma_{n=1}^{\infty}(\beta_{kn} + \gamma_{kn}) < \infty$ and $\mathcal{F} = \bigcap_{i=1}^{k} RF(T_i) \neq \emptyset$. If

(i) for all $1 \leq i \leq k$; $\lim_{n\to\infty} d(T_l(\omega, \zeta_n(\omega)), \zeta_n(\omega)) = 0$;
(ii) for some $1 \leq l \leq k$; T_l is semi-compact.

Then $\{\zeta_n(\omega)\}$ converges to a common fixed point of $\{T_i : i \in J\}$.

Proof

By conditions (i) and (ii), there exists a subsequence $\{\zeta_{n_j}(\omega)\} \subset \{\zeta_n(\omega)\}$ such that $\lim_{j\to\infty} \zeta_{n_j} = \zeta^*(\omega), \forall \omega \in \Omega$, where $\zeta^*(\omega) \in K$. Since T_i are continuous for $i \in J$. So, $\{\zeta_n(\omega)\}$ is a sequence of measurable mappings. Hence, $\zeta^* : \Omega \to K$ is also measurable. Since $\lim_{j\to\infty} d(T_i(\omega, \zeta_{n_j}(\omega)), \zeta_{n_j}(\omega)) = d(T_i(\omega, \zeta^*(\omega)), \zeta^*(\omega)) = 0$, we get, $\zeta^*(\omega) \in \mathcal{F}, \forall \omega \in \Omega$. By Lemma 6, we get

$$d(\zeta_{n+1}(\omega), \zeta^*(\omega)) \leq (1 + \Theta_n \mathcal{M}_0) d(\zeta_n(\omega), \zeta^*(\omega)) + \Theta_n \mathcal{M}_1.$$

Since $\Sigma_{n=1}^{\infty} \Theta_n < \infty$, by Lemma 2, there exists $\rho \geq 0$ such that

$$\lim_{n\to\infty} d(\zeta_n(\omega), \zeta^*(\omega)) = \rho. \tag{7}$$

Since $\liminf_{n\to\infty} d(\zeta_n(\omega), \zeta^*(\omega)) = 0$, we get

$$\lim_{n\to\infty} \zeta_n(\omega) = \zeta^*(\omega),$$

that is

$$\lim_{n\to\infty} d(\zeta_n(\omega) = \zeta^*(\omega)) = 0. \tag{8}$$

By (7) and (8), we get $\rho = 0$.

Hence, $\{\zeta_n(\omega)\}$ converges to common fixed point of $\{T_i : i \in J\}$. This completes the proof. □

Acknowledgements

The first author was supported by Rajamangala University of Technology Lanna (RMUTL) and the second author thank to RMUTT for financial support. Also, we would like to thank Dr. Poom Kumam for suggestions.

REFERENCES

1. A. Spacek, *Zufallige Gleichungen*, Czechoslov. Math. J. vol. 5, no. 80, pp. 462–466 1955.
2. O. Hans, *Random operator equations* In: Proceedings of 4th Berkeley Sympos. Math. Statist. Prob., vol. II, part I, pp. 185–202, University of California Press, Berkeley, 1961.
3. O. Hans, *Reduzierende zufallige transformationen*, Czechoslov. Math. J., vol. 7, no. 82, pp. 154–158, 1957.
4. A. Mukherjee, *Transformation aleatoires separable theorem all point fixed aleatoire*, C.R. Acad. Sci. Paris, Ser. A-B, vol. 263, pp. 393–395, 1966.
5. A.T., Bharucha-Reid, *Fixed point theorems in probabilistic analysis*, Bull. Amer. Math. Soc., vol. 82, pp. 641–657, 1976.
6. S. Itoh, *Random fixed-point theorems with an application to random differential equations in Banach spaces*, J. Math. Anal. Appl., vol. 67, pp. 261–273, 1979.
7. V.M. Sehgal, C. Waters, *Some random fixed point theorems for condensing operators*, Proc. Amer. Math. Soc., vol. 90, pp. 425–429, 1984.
8. E. Rothe, *Zur theorie der topologischen ordnung und der Vektorfelder in Banachschen Raumen*, Compos. Math., vol. 5, pp. 177–197, 1938.
9. I. Beg, N. Shahzad, *Random fixed points of random multivalued operator on Polish spaces*, to appear in Nonlinear Anal.
10. P. Kumam, W. Kumam, *Random fixed points of multivalued random operators with property (D)*, Random Oper.Stoch. Equat., vol. 15, pp. 127–136, 2007.
11. P. Kumam, S. Plubtieng, *Random fixed point theorems for asymptotically regular random operators*, Demonst. Math., vol. XLII, pp. 131–141, 2009.
12. P. Kumam, S. Plubtieng, *The characteristic of noncompact convexity and random fixed point theorem for set-valued operators*, Czechoslov. Math. J., vol. 57, no. 132, pp. 269–279, 2007.
13. P. Kumam, *Random common fixed points of single-valued and multivalued random operators in a uniformly convex Banach space*, J. Comput. Anal. Appl., vol. 13, pp. 368–375, 2011.
14. W. Kumam, P. Kumam, *Random fixed point theorems for multivalued subsequentially limit-contractive maps satisfying inwardness conditions*, J. Comput. Anal. Appl., vol. 14, pp. 239–251, 2012.
15. J.S. Jung, Y.J. Cho, S.M. Kang, B.S. Lee, B.S. Thakur, *Random fixed point theorems for a certain class of mappings in banach spaces*, Czechoslovak Mathematical Journal, Vol. 50, Issue 2, pp. 379–396, 2000.
16. M. Saha, *On some random fixed point of mappings over a Banach space with a probability measure*, Proc. Natl. Acad. Sci. India, Sect. A, vol. 76, pp. 219–224, 2006.
17. M. Saha, L. Debnath, *Random fixed point of mappings over a Hilbert space with a probability measure*, Adv. Stud. Contemp. Math., vol. 1, pp. 79–84, 2007.
18. W.J. Padgett, *On a nonlinear stochastic integral equation of the hammerstein type*, Proc. Amer. Soc., vol. 38, pp. 625–631, 1973.
19. J. Achari, *On a pair of random generalized nonlinear contractions*, Internat. J. Math. Math. Sci., vol. 6, pp. 467–475, 1983.

20. M. Saha, D. Dey, *Some random fixed point theorems for $(\theta, L)-$weak contractions* to appear in Hacet. J. Math. Stat.
21. M. Patriche, *Random fixed point theorems under mild continuity assumptions*, Fixed Point Theory and Applications, pp. 1–14, 2014.
22. V.N. Mishra, *Some Problems on Approximations of Functions in Banach Spaces*, Ph.D.Thesis (2007), Indian Institute of Technology, Roorkee 247 667, Uttarakhand, India.
23. L.N. Mishra, S.K. Tiwari and V.N. Mishra, *Fixed point theorems for generalized weakly S-contractive mappings in partial metric spaces*, Journal of Applied Analysis and Computation, vol. 5, pp. 600–612, 2015, doi:10.11948/2015047.
24. Deepmala, *Study on Fixed Point Theorems for Nonlinear Contractions and its Applications*, Ph.D. Thesis (2014), Pt. Ravishankar Shukla University, Raipur 492 010, Chhatisgarh, India.
25. H.K. Pathak and Deepmala, *Common fixed point theorems for PD-operator pairs under Relaxed conditions with applications*, Journal of Computational and Applied Mathematics, vol. 239, pp. 103–113, 2013.
26. V.N. Mishra, L.N. Mishra, *Trigonometric Approximation of Signals (Functions) in L_p $(p \geq 1)$ norm*, International Journal of Contemporary Mathematical Sciences, vol. 7, no. 19, pp. 909–918, 2012.
27. W. Takahashi, *A convexity in metric space and nonexpansive mapping*, Kodai.Math.Sem.Rep., vol. 22, pp. 142–149, 1970.
28. W.A. Kirk, *Krasnoselskfi's iteration process in hyperbohc space*, Number, Funet., Anal. Optm, vol. 4, pp. 371-381, 1982.
29. K. Goebel, W.A. Kirk, *Iteration processes for nonexpansive mappings*, In Contemporary Math, Amer Math. Soc, Providence, RI, vol. 1, pp. 115–123, 1983.
30. Q. Liu, *Iterative sequences for asymptotically quasi-nonexpansive mappings*, Journal of Mathematical Analysis and Applications, vol. 207, no. 1, pp. 96–103, 1997.
31. Q. Liu, *Iterative sequences for asymptotically quasi-nonexpansive mappings with error member*, Journal of Mathematical Analysis and Applications, vol. 259, no. 1, pp. 18–24, 2001.
32. Q. Liu, *Iterative sequences for asymptotically quasi-nonexpansive mapping with an error member of uniform convex Banach space*, Journal of Mathematical Analysis and Applications, vol. 266, no. 2, pp. 468–471, 2002.
33. Y.X. Tian, *Convergence of an Ishikawa type iterative scheme for asymptotically quasi-nonexpansive mappings*, Comput.Math.Appl., vol. 49, pp. 1905–1912, 2005.
34. G. Das, J.P. Debata, *Fixed points of quasi-nonexpansive mappings*, Indian J. Pure. Appl. Math., vol. 17, pp. 1263–1269, 1986.
35. A.R. Khan, M.A. Ahmed, *Convergence of a general iterative scheme for a finite family of asymptotically quasi-nonexpansive mappings in convex metric spaces and applications*, Comput.Math.Appl., vol. 59, pp. 2990–2995, 2010.
36. C. Wang, L. Liu, *Convergence theorems for fixed points of uniformly quasi-Lipschitzian mappings in convex metric spaces*, Nonlinear Analysis : TMA., vol. 70, pp. 2067–2071, 2009.
37. R.P. Agarwal, D. O'Regan and D.R. Sahu, *Iterative construction of fixed points of nearly asymptotically nonexpansive mappings*, J. Nonlinear Convex. Anal., vol. 8, no. 1, pp. 61–79, 2007.
38. P. Saipara, P. Chaipunya, Y.J. Cho and P. Kumam, *On Strong and Δ-convergence of modified S-iteration for uniformly continuous total asymptotically nonexpansive mappings in CAT(κ) spaces*, The Journal of Nonlinear Science and Applications, vol. 8, pp. 965–975, 2015.
39. R. Suparatulatorn and P. Cholumjiak, *The modified S-iteration process for nonexpansive mappings in $CAT(\kappa)$ spaces*, Fixed Point Theory and Applications, vol. 1, pp. 1–12, 2016.
40. V. Kumar, A. Latif, A. Rafiq and N. Hassain, *S-iteration process for quasi-contr active mappings*, Journal of Inequalities and Applications, vol. 206, 2013.
41. S. Kosol, *Rate of convergence of S-iteration and Ishikawa iteration for continuous functions on closed intervals*, Thai Journal of Mathematics, vol.11, pp. 703–709, 2013.
42. P. Kumam, G.S. Saluja and H.K. Nashine, *Convergence of modified S-iteration process for two asymptotically nonexpansive mappings in the intermediate sense in CAT(0) spaces*, Journal of Inequalities and Applications, vol. 368, 2014.
43. I. Beg, *Approximation of random fixed point in normed space*, Nonlinear Analysis: TMA., vol. 51, pp. 1363–1372, 2002.
44. K.K. Tan, X.Z. Yuan, *Some random fixed point theorems*, in: K.K.Tan (Ed.), Fixed Point Theory and Applications, World Scientific, Singapore, pp.334-345, 1992.

9

Bootstrap approach to the one-sample and two-sample test of variances of a fuzzy random variable

Jalal Chachi*

Department of Mathematics, Statistics and Computer Sciences, Semnan University, Iran

Abstract The aim of this paper is to present in a concise and integrated way of the bootstrap approach to statistical testing of hypotheses about the variance of fuzzy random variable. In this approach, first a notion of fuzzy random variables is recalled. Then, we will consider hypothesis-tests for the (crisp-valued) variance of fuzzy data in a population. For this purpose, the α-pessimistic values of the imprecise observations are used for defining a new notion of distance measure between fuzzy data, which is then used to make a procedure for testing the statistical hypotheses. Based on this argument, the application of bootstrap techniques in dealing with these testing problems will be introduced. The procedure develops a non-parametric approach to testing statistical hypotheses based on one-sample and two-sample fuzzy data.

Keywords α-Pessimistic, Bootstrap, Fuzzy Random Variable, Testing Hypotheses

1. Introduction

The bootstrap techniques contained in the literature are simple and straight-forward methods for calculating approximated biases, standard deviations, confidence intervals, and so forth, in almost any non-parametric estimation problem [8, 31]. As a common conclusion, we can state that bootstrap techniques are very valuable in dealing with testing about means and variances of imprecise concepts such as fuzzy random variables, since

1. there are not stochastic models in the literature which can be really widely applicable to describe imprecise quantities involved in practical situations,
2. most of the parameters in general asymptotic approaches could be unknown, the accuracy of bootstrap approaches is greater than that of asymptotic ones.

On the other hand, researchers in economics, engineering, environment science, the social science, medical science, business, management, and many other fields deal daily with the complexities of modeling uncertain data. Classical methods are not always successful dealing with uncertain data, because the uncertainties appearing in these domains may be of various types, such as probability theory [4], fuzzy set theory [36], intuitionistic fuzzy set theory [2], vague set theory [10], interval mathematics [14], and other mathematical tools are well know and often useful approaches to describing uncertainty. Therefore, in the studies proposed by Montenegro et al. [25], González-Rodríguez et al. [13], and Gil et al. [11] it has been stated that the well-known bootstrap techniques are a valuable tool in testing statistical hypotheses about the means and variances of fuzzy random variables, when these

*Correspondence to: Jalal Chachi (Email: jchachi@semnan.ac.ir). Department of Mathematics, Statistics and Computer Sciences, Semnan University, Semnan 35195-363, Iran.

variables are supposed to take on a finite number of different values and these values being fuzzy subsets of the one-dimensional Euclidean space (see also, [1, 7, 32, 37, 35]).

Therefore, the proposed paper provides the bootstrap approach to statistical testing of hypotheses about the variance of fuzzy random variable. For this purpose, the α-pessimistic values of the imprecise observations are used for defining the new notion of fuzzy random variables as well as a new notion of distance measure between fuzzy data. The procedure develops a non-parametric approach to testing statistical hypotheses based on one-sample and two-sample fuzzy data.

The paper is organized as follows: In the next section, some basic concepts of fuzzy set theory and uncertainty theory are recalled. In this section also a new definition of distance measure between fuzzy numbers is defined. In Section 3, a new definition of fuzzy random variable proposed by Hesamian and Chachi [15] is recalled. In Section 4, the non-parametric approach to testing statistical hypotheses based on one-sample and two-sample fuzzy data is provided. Numerical examples are also given in this section to clarify the proposed testing method. Finally, a brief conclusion is provided in Section 5.

2. Preliminary concepts

In this section, first, we shall review the basic definitions and terminologies of the fuzzy set theory and uncertainty theory which are necessary for our paper (for further details, the reader is referred to [21, 22, 33, 38]). Then, a new definition of distance measure between fuzzy numbers is defined.

A fuzzy set $\widetilde{A}$ of the universal set $\mathbb{X}$ is defined by its membership function $\widetilde{A}:\mathbb{X}\to[0,1]$. In this paper, we consider $\mathbb{R}$ (the real line) as the universal set. We denote by $\widetilde{A}[\alpha]=\{x\in\mathbb{R}:\widetilde{A}(x)\geq\alpha\}$ the α-level set (α-cut) of the fuzzy set $\widetilde{A}$ of $\mathbb{R}$, for every $\alpha\in(0,1]$, and $\widetilde{A}[0]$ is the closure of the set $\{x\in\mathbb{R}:\widetilde{A}(x)>0\}$. A fuzzy set $\widetilde{A}$ of $\mathbb{R}$ is called a fuzzy number if for every $\alpha\in[0,1]$, the set $\widetilde{A}[\alpha]$ is a non-empty compact interval. We denote by $\mathcal{F}(\mathbb{R})$, the set of all fuzzy numbers of $\mathbb{R}$.

The imprecision or vagueness can be treated by means of a particular kind (family) of fuzzy numbers, the LR-fuzzy numbers. These are very useful in practice since they can be characterized by means of three real numbers: the center, the left spread, and the right spread. The term LR is due to the left (L) and the right (R) shape of the membership function referred to the fuzzy set. Typically, the LR fuzzy number $\widetilde{N}=(n,l,r)_{LR}$ with central value $n\in\mathbb{R}$, left and right spreads $l\in\mathbb{R}^+$, $r\in\mathbb{R}^+$, decreasing left and right shape functions $L:\mathbb{R}^+\to[0,1]$, $R:\mathbb{R}^+\to[0,1]$, with $L(0)=R(0)=1$, has the following membership function

$$\widetilde{N}(x)=\begin{cases} L(\frac{n-x}{l}) & \text{if} \quad x\leq n,\\ R(\frac{x-n}{r}) & \text{if} \quad x\geq n.\end{cases}$$

We can easily obtain the α-cut of $\widetilde{N}$ as follows

$$\widetilde{N}[\alpha]=[n-L^{-1}(\alpha)l,n+R^{-1}(\alpha)r],\quad \alpha\in[0,1].$$

A special type of LR-fuzzy numbers is the so-called triangular fuzzy numbers. The membership function and the α-cut of triangular fuzzy number $\widetilde{A}=(a,l,r)_T$ is given by

$$\begin{aligned}\widetilde{A}(x) &= \begin{cases} 0 & x<a-l,\\ \frac{x-a+l}{l} & a-l\leq x<a,\\ \frac{a+r-x}{r} & a\leq x\leq a+r,\\ 0 & x>a+r.\end{cases}\\ \widetilde{A}[\alpha] &= [n-l(1-\alpha),n+r(1-\alpha)],\quad \alpha\in[0,1].\end{aligned}$$

For the algebraic operations of LR-fuzzy numbers, we have the following result on the basis of Zadeh's extension principle. Let $\widetilde{M}=(m,l_m,r_m)_{LR}$ and $\widetilde{N}=(n,l_n,r_n)_{LR}$ be two triangular fuzzy numbers and λ be a real number.

Then

$$\lambda \otimes \widetilde{M} = \begin{cases} (\lambda m, \lambda l_m, \lambda r_m)_{LR} & \text{if} \quad \lambda > 0, \\ \mathcal{I}_{\{0\}} & \text{if} \quad \lambda = 0, \\ (\lambda m, |\lambda| r_m, |\lambda| l_m)_{RL} & \text{if} \quad \lambda < 0, \end{cases}$$

$$\widetilde{M} \oplus \widetilde{N} = (m + n, l_m + l_n, r_m + r_n)_{LR},$$

where $\mathcal{I}_{\{0\}}$ stands for the indicator function of the crisp zero.

In the following, we introduce an index to compare fuzzy number $\widetilde{A} \in \mathcal{F}(\mathbb{R})$ and crisp value $x \in \mathbb{R}$. The index will be used for defining a new notion of fuzzy random variable.

Definition 1 ([23])
Let $\widetilde{A} \in \mathcal{F}(\mathbb{R})$ and $x \in \mathbb{R}$. The index $C : \mathcal{F}(\mathbb{R}) \times \mathbb{R} \longrightarrow [0, 1]$, which is defined by

$$C\{\widetilde{A} \leq x\} = \frac{\sup_{y \leq x} \widetilde{A}(y) + 1 - \sup_{y > x} \widetilde{A}(y)}{2},$$

shows the credibility degree that $\widetilde{A}$ is less than or equal to x. Similarly, $C\{\widetilde{A} > x\} = 1 - C\{\widetilde{A} \leq x\}$ shows the credibility degree that $\widetilde{A}$ is greater than x (see also [22]).

Definition 2
Let $\widetilde{A} \in \mathcal{F}(\mathbb{R})$ and $\alpha \in [0, 1]$, then $\widetilde{A}_\alpha = \inf\{x \in \widetilde{A}[0] : C\{\widetilde{A} \leq x\} \geq \alpha\}$, is called the α-pessimistic value of $\widetilde{A}$. It is clear that $\widetilde{A}_\alpha$ is a non-decreasing function of $\alpha \in (0, 1]$ (for more details, see [21, 26]).

Example 1
Suppose that $\widetilde{A} = (a, l, r)_{LR}$ is a LR-fuzzy number, and let $x \in \mathbb{R}$, then

$$C\{\widetilde{A} \leq x\} = \begin{cases} \frac{1}{2} L(\frac{a-x}{l}) & \text{if} \quad x \leq a, \\ 1 - \frac{1}{2} R(\frac{x-a}{r}) & \text{if} \quad x \geq a. \end{cases}$$

We can easily obtain the α-pessimistic values of $\widetilde{A}$ as follows

$$\widetilde{A}_\alpha = \begin{cases} a - lL^{-1}(2\alpha) & \text{for} \quad 0.0 < \alpha \leq 0.5, \\ a + rR^{-1}(2(1-\alpha)) & \text{for} \quad 0.5 \leq \alpha \leq 1.0. \end{cases}$$

In the following, based on the notion of α-pessimistic value of fuzzy sets, a new definition of metrics between fuzzy numbers is defined. In the literature one can find many useful metrics between fuzzy numbers and a few ones between fuzzy values. Valuable references on this point can be found in [5].

Definition 3
The distance measure is defined as the mapping $D : \mathcal{F}(\mathbb{R}) \otimes \mathcal{F}(\mathbb{R}) \to [0, \infty)$ such that it associates with two fuzzy numbers $\widetilde{A}, \widetilde{B} \in \mathcal{F}(\mathbb{R})$ the value $D(\widetilde{A}, \widetilde{B})$ such that

$$D(\widetilde{A}, \widetilde{B}) = \int_0^1 (\widetilde{A}_\alpha - \widetilde{B}_\alpha)^2 d\alpha.$$

One can conclude that the mapping $D : \mathcal{F}(\mathbb{R}) \otimes \mathcal{F}(\mathbb{R}) \to [0, \infty)$ satisfies the following conditions

1. For any $\widetilde{A}, \widetilde{B} \in \mathcal{F}(\mathbb{R})$, $D(\widetilde{A}, \widetilde{B}) = 0$ if and only if $\widetilde{A} = \widetilde{B}$.
2. For any $\widetilde{A}, \widetilde{B} \in \mathcal{F}(\mathbb{R})$, $D(\widetilde{A}, \widetilde{B}) = D(\widetilde{B}, \widetilde{A})$.
3. For any $\widetilde{A}, \widetilde{B}, \widetilde{C} \in \mathcal{F}(\mathbb{R})$, $D(\widetilde{A}, \widetilde{C}) \leq D(\widetilde{A}, \widetilde{B}) + D(\widetilde{B}, \widetilde{C})$.
4. For any $\widetilde{A}, \widetilde{B}, \widetilde{C} \in \mathcal{F}(\mathbb{R})$, such that $\widetilde{A} \subseteq \widetilde{B} \subseteq \widetilde{C}$, then $D(\widetilde{A}, \widetilde{C}) \geq \max\{D(\widetilde{A}, \widetilde{B}), D(\widetilde{B}, \widetilde{C})\}$.

As an example, we can easily obtain the distance between two LR-fuzzy numbers $\widetilde{A} = (a, l_1, r_1)_{LR}$ and $\widetilde{B} = (b, l_2, r_2)_{LR}$ as follows

$$\begin{aligned} D(\widetilde{A},\widetilde{B}) &= (a-b)^2 + \frac{(l_1-l_2)^2}{2}\int_0^1 \left(L^{-1}(\alpha)\right)^2 d\alpha + \frac{(r_1-r_2)^2}{2}\int_0^1 \left(R^{-1}(\alpha)\right)^2 d\alpha \\ & -(a-b)(l_1-l_2)\int_0^1 L^{-1}(\alpha)\,d\alpha + (a-b)(r_1-r_2)\int_0^1 R^{-1}(\alpha)\,d\alpha. \end{aligned}$$

For symmetric fuzzy numbers $\widetilde{A} = (a, l, l)_L$ and $\widetilde{B} = (b, r, r)_L$ we have

$$D(\widetilde{A},\widetilde{B}) = (a-b)^2 + (l-r)^2 \int_0^1 \left(L^{-1}(\alpha)\right)^2 d\alpha.$$

3. Fuzzy random variables

In the context of random experiments whose outcomes are not numbers (or vectors in $\mathbb{R}^p$) but they are expressed in inexact terms, the concept of fuzzy random variable turns out to be useful. Random fuzzy numbers (or, more generally, random fuzzy sets [5]) is a well-stated and supported model within the probabilistic setting for the random mechanisms generating fuzzy data. They integrate randomness and fuzziness, so that the first one affects the generation of experimental data, whereas the second one affects the nature of experimental data which are assumed to be intrinsically imprecise. The notion of random fuzzy set can be formalized in several equivalent ways. Thus, in this regard, different notions of fuzzy random variable have been introduced and investigated from different points of view in the literature, e.g. see, [5, 6, 12, 15, 17, 18, 19, 20, 24, 27, 28, 29, 30].

Definition 4 ([15])
Suppose that a random experiment is described by a probability space $(\Omega, \mathcal{A}, \mathbf{P})$, where Ω is a set of all possible outcomes of the experiment, $\mathcal{A}$ is a σ-algebra of subsets of Ω and $\mathbf{P}$ is a probability measure on the measurable space $(\Omega, \mathcal{A})$. The fuzzy-valued mapping $\widetilde{X} : \Omega \to \mathcal{F}(\mathbb{R})$ is called a fuzzy random variable if for any $\alpha \in [0, 1]$, the real-valued mapping $\widetilde{X}_\alpha : \Omega \to \mathbb{R}$ is a real-valued random variable on $(\Omega, \mathcal{A}, \mathbf{P})$. Throughout this paper, we assume that all random variables have the same probability space $(\Omega, \mathcal{A}, \mathbf{P})$.

Remark 1
Kwakernaak [19, 20] introduced the notion of fuzzy random variables which has been later formalized in a clear way by Kruse and Meyer [18]. In this approach, the fuzzy random variable is viewed as a fuzzy perception (observation or report) of a classical real-valued random variable which is equivalent as: given a probability space $(\Omega, \mathcal{A}, \mathbf{P})$, a mapping $\widetilde{X} : \Omega \to \mathcal{F}(\mathbb{R})$ is said to be a fuzzy random variable if for all $\alpha \in (0, 1]$ the two real-valued mappings $\widetilde{X}^L_\alpha : \Omega \to \mathbb{R}$ and $\widetilde{X}^U_\alpha : \Omega \to \mathbb{R}$ are real-valued random variables (see also [6, 27, 28]).

One can easily show that the following relations are also held between the definition of fuzzy random variable proposed in this paper and Kwakernaak and Kruse's definition (see also, Example 1)

$$\begin{aligned} \widetilde{X}_\alpha &= \begin{cases} \widetilde{X}^L_{2\alpha} & \text{for} \quad 0.0 < \alpha \le 0.5, \\ \widetilde{X}^U_{2(1-\alpha)} & \text{for} \quad 0.5 \le \alpha \le 1.0, \end{cases} \\ \widetilde{X}[\alpha] &= [\widetilde{X}_{\frac{\alpha}{2}}, \widetilde{X}_{1-\frac{\alpha}{2}}], \qquad \alpha \in (0, 1]. \end{aligned}$$

The first above relation shows that the information contained in the two-dimensional variable $(\widetilde{X}^L_\alpha, \widetilde{X}^U_\alpha)$ is summarized in the one-dimensional variable $\widetilde{X}_\alpha$ making the computational procedures in the problems more easier. These relations confirms that this kind of definition of fuzzy random variables is equivalent to that of Kwakernaak and Kruse's definition.

Using classical techniques in Probability Theory, some aspects and results associated to a random variable were extended to this new environment (see e.g. [6, 15, 29, 30, 37]).

Definition 5 ([15])
Two fuzzy random variables $\widetilde{X}$ and $\widetilde{Y}$ are said to be independent if $\widetilde{X}_\alpha$ and $\widetilde{Y}_\alpha$ are independent, for all $\alpha \in [0,1]$. In addition, we say that two fuzzy random variables $\widetilde{X}$ and $\widetilde{Y}$ are identically distributed if $\widetilde{X}_\alpha$ and $\widetilde{Y}_\alpha$ are identically distributed, for all $\alpha \in [0,1]$. Similar arguments can be used for more than two fuzzy random variables. We also say that $\widetilde{X}_1, \ldots, \widetilde{X}_n$ is a fuzzy random sample if $\widetilde{X}_i$'s are independent and identically distributed fuzzy random variables. We denote by $\widetilde{x}_1, \ldots, \widetilde{x}_n$ the observed values of fuzzy random sample $\widetilde{X}_1, \ldots, \widetilde{X}_n$.

In analyzing fuzzy data two main types of summary measures/parameters may be distinguished:

1. fuzzy-valued summary measures, like the mean value of a fuzzy random variable or the median of a fuzzy random variable as measures for the central tendency of their distributions [3, 9, 34];
2. real-valued summary measures, like the variance of a fuzzy random variable as a measure for the mean error/dispersion of the distributions of the fuzzy random variable, or the covariance as a measure of the (absolute) linear dependence/association of a fuzzy random variable [6, 16, 29].

The above summary measures for the fuzzy random variable as well as some properties of them are formalized as follows:

Definition 6
Given a probability space $(\Omega, \mathcal{A}, \mathbf{P})$ and an associated fuzzy random variable $\widetilde{X} : \Omega \to \mathcal{F}(\mathbb{R})$ such that for any $\alpha \in [0,1]$, the real-valued variable $\widetilde{X}_\alpha : \Omega \to \mathbb{R}$ on $(\Omega, \mathcal{A}, \mathbf{P})$ has finite mean (i.e. $\widetilde{X}_\alpha \in L^1(\Omega, \mathcal{A}, \mathbf{P})$), the mean value of $\widetilde{X}$ is the fuzzy value $\widetilde{E}(\widetilde{X}) \in \mathcal{F}(\mathbb{R})$ such that for all $\alpha \in [0,1]$

$$\widetilde{E}(\widetilde{X})_\alpha = E(\widetilde{X}_\alpha) = \int_\Omega \widetilde{X}_\alpha \, d\mathbf{P}.$$

Proposition 1
$\widetilde{E}$ is additive (i.e., equivariant under the sum of fuzzy random variables), that is, for fuzzy random variables $\widetilde{X}$ and $\widetilde{Y}$ associated with the same probability space $(\Omega, \mathcal{A}, \mathbf{P})$ and such that $\widetilde{X}_\alpha, \widetilde{Y}_\alpha \in L^1(\Omega, \mathcal{A}, \mathbf{P})$, we have that

$$\begin{aligned} \widetilde{E}(\widetilde{X} \oplus \widetilde{Y}) &= \widetilde{E}(\widetilde{X}) \oplus \widetilde{E}(\widetilde{Y}), \\ \widetilde{E}(\lambda \otimes \widetilde{X}) &= \lambda \otimes \widetilde{E}(\widetilde{X}), \qquad \lambda \in \mathbb{R}. \end{aligned}$$

Definition 7
The variance of a fuzzy random variable $\widetilde{X}$ is defined as

$$\begin{aligned} \nu(\widetilde{X}) &= E\left[D(\widetilde{X}, \widetilde{E}(\widetilde{X}))\right] \\ &= E\left(\int_0^1 \left(\widetilde{X}_\alpha - E(\widetilde{X}_\alpha)\right)^2 d\alpha\right) \\ &= \int_\Omega \int_0^1 \left(\widetilde{X}_\alpha - E(\widetilde{X}_\alpha)\right)^2 d\alpha \, d\mathbf{P} \\ &= \int_0^1 \int_\Omega \left(\widetilde{X}_\alpha - E(\widetilde{X}_\alpha)\right)^2 d\mathbf{P} \, d\alpha \\ &= \int_0^1 Var(\widetilde{X}_\alpha) \, d\alpha. \end{aligned}$$

Proposition 2
Let $\widetilde{\mathbf{X}} = (\widetilde{X}_1, \widetilde{X}_2, \ldots, \widetilde{X}_n)$ be a fuzzy random sample, and $S_n^2(\widetilde{\mathbf{X}}) = \frac{1}{n-1} \sum_{i=1}^n D\left(\widetilde{X}_i, \bar{\widetilde{\mathbf{X}}}\right)$ be the crisp variance value of the fuzzy sample $\widetilde{\mathbf{X}}$, where $\bar{\widetilde{\mathbf{X}}} = \frac{1}{n} \oplus_{i=1}^n \widetilde{X}_i$ is the fuzzy sample mean value. Then $S_n^2(\widetilde{\mathbf{X}})$ is an unbiased estimator of the parameter $\nu(\widetilde{X})$ (population variance), i.e. $E[S_n^2(\widetilde{\mathbf{X}})] = \nu(\widetilde{X})$; and $\lim_{n\to\infty} S_n^2(\widetilde{\mathbf{X}}) = \nu(\widetilde{X})$.

Proof

$$
\begin{aligned}
E[S_n^2] &= \frac{1}{n-1}\sum_{i=1}^{n}\int_0^1 E[\widetilde{X}_{i\alpha} - \bar{\widetilde{\mathbf{X}}}_\alpha]^2 d\alpha \\
&= \frac{1}{n-1}\sum_{i=1}^{n}\int_0^1 E[\widetilde{X}_{i\alpha} - E(\widetilde{X}_{i\alpha}) + E(\widetilde{X}_{i\alpha}) - \bar{\widetilde{\mathbf{X}}}_\alpha]^2\, d\alpha \\
&= \frac{1}{n-1}\sum_{i=1}^{n}\left[\nu(\widetilde{X}_i) + \int_0^1 Var(\bar{\widetilde{\mathbf{X}}}_\alpha)\, d\alpha - 2\int_0^1 \frac{1}{n} Var(\widetilde{X}_{i\alpha})\, d\alpha\right] \\
&= \frac{1}{n-1}\sum_{i=1}^{n}\left[\nu(\widetilde{X}_i) + \frac{1}{n}\nu(\widetilde{X}_i) - \frac{2}{n}\nu(\widetilde{X}_i)\right] \\
&= \nu(\widetilde{X}).
\end{aligned}
$$

□

4. Bootstrap methods for testing hypotheses about the variance

In the following, using the general bootstrap methodology, we follow the application of the bootstrap in fuzzy environment and propose an approach to testing hypotheses on the basis of the evidence supplied by a set of sample fuzzy data. The treatment of the bootstrap method described here comes from Efron and Tibshirani [8] (see also [31]). The interested reader is referred to that text for more information on the underlying theory behind the bootstrap.

4.1. One-sample test of hypotheses

In this section we describe on way to get bootstrap hypothesis testing about the variance of fuzzy random variables based on one-sample fuzzy data. Suppose that we have fuzzy random sample $\widetilde{\mathbf{X}} = (\widetilde{X}_1, \widetilde{X}_2, \ldots, \widetilde{X}_n)$, and we want to test the following hypotheses

$$
\begin{aligned}
H_0 &: \text{ the variance of population is } \sigma_0^2, \\
H_1 &: \text{ the variance of population is not } \sigma_0^2.
\end{aligned}
$$

The bootstrap is a method of Monte Carlo simulation where no parametric assumptions are made about the underlying population that generated the random sample $\widetilde{\mathbf{X}} = (\widetilde{X}_1, \widetilde{X}_2, \ldots, \widetilde{X}_n)$. Instead, we use the sample as an estimate of the population. This estimate is called the empirical distribution F_n where each $\widetilde{X}_i$ $(i = 1, \ldots, n)$ has probability mass $\frac{1}{n}$. Thus, each $\widetilde{X}_i$ $(i = 1, \ldots, n)$ has the same likelihood of being selected in a new sample taken from F_n. When we use F_n as our pseudo-population, then we resample with replacement from the original sample $\widetilde{\mathbf{X}} = (\widetilde{X}_1, \widetilde{X}_2, \ldots, \widetilde{X}_n)$. We denote the new sample obtained in this manner by $\widetilde{\mathbf{X}}^{*b} = (\widetilde{X}_1^{*b}, \widetilde{X}_2^{*b}, \ldots, \widetilde{X}_n^{*b})$. We use the notation $\widetilde{\mathbf{X}}^{*b}$ $(b = 1, \ldots, B)$ for the bth bootstrap data set. These B bootstrap replicates provide us with an estimate of the distribution of the test statistic which will be introduced later. But, we need a distribution that estimates the population of treatment times under H_0. Note first that the empirical distribution (i.e., putting probability $\frac{1}{n}$ on each member of $\widetilde{\mathbf{X}}$) is not an appropriate estimate for the population distribution because it does not obey H_0. Somehow we need to obtain an estimate of the population distribution that has variance σ_0^2. A simple way is to transfer the empirical distribution so that it has the desired variance. In other word, we use as our estimated null distribution the empirical distribution on the values

$$
\widetilde{\mathbf{X}}_c = \left(\frac{\sigma_0 \widetilde{X}_1}{S_n(\widetilde{\mathbf{X}})}, \ldots, \frac{\sigma_0 \widetilde{X}_n}{S_n(\widetilde{\mathbf{X}})}\right),
$$

because under the hypothesis H_0 one can easily show that the sample variance of $\widetilde{\mathbf{X}}_c$ is σ_0^2 (i.e., $S_n^2(\widetilde{\mathbf{X}}_c) = \sigma_0^2$). In testing the null hypothesis H_0 at the nominal significance level $\gamma \in [0,1]$, the following test statistic is proposed

$$T(\widetilde{\mathbf{X}}) = \frac{(n-1)S_n^2(\widetilde{\mathbf{X}})}{\sigma_0^2}.$$

To get estimate of the distribution of the test statistic $T(\widetilde{\mathbf{X}})$ under null hypothesis H_0, we obtain B bootstrap samples by sampling with replacement from the original sample $\widetilde{\mathbf{X}}_c = (\widetilde{X}_{c1}, \ldots, \widetilde{X}_{cn})$. We denote the new samples obtained in this manner by $\widetilde{\mathbf{X}}_c^{*1}, \widetilde{\mathbf{X}}_c^{*2}, \ldots, \widetilde{\mathbf{X}}_c^{*B}$. For every bootstrap sample $\widetilde{\mathbf{X}}_c^{*b}$ ($b = 1, \ldots, B$), we calculate the same statistic to obtain the bootstrap replications of $T(\widetilde{\mathbf{X}})$, as follows

$$T(\widetilde{\mathbf{X}}_c^{*b}) = \frac{(n-1)S_n^2(\widetilde{\mathbf{X}}_c^{*b})}{\sigma_0^2}, \qquad b = 1, \ldots, B.$$

Once we have the B bootstrapped values $T(\widetilde{\mathbf{X}}_c^{*b})$ from above equation, we can use them to understand the estimated distribution for $T(\widetilde{\mathbf{X}})$. We use the estimated distribution in the next step to estimate the quantiles needed for testing hypotheses. The $\frac{\gamma}{2}$-th quantile, denoted by $t_{\frac{\gamma}{2}}$ of the $T(\widetilde{\mathbf{X}})$, is estimated by

$$\frac{\gamma}{2} = \frac{\text{number of } (T(\widetilde{\mathbf{X}}_c^{*b}) \le \widehat{t}_{\frac{\gamma}{2}})}{B}.$$

We are now ready to test the hypotheses. This is given by "the null hypothesis H_0 is rejected whenever $T(\widetilde{\mathbf{X}}) < \widehat{t}_{\frac{\gamma}{2}}$ or $T(\widetilde{\mathbf{X}}) > \widehat{t}_{1-\frac{\gamma}{2}}$, otherwise, it is accepted".

Algorithm 1. The steps for the proposed bootstrap methodology of testing hypotheses for one-sample are given here:

1. Given a random sample $\widetilde{\mathbf{X}}=(\widetilde{X}_1, \widetilde{X}_2, \ldots, \widetilde{X}_n)$, calculate $\widetilde{\mathbf{X}}_c=(\widetilde{X}_{c1}, \widetilde{X}_{c2}, \ldots, \widetilde{X}_{cn})$, and $T(\widetilde{\mathbf{X}}) = \frac{(n-1)S_n^2(\widetilde{\mathbf{X}})}{\sigma_0^2}$.
2. Sample with replacement from the sample $\widetilde{\mathbf{X}}_c$ to get $\widetilde{\mathbf{X}}_c^{*1}, \widetilde{\mathbf{X}}_c^{*2}, \ldots, \widetilde{\mathbf{X}}_c^{*B}$.
3. Calculate the same statistic using the bootstrap samples in step 2 to get $T(\widetilde{\mathbf{X}}_c^{*b}) = \frac{(n-1)S_n^2(\widetilde{\mathbf{X}}_c^{*b})}{\sigma_0^2}$, $b = 1, \ldots, B$.
4. Use the bootstrap replicates $T(\widetilde{\mathbf{X}}_c^{*1}), T(\widetilde{\mathbf{X}}_c^{*2}), \ldots, T(\widetilde{\mathbf{X}}_c^{*B})$ to estimate the distribution of $T(\widetilde{\mathbf{X}})$.
5. Compute the estimated bootstrap quantiles $\widehat{t}_{\frac{\gamma}{2}}$ and $\widehat{t}_{1-\frac{\gamma}{2}}$.
6. The hypothesis H_0 is rejected whenever $T(\widetilde{\mathbf{X}}) < \widehat{t}_{\frac{\gamma}{2}}$ or $T(\widetilde{\mathbf{X}}) > \widehat{t}_{1-\frac{\gamma}{2}}$.

Example 2

Suppose that we have taken the fuzzy random sample given in Table 1 from a population. Based on this data set, we want to test the following hypotheses

$$\begin{aligned} H_0 &: \text{ the variance of population is } \sigma_0^2, \\ H_1 &: \text{ the variance of population is not } \sigma_0^2, \end{aligned}$$

at significance level $\gamma = 0.05$ for the different values of σ_0^2 given in Table 2.

By following the steps in Algorithm 1 with $B = 10000$ bootstrap replications, one can obtain the bootstrap replicates of $T(\widetilde{\mathbf{x}}_c^{*1}), T(\widetilde{\mathbf{x}}_c^{*2}), \ldots, T(\widetilde{\mathbf{x}}_c^{*B})$. By sorting these values we can easily estimate any γ-quantile of the estimated distribution of the test statistic $T(\widetilde{\mathbf{X}})$. This says that, for example, the estimated $\widehat{t}_{0.025}$ is the 250th largest value of the $T(\widetilde{\mathbf{x}}_c^{*b})$ (because $B \times \frac{\gamma}{2} = 10000 \times 0.025 = 250$), and the estimated $\widehat{t}_{0.975}$ is the 9750th largest value of the $T(\widetilde{\mathbf{x}}_c^{*b})$. The estimated quantiles $\widehat{t}_{0.025}$ and $\widehat{t}_{0.975}$ for testing different values of σ_0^2 are obtained as given in Table 2. The histograms in Figure 1 show the estimated distributions of the test statistic $T(\widetilde{\mathbf{X}})$ for different values of σ_0^2 given in Table 2. Finally, at the nominal significance level 0.05, the test result of the null hypothesis H_0 based on the fuzzy random sample given in Table 1 is given in Table 2.

Table 1. Data set in Example 2

$\widetilde{x}_1 = (0.23, 0.04, 0.07)_T$	$\widetilde{x}_{16} = (1.78, 0.04, 0.06)_T$
$\widetilde{x}_2 = (0.76, 0.05, 0.02)_T$	$\widetilde{x}_{17} = (1.99, 0.08, 0.09)_T$
$\widetilde{x}_3 = (0.98, 0.12, 0.09)_T$	$\widetilde{x}_{18} = (2.25, 0.04, 0.04)_T$
$\widetilde{x}_4 = (1.14, 0.06, 0.09)_T$	$\widetilde{x}_{19} = (2.45, 0.01, 0.08)_T$
$\widetilde{x}_5 = (1.46, 0.10, 0.07)_T$	$\widetilde{x}_{20} = (2.57, 0.07, 0.02)_T$
$\widetilde{x}_6 = (1.69, 0.05, 0.12)_T$	$\widetilde{x}_{21} = (0.64, 0.11, 0.07)_T$
$\widetilde{x}_7 = (1.95, 0.05, 0.11)_T$	$\widetilde{x}_{22} = (0.94, 0.09, 0.04)_T$
$\widetilde{x}_8 = (2.17, 0.03, 0.05)_T$	$\widetilde{x}_{23} = (1.08, 0.10, 0.06)_T$
$\widetilde{x}_9 = (2.40, 0.08, 0.12)_T$	$\widetilde{x}_{24} = (1.37, 0.08, 0.06)_T$
$\widetilde{x}_{10} = (2.51, 0.10, 0.14)_T$	$\widetilde{x}_{25} = (1.64, 0.02, 0.08)_T$
$\widetilde{x}_{11} = (0.41, 0.03, 0.08)_T$	$\widetilde{x}_{26} = (1.83, 0.09, 0.05)_T$
$\widetilde{x}_{12} = (0.86, 0.08, 0.04)_T$	$\widetilde{x}_{27} = (2.04, 0.11, 0.06)_T$
$\widetilde{x}_{13} = (1.02, 0.03, 0.10)_T$	$\widetilde{x}_{28} = (2.36, 0.05, 0.09)_T$
$\widetilde{x}_{14} = (1.23, 0.03, 0.14)_T$	$\widetilde{x}_{29} = (2.49, 0.13, 0.05)_T$
$\widetilde{x}_{15} = (1.53, 0.13, 0.15)_T$	$\widetilde{x}_{30} = (2.61, 0.08, 0.06)_T$

Table 2. Test result for testing different values of σ_0^2 in Example 2 (see also Figure 1)

σ_0^2	$T(\widetilde{\mathbf{x}})$	$\widetilde{t}_{0.025}$	$\widehat{t}_{0.025}$	Result
0.28	50.61	18.48	38.51	reject H_0
0.38	37.20	18.50	38.51	accept H_0
0.48	29.58	18.50	38.51	accept H_0
0.68	20.81	18.52	38.51	accept H_0
0.90	15.73	18.54	38.54	reject H_0
1.50	9.36	18.54	38.54	reject H_0

4.2. Two-sample test of hypotheses

In this section we present a bootstrap approach to the two-sample test of equality of variances. In this way, let $(\Omega_X, \mathcal{A}_X, \mathbf{P}_X)$ and $(\Omega_Y, \mathcal{A}_Y, \mathbf{P}_Y)$ be two probability spaces, and let $\widetilde{\mathbf{X}} = (\widetilde{X}_1, \widetilde{X}_2, \ldots, \widetilde{X}_n)$ and $\widetilde{\mathbf{Y}} = (\widetilde{Y}_1, \widetilde{Y}_2, \ldots, \widetilde{Y}_m)$ be two independent fuzzy random variables associated with these spaces, respectively. Also let σ_X^2 and σ_Y^2 be the corresponding population variances. The goal of this section is testing the hypotheses

$$H_0 \quad : \quad \text{the variances of the two populations are equal, (i.e. } \frac{\sigma_X^2}{\sigma_Y^2} = 1\text{)},$$

$$H_1 \quad : \quad \text{the variances of the two populations are not equal, (i.e. } \frac{\sigma_X^2}{\sigma_Y^2} \neq 1\text{)},$$

on the basis of the available sample fuzzy information. To get bootstrap populations with a common variance from the available sample information in this case, one can multiply to each sample $\widetilde{\mathbf{X}}$ and $\widetilde{\mathbf{Y}}$ the ratios $\frac{1}{S_n(\widetilde{\mathbf{X}})}$ and $\frac{1}{S_m(\widetilde{\mathbf{Y}})}$, respectively. In other words, one can define new fuzzy random variables

$$\widetilde{\mathbf{X}}_c = \left(\frac{\widetilde{X}_1}{S_n(\widetilde{\mathbf{X}})}, \ldots, \frac{\widetilde{X}_n}{S_n(\widetilde{\mathbf{X}})}\right),$$

$$\widetilde{\mathbf{Y}}_c = \left(\frac{\widetilde{Y}_1}{S_m(\widetilde{\mathbf{Y}})}, \ldots, \frac{\widetilde{Y}_m}{S_m(\widetilde{\mathbf{Y}})}\right).$$

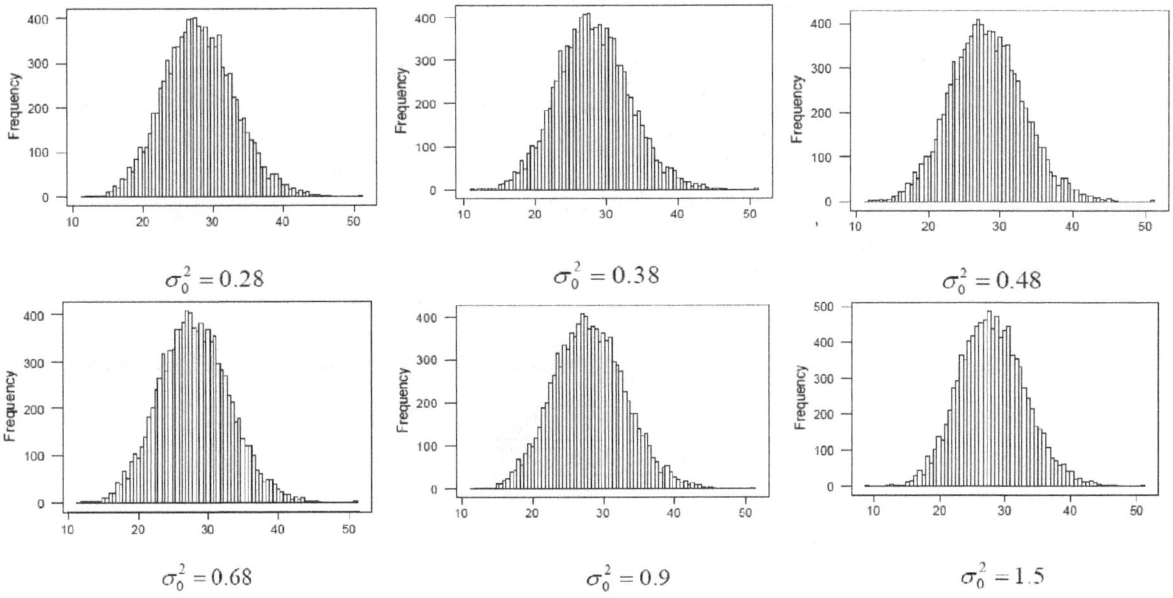

Figure 1. The histograms of the estimated distributions of the test statistic $T(\widetilde{\mathbf{X}})$ for different values of σ_0^2 given in Table 2 in Example 2.

Under the hypothesis H_0 one can easily show that the ratio of sample variances of $\widetilde{\mathbf{X}}_c$ and $\widetilde{\mathbf{Y}}_c$ is 1 (i.e., $\frac{S_n^2(\widetilde{\mathbf{X}}_c)}{S_m^2(\widetilde{\mathbf{Y}}_c)} = 1$). Therefore, resampling from these new populations is considered. In testing the null hypothesis $H_0 : \frac{\sigma_{\widetilde{X}}^2}{\sigma_{\widetilde{Y}}^2} = 1$ at the nominal significance level $\gamma \in [0, 1]$, the following test statistic is proposed

$$T(\widetilde{\mathbf{X}}, \widetilde{\mathbf{Y}}) = \frac{(n-1)S_n^2(\widetilde{\mathbf{X}})}{(m-1)S_m^2(\widetilde{\mathbf{Y}})}.$$

The null hypothesis H_0 should be rejected whenever $T(\widetilde{\mathbf{X}}, \widetilde{\mathbf{Y}}) < t_{\frac{\gamma}{2}}$ or $T(\widetilde{\mathbf{X}}, \widetilde{\mathbf{Y}}) > t_{1-\frac{\gamma}{2}}$ where t_γ is the $100(1-\gamma)\%$ quantile of the bootstrap distribution of the test statistic $T(\widetilde{\mathbf{X}}, \widetilde{\mathbf{Y}})$. Similar to the conclusions obtained for the one-sample test for the variance in previous section, the approach to be considered in the following will deal with the same. Therefore, to get estimate of the distribution of the test statistic $T(\widetilde{\mathbf{X}}, \widetilde{\mathbf{Y}})$ under null hypothesis H_0, we obtain B bootstrap samples by sampling with replacement from the original samples $\widetilde{\mathbf{X}}_c$, and $\widetilde{\mathbf{Y}}_c$. For every bootstrap sample $(\widetilde{\mathbf{X}}_c^{*b}, \widetilde{\mathbf{Y}}_c^{*b})$ $(b = 1, \ldots, B)$, we calculate the same statistic to obtain the bootstrap replications of $T(\widetilde{\mathbf{X}}, \widetilde{\mathbf{Y}})$, as follows

$$T(\widetilde{\mathbf{X}}_c^{*b}, \widetilde{\mathbf{Y}}_c^{*b}) = \frac{(n-1)S_n^2(\widetilde{\mathbf{X}}_c^{*b})}{(m-1)S_m^2(\widetilde{\mathbf{Y}}_c^{*b})}, \qquad b = 1, \ldots, B.$$

Once we have the B bootstrapped values $T(\widetilde{\mathbf{X}}_c^{*b}, \widetilde{\mathbf{Y}}_c^{*b})$ from above equation, we can use them to understand the estimated distribution for $T(\widetilde{\mathbf{X}}, \widetilde{\mathbf{Y}})$. We use the estimated distribution in the next step to estimate the quantiles needed for testing hypotheses. The $\frac{\gamma}{2}$-th quantile of $T(\widetilde{\mathbf{X}}, \widetilde{\mathbf{Y}})$ under null hypothesis H_0 is estimated by

$$\frac{\gamma}{2} = \frac{\text{number of } (T(\widetilde{\mathbf{X}}_c^{*b}, \widetilde{\mathbf{Y}}_c^{*b}) \leq \widehat{t}_{\frac{\gamma}{2}})}{B}.$$

Algorithm 2. The steps for the proposed bootstrap methodology of testing hypotheses for two-sample fuzzy data set are given here:

Table 3. Fuzzy random sample of size 15 from a population in Example 3

$\widetilde{y}_1 = (2.45, 0.04, 0.02)_T$	$\widetilde{y}_9 = (2.00, 0.08, 0.07)_T$
$\widetilde{y}_2 = (1.04, 0.10, 0.05)_T$	$\widetilde{y}_{10} = (2.02, 0.11, 0.02)_T$
$\widetilde{y}_3 = (1.09, 0.04, 0.01)_T$	$\widetilde{y}_{11} = (1.28, 0.10, 0.00)_T$
$\widetilde{y}_4 = (0.51, 0.07, 0.08)_T$	$\widetilde{y}_{12} = (0.95, 0.12, 0.09)_T$
$\widetilde{y}_5 = (2.40, 0.07, 0.06)_T$	$\widetilde{y}_{13} = (2.04, 0.01, 0.04)_T$
$\widetilde{y}_6 = (1.96, 0.11, 0.03)_T$	$\widetilde{y}_{14} = (0.68, 0.02, 0.05)_T$
$\widetilde{y}_7 = (1.34, 0.01, 0.08)_T$	$\widetilde{y}_{15} = (0.89, 0.09, 0.01)_T$
$\widetilde{y}_8 = (1.43, 0.08, 0.09)_T$	

Table 4. The estimated quantiles of the bootstrap distribution of $T(\widetilde{\mathbf{X}}, \widetilde{\mathbf{Y}})$ in Example 3 (see also Figure 2)

γ	0.005	0.025	0.05	0.10	0.90	0.95	0.975	0.995
$\widehat{t}_\gamma$	1.43	1.67	1.79	1.98	3.16	4.16	5.10	6.98

1. Given two random samples $\widetilde{\mathbf{X}}=(\widetilde{X}_1, \widetilde{X}_2, \ldots, \widetilde{X}_n)$, and $\widetilde{\mathbf{Y}}=(\widetilde{Y}_1, \widetilde{Y}_2, \ldots, \widetilde{Y}_m)$, calculate

$$\widetilde{\mathbf{X}}_c = \left(\frac{\widetilde{X}_1}{S_n(\widetilde{\mathbf{X}})}, \ldots, \frac{\widetilde{X}_n}{S_n(\widetilde{\mathbf{X}})}\right),$$

$$\widetilde{\mathbf{Y}}_c = \left(\frac{\widetilde{Y}_1}{S_m(\widetilde{\mathbf{Y}})}, \ldots, \frac{\widetilde{Y}_m}{S_m(\widetilde{\mathbf{Y}})}\right),$$

$$T(\widetilde{\mathbf{X}}, \widetilde{\mathbf{Y}}) = \frac{(n-1)S_n^2(\widetilde{\mathbf{X}})}{(m-1)S_m^2(\widetilde{\mathbf{Y}})}.$$

2. Sample with replacement from the samples $\widetilde{\mathbf{X}}_c$ and $\widetilde{\mathbf{Y}}_c$ to get $\widetilde{\mathbf{X}}_c^{*1}, \widetilde{\mathbf{X}}_c^{*2}, \ldots, \widetilde{\mathbf{X}}_c^{*B}$, and $\widetilde{\mathbf{Y}}_c^{*1}, \widetilde{\mathbf{Y}}_c^{*2}, \ldots, \widetilde{\mathbf{Y}}_c^{*B}$.
3. Calculate the same statistic using the bootstrap samples in step 2 to get $T(\widetilde{\mathbf{X}}_c^{*b}, \widetilde{\mathbf{Y}}_c^{*b}) = \frac{(n-1)S_n^2(\widetilde{\mathbf{X}}_c^{*b})}{(m-1)S_m^2(\widetilde{\mathbf{Y}}_c^{*b})}$, $b = 1, \ldots, B$.
4. Use the bootstrap replicates $T(\widetilde{\mathbf{X}}_c^{*1}, \widetilde{\mathbf{Y}}_c^{*1}), T(\widetilde{\mathbf{X}}_c^{*2}, \widetilde{\mathbf{Y}}_c^{*2}), \ldots, T(\widetilde{\mathbf{X}}_c^{*B}, \widetilde{\mathbf{Y}}_c^{*B})$, to estimate the distribution of $T(\widetilde{\mathbf{X}}, \widetilde{\mathbf{Y}})$ under null hypothesis H_0.
5. Compute the estimated bootstrap quantiles $\widehat{t}_{\frac{\gamma}{2}}$ and $\widehat{t}_{1-\frac{\gamma}{2}}$.
6. The hypothesis H_0 is rejected whenever $T(\widetilde{\mathbf{X}}, \widetilde{\mathbf{Y}}) < \widehat{t}_{\frac{\gamma}{2}}$, or $T(\widetilde{\mathbf{X}}, \widetilde{\mathbf{Y}}) > \widehat{t}_{1-\frac{\gamma}{2}}$.

Example 3
Suppose that we have taken two fuzzy random samples of size $n = 30$ (Table 1) and $m = 15$ (Table 3) from two populations and we want to test the equality of variances of the two populations at significance level $\gamma \in [0, 1]$.

By following the steps in Algorithm 2 with $B = 20000$ bootstrap replications, one can obtain the estimated quantiles of the bootstrap distribution of $T(\widetilde{\mathbf{X}}, \widetilde{\mathbf{Y}})$ under null hypothesis as given in Table 4. Now we can easily test the hypotheses of interest at significance level $\gamma \in [0, 1]$ by checking the inequalities $T(\widetilde{\mathbf{x}}, \widetilde{\mathbf{y}}) < \widehat{t}_{\frac{\gamma}{2}}$, and $T(\widetilde{\mathbf{x}}, \widetilde{\mathbf{y}}) > \widehat{t}_{1-\frac{\gamma}{2}}$. The histogram for the bootstrap replicate of $T(\widetilde{\mathbf{x}}_c^{*1}, \widetilde{\mathbf{y}}_c^{*1}), T(\widetilde{\mathbf{x}}_c^{*2}, \widetilde{\mathbf{y}}_c^{*2}), \ldots, T(\widetilde{\mathbf{x}}_c^{*B}, \widetilde{\mathbf{y}}_c^{*B})$ is shown in Figure 2, showing the estimated distributions of the test statistic $T(\widetilde{\mathbf{X}}, \widetilde{\mathbf{Y}})$.

5. Conclusions

In this paper we considered hypothesis-tests for the (crisp-valued) variance of a fuzzy random variable in a population. For this purpose, we made use of a new definition of fuzzy random variables and a new metric for

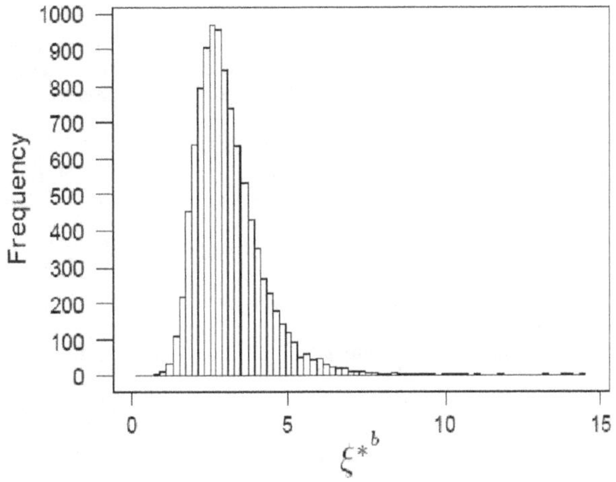

Figure 2. The estimated distributions of the test statistic $T(\widetilde{\mathbf{X}}, \widetilde{\mathbf{Y}})$ in Example 3.

fuzzy numbers. Using the general bootstrap methodology, we followed the application of the bootstrap in fuzzy environment and proposed an approach to testing hypotheses on the basis of the evidence supplied by a set of sample fuzzy data. It should be pointed out that the proposed procedure in this paper can be applied in practice with no limitations.

The distribution of the test statistic in this paper under the alternative hypothesis can be obtained by means of either asymptotic or bootstrap techniques, which seems that the study of the power function of the tests proposed in this paper can be a potential research for future studies. The bootstrap approach to test the multi-sample hypothesis of variances is another topic for further research in the future, as well as discussing the adequacy of the tests through a simulation comparative study.

REFERENCES

1. M. G. Akbari, and A. Rezaei, *Bootstrap testing fuzzy hypotheses and observations on fuzzy statistic*, Expert Systems with Applications, vol. 37, pp. 5782–5787, 2010.
2. K. Atanassov, *Intuitionistic Fuzzy Sets: Theory and Applications*, Physica-Verlag, Heidelberg, 1999.
3. R. J. Auman, *Integrals of set-valued functions*, Journal of Mathematical Analysis and Applications, vol. 12, pp. 1–12, 1965.
4. P. Billingsley, *Probability and Measure*, 3rd ed, Wiley, New York, 1995.
5. A., Blanco-Fernández, M.R., Casals, A., Colubi, N., Corral, M., García-Bárzana, M.A., Gil, G., González-Rodríguez, M.T., López, M.A., Lubiano, M., Montenegro, A.B., Ramos-Guajardo, S. De La Rosa De Sáa, and B. Sinova, *Random fuzzy sets: a mathematical tool to develop statistical fuzzy data analysis*, Iranian Journal of Fuzzy Systems, vol. 10, pp. 1–28, 2013.
6. J. Chachi, *On distribution characteristics of a fuzzy random variable*, Austrian Journal of Statistics, 2017, In Press.
7. J. Chachi, and S.M. Taheri, *Optimal statistical tests based on fuzzy random variables*, Iranian Journal of Fuzzy Systems, 2017, In Press.
8. B. Efron, and R. J. Tibshirani, *An Introduction to the Bootstrap*, Chapman and Hall, London, 1993.
9. M. Frechet, *Les elements aleatoires de natures quelconque dans une space distancie*, Ann. Inst. H. Poincare, vol. 10, no. 4, pp. 215–310, 1948. (In Ferench).
10. W. L. Gau, and D. J. Buehrer, *Vague sets*, IEEE Transactions on Systems Man and Cybernetics, vol. 23, pp. 610–614, 1993.
11. M.Á. Gil, M. Montenegro, G. González-Rodríguez, A. Colubi, and M. R. Casals, *Bootstrap approach to the multi-sample test of means with imprecise data*, Computational Statistics and Data Analysis, vol. 51, pp. 148–162, 2006.
12. M.Á. Gil, M. López-Díaz, and D. A. Ralescu, *Overview on the development of fuzzy random variables*, Fuzzy Sets and Systems, vol. 157, pp. 2546–2557, 2006.
13. G. González-Rodríguez, M. Montenegro, A. Colubi, and M. A. Gil, *Bootstrap techniques and fuzzy random variables: Synergy in hypothesis testing with fuzzy data*, Fuzzy Sets and Systems, vol. 157, pp. 2608–2613, 2006.
14. M. B. Gorzalzany, *A method of inference in approximate reasoning based on interval-valued fuzzy sets*, Fuzzy Sets and Systems, vol. 21, pp. 1–17, 1987.
15. G. Hesamian, and J. Chachi, *Two-sample Kolmogorov-Smirnov fuzzy test for fuzzy random variables*, Statistical Papers, vol. 56, pp. 61–82, 2013.
16. R. Korner, *On the variance of random fuzzy variables*, Fuzzy Sets and Systems, vol. 92, no. 1, pp. 83–93, 1997.

17. V. Krätschmer, *A unified approach to fuzzy random variables*, Fuzzy Sets and Systems, vol. 123, pp. 1–9, 2001.
18. R. Kruse, and K. D. Meyer, *Statistics with Vague Data*, Reidel Publishing Company, Dordrecht, 1987.
19. H. Kwakernaak, *Fuzzy random variables part I: Definitions and theorems*, Information Sciences, vol. 19, pp. 1–15, 1978.
20. H. Kwakernaak, *Fuzzy random variables part II: Algorithms and examples for the discrete case*, Information Sciences, vol. 17, pp. 253–278, 1979.
21. B. Liu, *Theory and Practice of Uncertain Programming*, Physica-Verlag, Heidelberg, 2002.
22. B. Liu, *Uncertainty Theory*, 5th ed, Springer-Verlag, Berlin (URL: http://orsc.edu.cn/liu/ut.pdf), 2016.
23. B. Liu, and Y. K. Liu, *Expected value of fuzzy variable and fuzzy expected value models*, IEEE Transactions on Fuzzy Systems, vol. 10, pp. 445–450, 2002.
24. Y. K. Liu, and B. Liu, *Fuzzy random variables: A scalar expected value operator*, Fuzzy Optimization and Decision Making, vol. 2, pp. 143–160, 2003.
25. M. Montenegro, A. Colubi, M. R. Casals, and M.Á. Gil, *Asymptotic and bootstrap techniques for testing the expected value of a fuzzy random variable*, Metrika, vol. 59, pp. 31–49, 2004.
26. J. Peng, and B. Liu, *Some properties of optimistic and pessimistic values of fuzzy*, IEEE International Conference on Fuzzy Systems, vol. 2, pp. 745–750, 2004.
27. M. L. Puri, and D. A. Ralescu, *The concept of normality for fuzzy random variables*, The Annals of Probability, vol. 13, pp. 1373–1379, 1985.
28. M. L. Puri, and D. A. Ralescu, D.A., *Fuzzy random variables*, Journal of Mathematical Analysis and Applications, vol. 114, pp. 409–422, 1986.
29. A. B. Ramos-Guajardo, A. Colubi, G. González-Rodríguez, and M. A. Gil, *One sample tests for a generalized Fréchet variance of a fuzzy random variable*, Metrika, vol. 71, no. 2, pp. 185–202, 2010.
30. A. B. Ramos-Guajardo, and M. A. Lubiano, *K-sample tests for equality of variances of random fuzzy sets*, Computational Statistics and Data Analysis, vol. 56, no. 4, pp. 956–966, 2012.
31. J. Shao, and D. Tu, *The Jackknife and Bootstrap*. Springer, NewYork, 1995.
32. S. M. Taheri, and G. Hesamian, *A generalization of the Wilcoxon signed-rank test and its applications,* Stat Papers, vol. 54, pp. 457–470, 2013.
33. R. Viertl, *Statistical Methods for Fuzzy Data*, Wiley, Chichester, 2011.
34. R. A. Vitale, *An alternate formulation of mean value for random geometric figures*, J. Microscopy, vol. 151, no. 3, pp. 197–204, 1998.
35. S. Yosefi, M. Arefi, and M. G. Akbari, *A new approach for testing fuzzy hypotheses based on likelihood ratio statistic*, Stat Papers, vol. 57, pp. 665–688, 2016.
36. L. A. Zadeh, *Fuzzy sets*, Information and Control, vol. 8, pp. 338–353, 1965.
37. Z. Zainali, M. G. Akbari, and H. Alizadeh Noughabi, *Intuitionistic fuzzy random variable and testing hypothesis about its variance*, Soft Computing, vol. 19, no. 9, pp. 2681–2689, 2015.
38. H. J. Zimmermann, *Fuzzy Set Theory and Its Applications*, 4th ed., Kluwer Nihoff, Boston, 2001.

10

Robust Bayesian analysis of generalized half logistic distribution

Ajit Chaturvedi, Taruna Kumari*

Department of Statistics, University of Delhi, India

Abstract In this paper, Robust Bayesian analysis of the generalized half logistic distribution (GHLD) under an ϵ-contamination class of priors for the shape parameter λ is considered. ML-II Bayes estimators of the parameters, reliability function and hazard function are derived under the squared-error loss function (SELF) and linear exponential (LINEX) loss function by considering the Type II censoring and the sampling scheme of Bartholomew(1963). Both the cases when scale parameter is known and unknown is considered under Type II censoring and under the sampling scheme of Bartholomew. Simulation study and analysis of a real data set are presented.

Keywords GHLD; ϵ-contamination class of prior distributions, ML-II posterior density, Reliability function, Hazard function, SELF, LINEX loss function, Type II censoring, Sampling scheme by Bartholomew, Markov Chain Monte Carlo (MCMC) procedure, Metropolis-Hastings algorithm, Gibbs sampling technique

1. Introduction and preliminaries

Half logistic has been used in many reliability and survival analysis (especially when the data is censored). Inferences for the half logistic distribution (HLD) have been discussed by several authors. Balakrishnan (1985) discussed on order statistics from the HLD. Balakrishnan and Puthenpura (1986) introduced the best linear unbiased estimators of location and scale parameters of the HLD through linear functions of order statistics. Balakrishnan and Asgharzadeh (2005) gave the inferences for the scaled HLD based on progressively Type II censored samples. Balakrishnan and Hossain (2007) considered generalized (Type II) version of logistic distribution and derived some interesting properties of the distribution. Arora, Bhimani and Patel (2010) obtained the maximum likelihood estimator of the shape parameter in a GHLD based on Type I progressive censoring with varying failure rates. Asgharzadeh, Rezaie and Abdi (2011) compared the methods of estimation for the HLD. Kim, Kang and Seo (2011) proposed the Bayes estimators of the shape parameter and reliability function for the GHLD based on progressively Type II censored data under various loss functions. Seo, Lee and Kang (2012) developed an entropy estimation method for the upper record values from the GHLD. Azimi (2013) derived the Bayes estimators of the shape parameter and the reliability function for the GHLD based on Type II doubly censored samples. Seo and Kang (2014) derived the entropy of GHLD by using the Bayes estimators of an unknown parameter based on Type II censored samples. They also compared these estimators in terms of the mean square error and the bias. Recently, Chaturvedi, Kang and Pathak (2016) developed estimation and testing procedures for the powers of the parameter and the reliability function under Type II censoring and under the sampling scheme of Bartholomew (1963).

*Correspondence to: Taruna Kumari (Email: taruna1986@yahoo.com). Department of Statistics, University of Delhi, Delhi 110007, India.

The robust Bayesian viewpoint, since Good (1950), assumed that the subjective information can be quantified in terms of a class Γ of possible distributions. The goal is to make inferences or decisions which are robust or flexible over Γ, i.e., which are relatively insensitive to derivations as the prior distribution varies over Γ. For a brief review of the literature on robustness in Bayesian model, one may refer to Jeffreys (1961), Box and Tiao (1962), Good (1965, 1967, 1980), Lindley (1961), Dempster (1976), Box (1980).

Bayesian analysis based on ϵ-contamination class of priors has been extensively studied by Berger (1982, 1983), Berger and Berliner (1983), Chaturvedi (1993, 1998), Chaturvedi, Pati and Tomer (2013). This class is a sensible class in the light of prior elicitation process for removing prior judgment error. Betro (2009) provides some numerical procedures for robust Bayesian analysis. For a detailed review on Bayesian robustness one may refer to Berger et al. (2000).

The randome variable (rv) X is said to follow GHLD if its probability density function (pdf) is given by

$$f(x;\lambda,\sigma)=\frac{\lambda}{\sigma}\left(\frac{2e^{-x/\sigma}}{1+e^{-x/\sigma}}\right)^{\lambda}\frac{1}{1+e^{-x/\sigma}},\quad x>0,\ \lambda,\sigma>0. \tag{1.1}$$

The reliability function is

$$\begin{aligned}R(t)&=P(X>t)\\&=\left(\frac{1+e^{t/\sigma}}{2}\right)^{-\lambda}\end{aligned} \tag{1.2}$$

and the hazard-rate comes out to be

$$\begin{aligned}h(t)&=\frac{f(t;\lambda)}{R(t)}\\&=\frac{\lambda}{\sigma\left(1+e^{-t/\sigma}\right)}.\end{aligned} \tag{1.3}$$

Form (1.3), we conclude that the GHLD is fit for the situation when we have increasing failure-rate.

The purpose of the present paper is to give robust Bayesian analysis of parameters, hazard function and reliability function of GHLD. In Section 2 and 4 respectively, for known scale parameter, we obtained the robust Bayes estimators of these parametric functions under SELF and LINEX loss functions considering Type II censoring and the sampling scheme of Bartholomew (1963). In Section3 and 5 respectively, we consider the case when both shape and scale parameters are unknown. In Section 6, simulation study is performed by using Monte Carlo simulation method, when scale parameter is known. In Section 7, analysis of a real data set using MCMC and Gibbs sampling technique is performed, for the case when both the parameters are unknown. Finally, in Section 8, discussion and some concluding remarks are presented.

2. Robust Bayesian analysis under Type II censoring with known scale parameter

Let n items are put on a life test and the test is terminated after r^{th} failure, so that, $(n-r)$ items survived. Let us denote by $0<X_{(1)}\le X_{(2)}\le\ldots\le X_{(r)}$, $0<r<n$, the lifetimes of first r failures. Denoting by $S_\sigma=\sum_{i=1}^{r}\ln\left(\frac{1+e^{X_{(i)}/\sigma}}{2}\right)+(n-r)\ln\left(\frac{1+e^{X_{(r)}/\sigma}}{2}\right)$ and $\underline{x}=(x_{(1)},x_{(2)},\ldots,x_{(r)})$, the likelihood function is [Seo, Kim and Kang (2013)].

$$l(\underline{x}\mid\lambda,\sigma)=\frac{n!}{(n-r)!}\left(\frac{\lambda}{\sigma}\right)^{r}\left(\prod_{i=1}^{r}\frac{1}{1+e^{-x_{(i)}/\sigma}}\right)\exp(-\lambda S_\sigma). \tag{2.1}$$

For the prior density, we consider the following ϵ-contamination class of prior distribution for λ [Berger (1985)].

$$\Gamma=\{\pi(\lambda):\pi(\lambda)=(1-\epsilon)g_o(\lambda\mid\mu_o)+\epsilon g(\lambda\mid\mu);\ g(\lambda\mid\mu)\in G\} \tag{2.2}$$

where ϵ $(0 \leq \epsilon \leq 1)$ is pre-assigned and represents the probability of error in the elicitation of the prior $g_o(\lambda \mid \mu_o)$. We consider the base prior, a natural conjugate prior, given by the pdf

$$g_o(\lambda \mid \mu_o) = \frac{\mu_o^\nu}{\Gamma(\nu)}\lambda^{\nu-1}\exp(-\mu_o\lambda); \quad \mu_o > 0,\ \nu > 0, \tag{2.3}$$

where (μ_o, ν) represents the vector of hyper parameters. The contamination class G is the class of all natural conjugate priors with the vector of hyper parameters (μ, ν), which is given as

$$G = \left\{g(\lambda \mid \mu) = \frac{\mu^\nu}{\Gamma(\nu)}\lambda^{\nu-1}\exp(-\mu\lambda);\ \mu \in (\mu_o, \infty)\right\}. \tag{2.4}$$

The predictive density corresponding to the prior $g(\lambda \mid \mu)$ is

$$m(\underset{\sim}{x} \mid g) = \int_0^\infty l(\lambda \mid \underset{\sim}{x})g(\lambda \mid \mu)d\lambda. \tag{2.5}$$

Using (2.1) and (2.4), we obtain from (2.5) that

$$m(\underset{\sim}{x} \mid g) = \frac{n!}{(n-r)!\sigma^r}\left(\prod_{i=1}^{r}\frac{1}{1+e^{-x_{(i)}/\sigma}}\right)\frac{\mu^\nu}{\Gamma(\nu)}\frac{\Gamma(r+\nu)}{(S_\sigma+\mu)^{r+\nu}}. \tag{2.6}$$

The predictive density corresponding to the base prior is obtained from (2.6) when $\mu = \mu_o$. Now, using (2.2), the predictive density corresponding to the generic prior $\pi \in \Gamma$ is

$$m(\underset{\sim}{x} \mid \pi) = (1-\epsilon)m(\underset{\sim}{x} \mid g_o) + \epsilon m(\underset{\sim}{x} \mid g). \tag{2.7}$$

In the ML-II process, we choose the value of the unknown hyper parameter μ in a data dependent fashion by maximizing the predictive density $m(\underset{\sim}{x} \mid \pi)$ over the class of all prior $\pi \in \Gamma$. Since g_o is fixed, we have

$$\sup_{\pi\in\Gamma} m(\underset{\sim}{x} \mid \pi) = (1-\epsilon)m(\underset{\sim}{x} \mid g_o) + \epsilon \sup_{g\in G} m(\underset{\sim}{x} \mid g) \tag{2.8}$$

and $m(\underset{\sim}{x} \mid g)$ is maximized when we replace μ by its maximum likelihood estimator (MLE), which is given by

$$\hat{\mu} = \max\left\{\mu_o, \frac{\nu S_\sigma}{r}\right\}$$

in $g(\lambda \mid \mu)$. Then we have

$$g(\lambda \mid \hat{\mu}) = \begin{cases} \frac{\lambda^{\nu-1}}{\Gamma(\nu)}\left(\frac{\nu S_\sigma}{r}\right)^\nu \exp\left(\frac{-\nu S_\sigma}{r}\lambda\right) = \hat{g}, & \mu_o < \frac{\nu S_\sigma}{r}, \\ g(\lambda \mid \mu_o) = g_o, & \mu_o \geq \frac{\nu S_\sigma}{r}. \end{cases}$$

Thus, the ML-II prior density is given by

$$\hat{\pi}(\lambda) = (1-\epsilon)g_o(\lambda \mid \mu_o) + \epsilon\hat{g}(\lambda \mid \hat{\mu}).$$

From Berger and Berliner (1983), the ML-II posterior of λ is obtained as

$$\hat{\pi}^*(\lambda) = \hat{\eta}g_o^*(\lambda) + (1-\hat{\eta})\hat{g}^*(\lambda), \tag{2.9}$$

where

$$\begin{aligned} g_o^*(\lambda) &= \frac{l(\lambda \mid \underset{\sim}{x})g_o(\lambda \mid \mu_o)}{\int_0^\infty l(\lambda \mid \underset{\sim}{x})g_o(\lambda \mid \mu_o)d\lambda} \\ &= \frac{(S_\sigma+\mu_o)^{r+\nu}}{\Gamma(r+\nu)}\lambda^{r+\nu-1}\exp(-\lambda(S_\sigma+\mu_o)), \quad \mu_o \geq \frac{\nu S_\sigma}{r}. \end{aligned} \tag{2.10}$$

Similarly, we get

$$\hat{g}^*(\lambda) = \begin{cases} \frac{\lambda^{r+\nu-1}}{\Gamma(r+\nu)}\left(1+\frac{\nu}{r}\right)^{r+\nu} S_\sigma^{r+\nu}\exp\left(-\lambda S_\sigma(1+\frac{\nu}{r})\right), & \mu_o < \frac{\nu S_\sigma}{r}, \\ g_o^*(\lambda), & \mu_o \geq \frac{\nu S_\sigma}{r} \end{cases} \tag{2.11}$$

and

$$\begin{aligned} \hat{\eta} &= \frac{(1-\epsilon)m(\underline{\mathrm{x}} \mid g_o)}{(1-\epsilon)m(\underline{\mathrm{x}} \mid g_o) + \epsilon m(\underline{\mathrm{x}} \mid \hat{g})} \\ &= \begin{cases} \left[1+\frac{\epsilon}{(1-\epsilon)}\frac{\nu^\nu r^r}{S_\sigma^r \mu_o^\nu}\frac{(S_\sigma+\mu_o)^{r+\nu}}{(r+\nu)^{(r+\nu)}}\right]^{-1}, & \mu_o < \frac{\nu S_\sigma}{r}, \\ (1-\epsilon), & \mu_o \geq \frac{\nu S_\sigma}{r}. \end{cases} \end{aligned} \tag{2.12}$$

2.1. Under SELF

Now we derive the ML-II estimators of the powers of λ, hazard-rate and reliability function under SELF. We also derive the expression for posterior variance. It is well known that the Bayes estimator of any parametric function under SELF is its posterior mean. Also the well known form of SELF for estimating λ through its estimator $\hat{\lambda}$ is given by

$$L(\hat{\lambda}) = (\hat{\lambda} - \lambda)^2$$

Let us first obtain the Bayes estimator λ^p (p is finite and non zero) and its posterior variance in the following theorem.

Theorem 1
The ML-II posterior mean and variance of λ^p are given, respectively by

$$E_{\hat{\pi}^*}(\lambda^p) = \begin{cases} \frac{\Gamma(r+\nu+p)}{(r+\nu)}\left\{\hat{\eta}(S_\sigma+\mu_o)^{-p} + (1-\hat{\eta})\left(1+\frac{v}{r}\right)^{-p} S_\sigma^{-p}\right\}, & \mu_o < \frac{\nu S_\sigma}{r}, \\ \frac{\Gamma(r+\nu+p)}{(r+\nu)}(S_\sigma+\mu_o)^{-p}, & \mu_o \geq \frac{\nu S_\sigma}{r} \end{cases} \tag{2.13}$$

and

$$V_{\hat{\pi}^*}(\lambda^p) = \begin{cases} \left\{\frac{\Gamma(r+\nu+2p)}{(r+\nu)} - \left(\frac{\Gamma(r+\nu+p)}{(r+\nu)}\right)^2\right\} & \\ \cdot\left\{\hat{\eta}(S_\sigma+\mu_o)^{-2p} + (1-\hat{\eta})\left(1+\frac{v}{r}\right)^{-2p} S_\sigma^{-2p}\right\} & \\ +\hat{\eta}(1-\hat{\eta})\left(\frac{\Gamma(r+\nu+p)}{(r+\nu)}\right)^2\left\{(S_\sigma+\mu_o)^{-p} - \left(1+\frac{v}{r}\right)^{-p} S_\sigma^{-p}\right\}^2, & \mu_o < \frac{\nu S_\sigma}{r}, \\ \left\{\frac{\Gamma(r+\nu+2p)}{(r+\nu)} - \left(\frac{\Gamma(r+\nu+p)}{(r+\nu)}\right)^2\right\}(S_\sigma+\mu_o)^{-2p}, & \mu_o \geq \frac{\nu S_\sigma}{r}. \end{cases} \tag{2.14}$$

Proof
Using (2.9), we have

$$E_{\hat{\pi}^*}(\lambda^p) = \hat{\eta} E_{g_o^*}(\lambda^p) + (1-\hat{\eta})E_{\hat{g}^*}(\lambda^p). \tag{2.15}$$

For $\mu_o \geq \frac{\nu S_\sigma}{r}$, we obtain

$$\begin{aligned} E_{g_o^*}(\lambda^p) &= \frac{(S_\sigma+\mu_o)^{(r+\nu)}}{\Gamma(r+\nu)}\int_0^\infty \lambda^{r+p+\nu-1} e^{-\lambda(S_\sigma+\mu_o)} d\lambda \\ &= \frac{\Gamma(r+\nu+p)}{\Gamma(r+\nu)}(S_\sigma+\mu_o)^{-p}. \end{aligned} \tag{2.16}$$

Similarly, for $\mu_o < \frac{\nu S_r}{r}$, it immediately follows that

$$E_{\hat{g}^*}(\lambda^p) = \frac{\Gamma(r+\nu+p)}{\Gamma(r+\nu)}\left(1+\frac{\nu}{r}\right)^{-p} S_\sigma^{-p}, \tag{2.17}$$

thus (2.13) follows on substituting (2.16) and (2.17) in (2.15).

In order to obtain the posterior variance of λ^p, we use the following expression from Berger(1985)

$$V_{\hat{\pi}^*}(\lambda^p) = \hat{\eta} V_{g_o^*}(\lambda^p) + (1-\hat{\eta})V_{\hat{g}^*}(\lambda^p) + \hat{\eta}(1-\hat{\eta})\left[E_{g_o^*}(\lambda^p) - E_{\hat{g}^*}(\lambda^p)\right]^2, \tag{2.18}$$

where

$$\begin{aligned} V_{g_o^*}(\lambda^p) &= E_{g_o^*}(\lambda^{2p}) - \left[E_{g_o^*}(\lambda^p)\right]^2 \\ &= \left[\frac{\Gamma(r+\nu+2p)}{(r+\nu)} - \left(\frac{\Gamma(r+\nu+p)}{(r+\nu)}\right)^2\right](S_\sigma+\mu_o)^{-2p} \end{aligned} \tag{2.19}$$

and

$$\begin{aligned} V_{\hat{g}^*}(\lambda^p) &= E_{\hat{g}^*}(\lambda^{2p}) - \left[E_{\hat{g}^*}(\lambda^p)\right]^2 \\ &= \left[\frac{\Gamma(r+\nu+2p)}{(r+\nu)} - \left(\frac{\Gamma(r+\nu+p)}{(r+\nu)}\right)^2\right] S_\sigma^{-2p}\left(1+\frac{\nu}{r}\right)^{-2p}, \end{aligned} \tag{2.20}$$

now (2.14) follows on using (2.16), (2.17), (2.19) and (2.20) in (2.18). □

Corollary 1
The Bayes estimator of λ when p=1 is given by

$$\hat{\lambda} = \begin{cases} \hat{\lambda}_o + (1-\hat{\eta})(\hat{\lambda}_M - \hat{\lambda}_o)^+, & \mu_o < \frac{\nu S_\sigma}{r}, \\ \hat{\lambda}_o, & \mu_o \geq \frac{\nu S_\sigma}{r}, \end{cases}$$

where $\hat{\lambda}_o = \frac{(r+\nu)}{(S_\sigma+\nu_o)}$, $\hat{\lambda}_M = \frac{r}{S_\sigma}$ and $(a_M - a_o)^+ = \begin{cases} a_M - a_o, & a_M > a_o, \\ 0, & \text{otherwise.} \end{cases}$

Corollary 2
The posterior variance of λ when p=1 is given by

$$V_{\hat{\pi}^*}(\lambda) = \begin{cases} (r+\nu)\Big\{\hat{\eta}(S_\sigma+\mu_o)^{-2} + (1-\hat{\eta})(1+\frac{v}{r})^{-2}S_\sigma^{-2} \\ \quad +\hat{\eta}(1-\hat{\eta})(r+\nu)\left[(S_\sigma+\mu_o)^{-1} - (1+\frac{v}{r})^{-1}S_\sigma^{-1}\right]^2\Big\}, & \mu_o < \frac{\nu S_\sigma}{r}, \\ (r+\nu)(S_\sigma+\mu_o)^{-2}, & \mu_o \geq \frac{\nu S_\sigma}{r}. \end{cases} \tag{2.21}$$

Now in the following Theorem, we obtain the ML-II estimator and posterior variance of the hazard rate function, as follows

Theorem 2
The ML-II estimator of the hazard rate function $h(t)$ is given by

$$\widehat{h(t)} = \begin{cases} \hat{h}_o(t) + (1-\hat{\eta})(\hat{h}_M(t) - \hat{h}_o(t))^+, & \mu_o < \frac{\nu S_\sigma}{r}, \\ \hat{h}_o(t), & \mu_o \geq \frac{\nu S_\sigma}{r}, \end{cases} \tag{2.22}$$

where $\hat{h}_o(t) = \frac{(r+\nu)}{\sigma(1+e^{-t/\sigma})(S_\sigma+\mu_o)}$ and $\hat{h}_M(t) = \frac{r}{\sigma(1+e^{-t/\sigma})S_\sigma}$.

Proof
It follows from the fact that

$$\begin{aligned}\widehat{h(t)} &= E_{\hat{\pi}^*}(h(t)) \\ &= \frac{1}{\sigma(1+e^{-t/\sigma})} E_{\hat{\pi}^*}(\lambda) \\ &= \frac{1}{\sigma(1+e^{-t/\sigma})} \left\{ \hat{\eta} E_{g_o^*}(\lambda) + (1-\hat{\eta}) E_{\hat{g}^*}(\lambda) \right\}. \end{aligned} \tag{2.23}$$

Upon using (2.16) and (2.17) with $p = 1$ in (2.23), we get (2.22). □

Theorem 3
The posterior variance of the hazard rate function $h(t)$ is given by

$$V_{\hat{\pi}^*}(h(t)) = \begin{cases} \frac{(r+\nu)}{\sigma^2(1+e^{-t/\sigma})^2} \Big\{ \hat{\eta}(S_\sigma + \mu_o)^{-2} + (1-\hat{\eta})(1+\frac{v}{r})^{-2} S_\sigma^{-2} \\ \quad + \hat{\eta}(1-\hat{\eta})(r+\nu)\left[(S_\sigma+\mu_o)^{-1} - (1+\frac{v}{r})^{-1} S_\sigma^{-1}\right]^2 \Big\}, & \mu_o < \frac{\nu S_\sigma}{r}, \\ \frac{(r+\nu)}{\sigma^2(1+e^{-t/\sigma})^2} (S_\sigma + \mu_o)^{-2}, & \mu_o \geq \frac{\nu S_\sigma}{r}. \end{cases} \tag{2.24}$$

Proof
Using (1.3), we have

$$\begin{aligned} V_{\hat{\pi}^*}(h(t)) &= V_{\hat{\pi}^*}\left(\frac{\lambda}{\sigma(1+e^{-t/\sigma})} \right) \\ &= \frac{1}{\sigma^2(1+e^{-t/\sigma})^2} V_{\hat{\pi}^*}(\lambda). \end{aligned} \tag{2.25}$$

Now, applying (2.21) in (2.25), we get (2.24). □

In Theorem 4 and Theorem 5, we will obtain the ML-II estimator and the posterior variance of the reliability function R(t), respectively.

Theorem 4
The ML-II estimator of the reliability function R(t) is given by

$$\widehat{R(t)} = \begin{cases} \hat{R}_o(t) + (1-\hat{\eta})(\hat{R}_M(t) - \hat{R}_o(t))^+, & \mu_o < \frac{\nu S_\sigma}{r}, \\ \hat{R}_o(t), & \mu_o \geq \frac{\nu S_\sigma}{r}, \end{cases}$$

where $\hat{R}_o(t) = \left[1 + \frac{\ln\left((1+e^{t/\sigma})/2\right)}{(S_\sigma + \mu_o)}\right]^{-(r+\nu)}$ and $\hat{R}_M(t) = \left[1 + \frac{r\ln\left((1+e^{t/\sigma})/2\right)}{S_\sigma(r+\nu)}\right]^{-(r+\nu)}$.

Proof
Proceeding in a similar manner as in Theorem 2, we get

$$\begin{aligned} \widehat{R(t)} &= E_{\hat{\pi}^*}(R(t)) \\ &= \hat{\eta} E_{g_o^*}(R(t)) + (1-\hat{\eta}) E_{\hat{g}^*}(R(t)). \end{aligned}$$

Let us denote $\hat{R}_o(t) = E_{g_o^*}(R(t))$ and $\hat{R}_M(t) = E_{\hat{g}^*}(R(t))$, then the above expression can be written as

$$\widehat{R(t)} = \hat{\eta}\hat{R}_o(t) + (1-\hat{\eta})\hat{R}_M(t). \tag{2.26}$$

Upon solving

$$\hat{R}_o(t) = \frac{(S_\sigma + \mu_o)^{(r+\nu)}}{\Gamma(r+\nu)} \int_0^\infty \lambda^{r+\nu-1} \exp\left[-\lambda \left\{ \ln((1+e^{t/\sigma})/2) + S_\sigma + \mu_o \right\}\right] d\lambda$$

$$= \left[1 + \frac{\ln((1+e^{t/\sigma})/2)}{(S_\sigma + \mu_o)}\right]^{-(r+\nu)} \quad (2.27)$$

and similarly,

$$\hat{R}_M(t) = \left[1 + \frac{r \ln((1+e^{t/\sigma})/2)}{S_\sigma (r+\nu)}\right]^{-(r+\nu)}, \quad (2.28)$$

on using (2.27) and (2.28) in (2.26) yield the required result. □

Theorem 5
The posterior variance of the reliability function $R(t)$ is given by

$$V_{\hat{\pi}^*}(R(t)) = \begin{cases} \hat{\eta}\left[\hat{R}_o^2(t) - (\hat{R}_o(t))^2\right] + (1-\hat{\eta})\left[\hat{R}_M^2(t) - (\hat{R}_M(t))^2\right] \\ \quad +\hat{\eta}(1-\hat{\eta})\left[\hat{R}_o(t) - \hat{R}_M(t)\right]^2, & \mu_o < \frac{\nu S_\sigma}{r}, \\ \hat{R}_o^2(t) - (\hat{R}_o(t))^2, & \mu_o \geq \frac{\nu S_\sigma}{r}, \end{cases} \quad (2.29)$$

where $\hat{R}_o^2(t) = \left[1 + \frac{2\ln((1+e^{t/\sigma})/2)}{(S_\sigma+\mu_o)}\right]^{-(r+\nu)}$ and $\hat{R}_M^2(t) = \left[1 + \frac{2r\ln((1+e^{t/\sigma})/2)}{S_\sigma(r+\nu)}\right]^{-(r+\nu)}$.

Proof
Let us denote $\hat{R}_o^2(t) = E_{g_o^*}(R^2(t))$ and $\hat{R}_M^2(t) = E_{\hat{g}^*}(R^2(t))$. By using (2.27) and (2.28) and proceeding in a similar manner as in the second part of Theorem 1, we get (2.29). □

2.2. *Under LINEX*

Let us now derive the ML-II estimators of λ, hazard-rate and reliability function under LINEX loss function. The expression for the LINEX loss function is given by

$$L(\Delta) = \left[\exp(a\Delta) - a\Delta - 1\right], \quad a \neq 0, \quad (2.30)$$

where $\Delta = \hat{\lambda} - \lambda$. For $a = 1$, the function is quite asymmetric with overestimation being more serious than underestimation. When $a < 0$, (2.30) behaves almost exponential when $\Delta < 0$ and almost linear when $\Delta > 0$. For small values of $|a|$, the function is almost symmetric and close to SELF, viz.,

$$L(\Delta) = \frac{a^2\Delta^2}{2}.$$

Under the LINEX loss function (2.30), he ML-II estimator λ is given by

$$\tilde{\lambda} = -\frac{1}{a} \ln E_{\hat{\pi}^*}\left[\exp(-a\lambda)\right]. \quad (2.31)$$

Theorem 6
The ML-II estimator of λ is given by

$$\tilde{\lambda} = \begin{cases} -\frac{1}{a}\left\{\ln \tilde{\tau}_o + \ln\left[1 + (1-\hat{\eta})\left(\frac{\tilde{\tau}_M - \tilde{\tau}_o}{\tilde{\tau}_o}\right)\right]\right\}, & \mu_o < \frac{\nu S_\sigma}{r}, \\ -\frac{1}{a}\ln \tilde{\tau}_o, & \mu_o \geq \frac{\nu S_\sigma}{r}, \end{cases} \quad (2.32)$$

where $\tilde{\tau}_o = \left[1 + \frac{a}{(S_\sigma+\mu_o)}\right]^{-(r+\nu)}$ and $\tilde{\tau}_M = \left[1 + \frac{ar}{S_\sigma(\nu+r)}\right]^{-(r+\nu)}$.

Proof
Using (2.31),

$$\tilde{\lambda} = -\frac{1}{a}\ln\left[\hat{\eta}E_{g_o^*}\left[\exp(-a\lambda)\right] + (1-\hat{\eta})E_{\hat{g}^*}\left[\exp(-a\lambda)\right]\right]. \tag{2.33}$$

Let us denote $\tilde{\tau}_o = E_{g_o^*}\left[\exp(-a\lambda)\right]$ and $\tilde{\tau}_M = E_{\hat{g}^*}\left[\exp(-a\lambda)\right]$, then the above expression can be rewritten as

$$\tilde{\lambda} = -\frac{1}{a}\ln\left[\hat{\eta}\tilde{\tau}_o + (1-\hat{\eta})\tilde{\tau}_M\right], \tag{2.34}$$

where

$$\begin{aligned}\tilde{\tau}_o &= \frac{(S_\sigma+\mu_o)^{(r+\nu)}}{\Gamma(r+\nu)}\int_0^\infty \lambda^{r+\nu-1}\exp\left[-\lambda\left(S_\sigma+\mu_o+a\right)\right]d\lambda \\ &= \left[1+\frac{a}{(S_\sigma+\mu_o)}\right]^{-(r+\nu)}.\end{aligned} \tag{2.35}$$

Similarly,

$$\tilde{\tau}_M = \left\{1+\frac{ar}{S_\sigma(\nu+r)}\right\}^{-(r+\nu)}. \tag{2.36}$$

Upon rewriting (2.34) with (2.35) and (2.36) yield (2.32). □

The expectation of the LINEX loss function for $\tilde{\lambda}$ with respect to ML-II posterior distribution of λ is

$$\begin{aligned}aE_{\hat{\pi}^*}(\lambda-\tilde{\lambda}) &= a\left[E_{\hat{\pi}^*}(\lambda)-\tilde{\lambda}\right] \\ &= \begin{cases} a\left[\hat{\lambda}_o + (1-\hat{\eta})(\hat{\lambda}_M-\hat{\lambda}_o)^+ - \tilde{\lambda}\right], & \mu_o < \frac{\nu S_\sigma}{r}, \\ a\left[\hat{\lambda}_o - \tilde{\lambda}\right], & \mu_o \ge \frac{\nu S_\sigma}{r}.\end{cases}\end{aligned} \tag{2.37}$$

Next we obtain the ML-II estimator of hazard rate function h(t) respectively. Proceeding in a similar manner as in Theorem 6 and denoting $\tilde{\phi}_o(t) = E_{g_o^*}\left[\exp(-ah(t))\right]$ and $\tilde{\phi}_M(t) = E_{\hat{g}^*}\left[\exp(-ah(t))\right]$, the expressions for ML-II estimator of $h(t)$ under LINEX can be derived as follows.

Theorem 7
The ML-II estimator of the hazard rate function $h(t)$ is given by

$$\widetilde{h(t)} = \begin{cases} -\frac{1}{a}\left\{\ln\tilde{\phi}_o(t) + \ln\left[1+(1-\hat{\eta})\left(\frac{\tilde{\phi}_M(t)-\tilde{\phi}_o(t)}{\tilde{\phi}_o(t)}\right)\right]\right\}, & \mu_o < \frac{\nu S_\sigma}{r}, \\ -\frac{1}{a}\ln\phi_o(t), & \mu_o \ge \frac{\nu S_\sigma}{r},\end{cases} \tag{2.38}$$

where $\tilde{\phi}_o(t) = \left[1+\frac{a}{\sigma(1+e^{-t/\sigma})(S_\sigma+\mu_o)}\right]^{-(r+\nu)}$ and $\tilde{\phi}_M(t) = \left[1+\frac{ar}{\sigma S_\sigma(\nu+r)(1+e^{-t/\sigma})}\right]^{-(r+\nu)}$.

The expectation of the LINEX loss function for $\widetilde{h(t)}$ is obtained as

$$\begin{aligned}aE_{\hat{\pi}^*}(h(t)-\tilde{h}(t)) &= a\left[E_{\hat{\pi}^*}(h(t))-\widetilde{h(t)}\right] \\ &= \begin{cases} a\left[\hat{h}_o(t) + (1-\hat{\eta})(\hat{h}_M(t)-\hat{h}_o(t))^+ - \widetilde{h(t)}\right], & \mu_o < \frac{\nu S_\sigma}{r}, \\ a\left[\hat{h}_o(t) - \widetilde{h(t)}\right], & \mu_o \ge \frac{\nu S_\sigma}{r}.\end{cases}\end{aligned}$$

In order to derive the ML-II estimator of the reliability function $R(t)$, let us denote $\tilde{\psi}_o(t) = E_{g_o^*}\left[\exp(-aR(t))\right]$ and $\tilde{\psi}_M(t) = E_{\hat{g}^*}\left[\exp(-aR(t))\right]$, then the expressions for ML-II estimator of $R(t)$ under LINEX can be derived as follows.

Theorem 8
The ML-II estimator of the reliability function $R(t)$ under LINEX is given by

$$\widetilde{R(t)} = \begin{cases} -\frac{1}{a}\left\{\ln\tilde{\psi}_o(t) + \ln\left[1+(1-\hat{\eta})\left(\frac{\tilde{\psi}_M(t)-\tilde{\psi}_o(t)}{\tilde{\psi}_o(t)}\right)\right]\right\}, & \mu_o < \frac{\nu S_\sigma}{r}, \\ -\frac{1}{a}\ln\tilde{\psi}_o(t), & \mu_o \geq \frac{\nu S_\sigma}{r}, \end{cases}$$

where
$\tilde{\psi}_o(t) = \sum_{k=0}^{\infty}\frac{(-a)^k}{k!}\left[1+\frac{k\ln((1+e^{t/\sigma})/2)}{(S_\sigma+\mu_o)}\right]^{-(r+\nu)}$ and $\tilde{\psi}_M(t) = \sum_{k=0}^{\infty}\frac{(-a)^k}{k!}\left[1+\frac{kr\ln((1+e^{t/\sigma})/2)}{S_\sigma(r+\nu)}\right]^{-(r+\nu)}$.

Further, the posterior expectation for the LINEX loss function of $\widetilde{R(t)}$ is given by

$$aE_{\hat{\pi}^*}(R(t)-\widetilde{R(t)}) = a\left[E_{\hat{\pi}^*}(R(t)) - \widetilde{R(t)}\right] \tag{2.39}$$

$$= \begin{cases} a\left[\hat{R}_o(t) + (1-\hat{\eta})(\hat{R}_M(t)-\hat{R}_o(t))^+ - \widetilde{R(t)}\right], & \mu_o < \frac{\nu S_\sigma}{r}, \\ a\left[\hat{R}_o(t) - \widetilde{R(t)}\right], & \mu_o \geq \frac{\nu S_\sigma}{r}. \end{cases} \tag{2.40}$$

3. Robust Bayesian analysis under Type II censoring with unknown shape and scale parameters

In this section we assume that both the parameters of the generalized half logistic distribution are unknown. We assume the following prior distribution to carry out the robust Bayesian analysis for the generalized half logistic distribution by assuming independent priors for both the parameters.

$$\Gamma = \{\pi(\sigma,\lambda):\ \pi(\sigma,\lambda) = (1-\epsilon)q_o + \epsilon q\}, \tag{3.1}$$

where $q_o = g_o(\lambda \mid \mu_o)\,P(\sigma)$ and $q = g(\lambda \mid \mu)\,P(\sigma)$ with g_o and g as defined in (2.3) and (2.4), respectively. The density of $P(\sigma)$, the prior for the scale parameter σ, is assumed to be inverted gamma

$$P(\sigma) = \frac{\beta^\alpha}{\Gamma(\alpha)\sigma^{\alpha+1}}\exp(-\beta/\sigma), \quad \beta > 0,\ \alpha > 0. \tag{3.2}$$

The predictive density corresponding to the prior $\pi(\sigma,\lambda)$ comes out to be

$$m(\underline{x} \mid \pi) = (1-\epsilon)m(\underline{x} \mid q_o) + \epsilon m(\underline{x} \mid q), \quad 0 < \lambda < \infty, \tag{3.3}$$

where

$$m(\underline{x} \mid q) = \int_{\sigma=0}^{\infty}\int_{\lambda=0}^{\infty} l(\lambda,\sigma \mid \underline{x})\, g(\lambda \mid \mu)\, P(\sigma)d\lambda d\sigma$$
$$= \frac{n!}{(n-r)!}\frac{\Gamma(r+\nu)}{\Gamma(\nu)}\frac{\mu^\nu\beta^\alpha}{\Gamma(\alpha)}\int_0^\infty \frac{\sigma^{-\alpha-r-1}\exp(-\beta/\sigma)}{(S_\sigma+\mu)^{r+\nu}}\left(\prod_{i=1}^{r}\frac{1}{1+e^{-x_{(i)}/\sigma}}\right)d\sigma. \tag{3.4}$$

In order to obtain, the ML-II Bayes estimate of λ and σ under Type II censoring, we need their marginal posterior under the base prior q_o and the generic prior q. But, in the present situation it is not possible to get explicit expressions for them. We, therefore, apply Gibbs sampler to get the samples from the marginal distributions of these parameters. For this the full conditionals of λ given σ is given by

$$\hat{\pi}^*(\lambda \mid \sigma) = \hat{\eta}g_o^*(\lambda \mid \sigma) + (1-\hat{\eta})\hat{g}^*(\lambda \mid \sigma), \quad 0 < \lambda < \infty. \tag{3.5}$$

Similarly, by combining the likelihood (2.1) with the prior (3.2), we get the conditional posterior density of $z^*(\sigma \mid \lambda)$ as

$$z^*(\sigma \mid \lambda) \propto \sigma^{-\alpha-r-1}\exp(-\lambda S_\sigma - \beta/\sigma)\left(\prod_{i=1}^{r}\frac{1}{1+e^{-x_{(i)}/\sigma}}\right). \tag{3.6}$$

4. Robust Bayesian analysis under the sampling scheme of Bartholomew with known scale parameter

Let $X_{(1)} \leq X_{(2)} \leq \ldots \leq X_{(n)}$ be the failure times of n items under test from (1.3). Then test begins at time $X_{(0)} = 0$ and the system operates till $X_{(1)} = x_1$ when the first failure occurs. The failed item is replaced by a new one and the system operates till the second failure occurs at time $X_{(2)} = x_2$ and so on. The experiment is terminated at time t_o. Here, $X_{(i)}$ is the time until i^{th} failure measured from time 0.

Lemma 1
If $N(t_o)$ be the number of failures during the interval $[0, t_o]$, then

$$P\left[N(t_o) = r \mid t_o\right] = \frac{\left[n\lambda \ln((1+e^{t_o/\sigma})/2)\right]^r}{r!} \exp\left[-n\lambda \ln((1+e^{t_o/\sigma})/2)\right], \tag{4.1}$$

Let us consider for the prior density, the same ϵ-contamination class of prior distribution as in Section 2. Then the predictive density corresponding to the prior $g(\lambda \mid \mu)$ in the sampling scheme of Bartholomew is given by

$$m(\underset{\sim}{x} \mid g) = \int_0^\infty P\left[N(t_o) = r \mid t_o\right] g(\lambda \mid \mu) d\lambda. \tag{4.2}$$

Using (4.1) and (2.4), we obtain from (4.2) that

$$m(\underset{\sim}{x} \mid g) = \binom{r+\nu-1}{r} \left[\frac{\mu}{n \ln((1+e^{t_o/\sigma})/2)}\right]^\nu \left[1 + \frac{\mu}{n \ln((1+e^{t_o/\sigma})/2)}\right]^{-(r+\nu)}. \tag{4.3}$$

The predictive density corresponding to the base prior is obtained from (4.3) when $\mu = \mu_o$. Now, as earlier, in the sampling scheme of Bartholomew the predictive density corresponding to the generic prior $\pi \in \Gamma$ is given by (2.7).

As we have already discussed here also we choose the value of the unknown hyper parameter μ in a data dependent fashion by maximizing the predictive density $m(\underset{\sim}{x} \mid \pi)$ over the class of all prior $\pi \in \Gamma$. Since g_o is fixed, we have

$$\sup_{\pi \in \Gamma} m(\underset{\sim}{x} \mid \pi) = (1-\epsilon) m(\underset{\sim}{x} \mid g_o) + \epsilon \sup_{g \in G} m(\underset{\sim}{x} \mid g) \tag{4.4}$$

and $m(\underset{\sim}{x} \mid g)$ is maximized when we replace μ by its maximum likelihood estimator (MLE), which is given by

$$\hat{\mu} = \max\left\{\mu_o, \frac{\nu n \ln\left((1+e^{t_o/\sigma})/2\right)}{r}\right\} \tag{4.5}$$

in $g(\lambda \mid \mu)$. Then we have

$$g(\lambda \mid \hat{\mu}) = \begin{cases} \frac{\lambda^{\nu-1}}{\Gamma(\nu)} \left(\frac{\nu n \ln((1+e^{t_o/\sigma})/2)}{r}\right)^\nu \exp\left(\frac{-\nu n \ln((1+e^{t_o/\sigma})/2)}{r} \lambda\right) = \hat{g}, & \mu_o < \frac{\nu n \ln\left((1+e^{t_o/\sigma})/2\right)}{r}, \\ g(\lambda \mid \mu_o) = g_o, & \mu_o \geq \frac{\nu n \ln\left((1+e^{t_o/\sigma})/2\right)}{r}. \end{cases}$$

Thus, the ML-II prior density is given by

$$\hat{\pi}(\lambda) = (1-\epsilon) g_o(\lambda \mid \mu_o) + \epsilon \hat{g}(\lambda \mid \hat{\mu}).$$

From Berger and Berliner (1983), the ML-II posterior of λ is obtained as

$$\hat{\pi}^*(\lambda) = \hat{\eta} g_o^*(\lambda) + (1-\hat{\eta}) \hat{g}^*(\lambda), \tag{4.6}$$

where for $\mu_o \geq \frac{\nu n \ln\left((1+e^{t_o/\sigma})/2\right)}{r}$

$$
\begin{aligned}
g_o^*(\lambda) &= \frac{P\left[N(t_o)=r \mid t_o\right] g_o(\lambda \mid \mu_o)}{\int_0^\infty P\left[N(t_o)=r \mid t_o\right] g_o(\lambda \mid \mu_o) d\lambda} \\
&= \frac{\left[\mu_o + n\ln((1+e^{t_o/\sigma})/2)\right]^{r+\nu}}{\Gamma(r+\nu)} \lambda^{r+\nu-1} \exp\left[-\lambda\left(\mu_o + n\ln\left((1+e^{t_o/\sigma})/2\right)\right)\right]. \qquad (4.7)
\end{aligned}
$$

Similarly, we get

$$
\hat{g}^*(\lambda) = \begin{cases} \frac{\lambda^{r+\nu-1}}{\Gamma(r+\nu)} \left(1+\frac{\nu}{r}\right)^{r+\nu} \left[n\ln\left((1+e^{t_o/\sigma})/2\right)\right]^{r+\nu} \\ \quad \cdot \exp\left[-\lambda n\ln((1+e^{t_o/\sigma})/2)(1+\frac{\nu}{r})\right], & \mu_o < \frac{\nu n \ln\left((1+e^{t_o/\sigma})/2\right)}{r}, \\ g_o^*(\lambda), & \mu_o \geq \frac{\nu n \ln\left((1+e^{t_o/\sigma})/2\right)}{r} \end{cases}
$$

and

$$
\begin{aligned}
\hat{\eta} &= \frac{(1-\epsilon) m(\underset{\sim}{x} \mid g_o)}{(1-\epsilon) m(\underset{\sim}{x} \mid g_o) + \epsilon m(\underset{\sim}{x} \mid \hat{g})} \\
&= \begin{cases} \left[1 + \frac{\epsilon}{(1-\epsilon)} \frac{\nu^\nu r^r}{\left[n\ln((1+e^{t_o/\sigma})/2)\right]^r \mu_o^\nu} \frac{\left[n\ln((1+e^{t_o/\sigma})/2)+\mu_o\right]^{r+\nu}}{(r+\nu)^{(r+\nu)}}\right]^{-1}, & \mu_o < \frac{\nu n \ln\left((1+e^{t_o/\sigma})/2\right)}{r}, \\ (1-\epsilon), & \mu_o \geq \frac{\nu n \ln\left((1+e^{t_o/\sigma})/2\right)}{r}. \end{cases} \qquad (4.8)
\end{aligned}
$$

4.1. Under SELF

Now we state without proofs some results concerning point estimators under the sampling scheme of Bartholomew. The proofs are on similar lines as in Type II censoring case.

Theorem 9

The ML-II posterior mean and variance of λ^p under SELF are given, respectively by

$$
E_{\hat{\pi}^*}(\lambda^p) = \begin{cases} \frac{\Gamma(r+\nu+p)}{(r+\nu)} \left\{\hat{\eta}\left[n\ln((1+e^{t_o/\sigma})/2)+\mu_o\right]^{-p} + (1-\hat{\eta}) \right. \\ \quad \left. \cdot(1+\frac{v}{r})^{-p}\left[n\ln((1+e^{t_o/\sigma})/2)\right]^{-p}\right\}, & \mu_o < \frac{\nu n \ln\left((1+e^{t_o/\sigma})/2\right)}{r}, \\ \frac{\Gamma(r+\nu+p)}{(r+\nu)} \left[n\ln((1+e^{t_o/\sigma})/2)+\mu_o\right]^{-p}, & \mu_o \geq \frac{\nu n \ln\left((1+e^{t_o/\sigma})/2\right)}{r} \end{cases}
$$

and

$$
V_{\hat{\pi}^*}(\lambda^p) = \begin{cases} \left\{\frac{\Gamma(r+\nu+2p)}{(r+\nu)} - \left(\frac{\Gamma(r+\nu+p)}{(r+\nu)}\right)^2\right\} \left\{\hat{\eta}\left[n\ln((1+e^{t_o/\sigma})/2)+\mu_o\right]^{-2p} \right. \\ \quad \left. +(1-\hat{\eta})\cdot\left(1+\frac{v}{r}\right)^{-2p}\left[n\ln\left((1+e^{t_o/\sigma})/2\right)\right]^{-2p}\right\} \\ \quad +\hat{\eta}(1-\hat{\eta})\left(\frac{\Gamma(r+\nu+p)}{(r+\nu)}\right)^2 \left\{\left[n\ln((1+e^{t_o/\sigma})/2)+\mu_o\right]^{-p} \right. \\ \quad \left. -(1+\frac{v}{r})^{-p}\left[n\ln\left((1+e^{t_o/\sigma})/2\right)\right]^{-p}\right\}^2, & \mu_o < \frac{\nu n ln((1+e^{t_o})/2)}{r}, \\ \left\{\frac{\Gamma(r+\nu+2p)}{(r+\nu)} - \left(\frac{\Gamma(r+\nu+p)}{(r+\nu)}\right)^2\right\}\left[n\ln\left((1+e^{t_o/\sigma})/2\right)+\mu_o\right]^{-2p}, & \mu_o \geq \frac{\nu n \ln\left((1+e^{t_o/\sigma})/2\right)}{r}. \end{cases}
$$

Theorem 10

The ML-II estimator of the hazard rate function h(t) under SELF is given by

$$
\widehat{h(t)} = \begin{cases} \hat{h}_o(t) + (1-\hat{\eta})(\hat{h}_M(t) - \hat{h}_o(t))^+, & \mu_o < \frac{\nu n \ln\left((1+e^{t_o/\sigma})/2\right)}{r}, \\ \hat{h}_o(t), & \mu_o \geq \frac{\nu n \ln\left((1+e^{t_o/\sigma})/2\right)}{r}, \end{cases}
$$

where $\hat{h}_o(t) = \frac{(r+\nu)}{\sigma\left(1+e^{-t/\sigma}\right)\left[n\ln\left((1+e^{t_o/\sigma})/2\right)+\mu_o\right]}$ and $\hat{h}_M(t) = \frac{r}{\sigma\left(1+e^{-t/\sigma}\right)\left[n\ln\left((1+e^{t_o/\sigma})/2\right)\right]}$.

Theorem 11

The posterior variance of the hazard rate function h(t) under SELF is given by

$$V_{\hat{\pi}^*}(h(t)) = \begin{cases} \frac{(r+\nu)}{\sigma^2\left(1+e^{-t/\sigma}\right)^2}\Big\{\hat{\eta}\left[n\ln\left((1+e^{t_o/\sigma})/2\right)+\mu_o\right]^{-2}+(1-\hat{\eta}) \\ \cdot\left(1+\frac{v}{r}\right)^{-2}\left[n\ln\left((1+e^{t_o/\sigma})/2\right)\right]^{-2}+\hat{\eta}(1-\hat{\eta})(r+\nu) \\ \cdot\Big[\left[n\ln\left((1+e^{t_o/\sigma})/2\right)+\mu_o\right]^{-1} \\ -\left(1+\frac{v}{r}\right)^{-1}\left[n\ln\left((1+e^{t_o/\sigma})/2\right)\right]^{-1}\Big]^2\Big\}, & \mu_o < \frac{\nu n\ln\left((1+e^{t_o/\sigma})/2\right)}{r}, \\ \frac{(r+\nu)}{\left(1+e^{-t/\sigma}\right)^2}\left[n\ln\left((1+e^{t_o/\sigma})/2\right)+\mu_o\right]^{-2}, & \mu_o \geq \frac{\nu n\ln\left((1+e^{t_o/\sigma})/2\right)}{r}. \end{cases}$$

Theorem 12

The ML-II estimator of the reliability function $R(t)$ under SELF is given by

$$\widehat{R(t)} = \begin{cases} \hat{R}_o(t)+(1-\hat{\eta})(\hat{R}_M(t)-\hat{R}_o(t))^+, & \mu_o < \frac{\nu n\ln\left((1+e^{t_o/\sigma})/2\right)}{r}, \\ \hat{R}_o(t), & \mu_o \geq \frac{\nu n\ln\left((1+e^{t_o/\sigma})/2\right)}{r}, \end{cases}$$

where $\hat{R}_o(t) = \left[1+\frac{\ln\left((1+e^{t/\sigma})/2\right)}{\left[n\ln\left((1+e^{t_o/\sigma})/2\right)+\mu_o\right]}\right]^{-(r+\nu)}$ and $\hat{R}_M(t) = \left[1+\frac{r\ln\left((1+e^{t/\sigma})/2\right)}{[n\ln((1+e^{t_o})/2)](r+\nu)}\right]^{-(r+\nu)}$.

Theorem 13

The posterior variance of the reliability function $R(t)$ under SELF is given by

$$V_{\hat{\pi}^*}(R(t)) = \begin{cases} \hat{\eta}\left[\hat{R}_o^2(t)-(\hat{R}_o(t))^2\right]+(1-\hat{\eta})\left[\hat{R}_M^2(t)-(\hat{R}_M(t))^2\right] \\ +\hat{\eta}(1-\hat{\eta})\left[\hat{R}_o(t)-\hat{R}_M(t)\right]^2, & \mu_o < \frac{\nu n\ln\left((1+e^{t_o/\sigma})/2\right)}{r}, \\ \hat{R}_o^2(t)-(\hat{R}_o(t))^2, & \mu_o \geq \frac{\nu n\ln\left((1+e^{t_o/\sigma})/2\right)}{r}, \end{cases}$$

where $\hat{R}_o^2(t) = \left[1+\frac{2\ln\left((1+e^{t/\sigma})/2\right)}{\left[n\ln\left((1+e^{t_o/\sigma})/2\right)+\mu_o\right]}\right]^{-(r+\nu)}$ and $\hat{R}_M^2(t) = \left[1+\frac{2r\ln\left((1+e^{t/\sigma})/2\right)}{\left[n\ln\left((1+e^{t_o/\sigma})/2\right)\right](r+\nu)}\right]^{-(r+\nu)}$.

4.2. Under LINEX

Theorem 14

The ML-II estimator of λ under LINEX is given by

$$\tilde{\lambda} = \begin{cases} -\frac{1}{a}\Big\{\ln\tilde{\tau}_o+\ln\left[1+(1-\hat{\eta})\left(\frac{\tilde{\tau}_M-\tilde{\tau}_o}{\tilde{\tau}_o}\right)\right]\Big\}, & \mu_o < \frac{\nu n\ln\left((1+e^{t_o/\sigma})/2\right)}{r}, \\ -\frac{1}{a}\ln\tilde{\tau}_o, & \mu_o \geq \frac{\nu n\ln\left((1+e^{t_o/\sigma})/2\right)}{r}, \end{cases}$$

where $\tilde{\tau}_o = \left[1+\frac{a}{\left[n\ln\left((1+e^{t_o/\sigma})/2\right)+\mu_o\right]}\right]^{-(r+\nu)}$ and $\tilde{\tau}_M = \left[1+\frac{ar}{\left[n\ln\left((1+e^{t_o/\sigma})/2\right)\right](\nu+r)}\right]^{-(r+\nu)}$.

Theorem 15

The ML-II estimator of the hazard rate function $h(t)$ under LINEX is given by

$$\widetilde{h(t)} = \begin{cases} -\frac{1}{a}\Big\{\ln\tilde{\phi}_o(t)+\ln\left[1+(1-\hat{\eta})\left(\frac{\tilde{\phi}_M(t)-\tilde{\phi}_o(t)}{\tilde{\phi}_o(t)}\right)\right]\Big\}, & \mu_o < \frac{\nu n\ln\left((1+e^{t_o/\sigma})/2\right)}{r}, \\ -\frac{1}{a}\ln\tilde{\phi}_o(t), & \mu_o \geq \frac{\nu n\ln\left((1+e^{t_o/\sigma})/2\right)}{r}, \end{cases}$$

where
$\tilde{\phi}_o(t) = \left[1 + \frac{a}{\sigma\left(1+e^{-t/\sigma}\right)\left[n\ln((1+e^{t_o})/2)+\mu_o\right]}\right]^{-(r+\nu)}$ and $\tilde{\phi}_M(t) = \left[1 + \frac{ar}{\sigma\left[n\ln\left((1+e^{t_o/\sigma})/2\right)\right](\nu+r)\left(1+e^{-t/\sigma}\right)}\right]^{-(r+\nu)}$.

Theorem 16
The ML-II estimator of the reliability function $R(t)$ under LINEX is given by

$$\widetilde{R(t)} = \begin{cases} -\frac{1}{a}\left\{\ln\tilde{\psi}_o(t) + \ln\left[1 + (1-\hat{\eta})\left(\frac{\tilde{\psi}_M(t)-\tilde{\psi}_o(t)}{\tilde{\psi}_o(t)}\right)\right]\right\}, & \mu_o < \frac{\nu n\ln\left((1+e^{t_o/\sigma})/2\right)}{r}, \\ -\frac{1}{a}\ln\tilde{\psi}_o(t), & \mu_o \geq \frac{\nu n\ln\left((1+e^{t_o/\sigma})/2\right)}{r}, \end{cases}$$

where $\tilde{\psi}_o(t) = \sum_{k=0}^{\infty}\frac{(-a)^k}{k!}\left[1 + \frac{k\ln\left((1+e^{t/\sigma})/2\right)}{\left[n\ln\left((1+e^{t_o/\sigma})/2\right)+\mu_o\right]}\right]^{-(r+\nu)}$

and $\tilde{\psi}_M(t) = \sum_{k=0}^{\infty}\frac{(-a)^k}{k!}\left[1 + \frac{kr\ln\left((1+e^{t/\sigma})/2\right)}{\left[n\ln\left((1+e^{t_o/\sigma})/2\right)\right](r+\nu)}\right]^{-(r+\nu)}$.

5. Robust Bayesian analysis under the sampling scheme of Bartholomew censoring with unknown shape and scale parameters

Under the same set up as in Section 3, the predictive density corresponding to the prior $\pi(\sigma,\lambda)$ under the sampling scheme of Bartholomew comes out to be

$$m(\underset{\sim}{x} \mid \pi) = (1-\epsilon)m(\underset{\sim}{x} \mid q_o) + \epsilon m(\underset{\sim}{x} \mid q), \quad 0 < \lambda < \infty, \tag{5.1}$$

where

$$\begin{aligned} m(\underset{\sim}{x} \mid q) &= \int_{\sigma=0}^{\infty}\int_{\lambda=0}^{\infty} P\left[N(t_o) = r \mid t_o\right] g\left(\lambda \mid \mu\right) P(\sigma) d\lambda d\sigma \\ &= \frac{\Gamma(r+\nu)}{\Gamma(\nu)r!}\frac{\mu^\nu\beta^\alpha}{\Gamma(\alpha)}\int_0^\infty \frac{\left[n\ln\left((1+e^{to/\sigma})/2\right)\right]^r\sigma^{-\alpha-1}\exp(-\beta/\sigma)}{\left[n\ln\left((1+e^{to/\sigma})/2\right)+\mu\right]^{r+\nu}}d\sigma. \end{aligned} \tag{5.2}$$

Again, to obtain, the ML-II Bayes estimate of λ and σ under the sampling scheme of Bartholomew, we need their marginal posterior under the base prior q_o and the generic prior q. But, it is not possible to get explicit expressions for them. We, therefore, apply Gibbs sampler to get the samples from the marginal distributions of these parameters. For this the full conditionals of λ given σ is given by

$$\hat{\pi}^*(\lambda \mid \sigma) = \hat{\eta}g_o^*(\lambda \mid \sigma) + (1-\hat{\eta})\hat{g}^*(\lambda \mid \sigma), \quad 0 < \lambda < \infty. \tag{5.3}$$

Similarly, by combining the likelihood (2.1) with the prior (3.2), we get the conditional posterior density of $z^*(\sigma \mid \lambda)$ as

$$z^*(\sigma \mid \lambda) \propto \sigma^{-\alpha-1}\exp\left[-\beta/\sigma - \lambda n\ln\left((1+e^{to/\sigma})/2\right)\right]\left[n\lambda\ln\left((1+e^{to/\sigma})/2\right)\right]^r. \tag{5.4}$$

6. Simulation study

6.1. When scale parameter is known

In order to obtain the ML-II Bayes estimates of λ under Type II censoring, we have generated 1000 samples using Monte Carlo simulation method, each of size 50 from GHLD with $\lambda = 2$ and $\sigma = 1.2$. For each sample we arranged the data in ascending order and considered a sample of first 25 and 30 observations, respectively. We have considered the values of the hyper-parameters of the base prior to be $(\mu_o = 0.5, \nu = 4)$ and $(\mu_o = 1.5, \nu = 4)$ and obtained the ML-II Bayes estimates of λ under SELF and LINEX loss functions. We used the software R (`www.r-project.org`) for the computations of functions in the various expressions.

For various values of ϵ the average ML-II posterior mean, posterior risk and average ML-II posterior variance under SELF and LINEX for λ by considering Type II censoring for $r = 25$ and 30 is presented in Table 1.

Table 1. Average ML-II posterior mean (posterior risk) under SELF and LINEX $(a = 0.01)$ of λ.

		SELF		LINEX	
$r\downarrow$	$\epsilon\downarrow$ $\mu_o \rightarrow$	0.5	1.5	0.5	1.5
25	0.0	2.3441(0.2211)	2.1452 (0.16)	2.3431(1.1e-05)	2.1446 (8e-06)
		0.1971	0.1642		
	0.2	2.1824 (0.2297)	2.1306 (0.1669)	2.1815 (1.1e-05)	2.13 (8e-06)
		0.184	0.1632		
	0.4	2.1449 (0.2159)	2.1177 (0.1725)	2.1441 (1.1e-05)	2.117 (9e-06)
		0.1735	0.1617		
	0.6	2.1272 (0.207)	2.1059 (0.1771)	2.1263 (1e-05)	2.1053 (9e-06)
		0.1673	0.1601		
	0.8	2.1166 (0.2009)	2.0952 (0.181)	2.1158 (1e-05)	2.0946 (9e-06)
		0.1632	0.1582		
	1	2.1096 (0.1966)	2.0853 (0.1843)	2.1088 (1e-05)	2.0847 (9e-06)
		0.1602	0.1563		
30	0.0	2.2639 (0.1542)	2.1267 (0.1301)	2.2631 (8e-06)	2.1262 (7e-06)
		0.1553	0.1369		
	0.2	2.1258 (0.1584)	2.1143 (0.135)	2.125 (8e-06)	2.1137 (7e-06)
		0.1456	0.1361		
	0.4	2.0957 (0.1498)	2.1031 (0.1391)	2.095 (7e-06)	2.1025 (7e-06)
		0.1384	0.135		
	0.6	2.0819 (0.1446)	2.0928 (0.1425)	2.0812 (7e-06)	2.0923 (7e-06)
		0.1343	0.1338		
	0.8	2.0738 (0.1411)	2.0834 (0.1454)	2.0732 (7e-06)	2.0829 (7e-06)
		0.1317	0.1324		
	1	2.0686 (0.1387)	2.0747 (0.1479)	2.0679 (7e-06)	2.0742 (7e-06)
		0.1299	0.131		

Note: The second row represents the average ML-II posterior variance.

To investigate $R(t)$ and $h(t)$ under Type II censoring, we have generated 1000 samples using Monte Carlo simulation method, each of size 50 from GHLD with $\lambda = 2$ and $\sigma = 1$. For $r = 25$, we obtained the ML-II Bayes estimates of $R(t)$ and $h(t)$ under SELF and LINEX loss functions, with the same values of the hyper-parameters as considered earlier.

In Table 2, $R(t)$, the average ML-II posterior mean and average variance for different values of t, u_o and ϵ under SELF and Type II censoring are presented and in Table 3, $h(t)$, the average ML-II posterior mean and average ML-II posterior variance for different values of t, u_o and ϵ under SELF and Type II censoring are presented. Also the average ML-II posterior mean for different values of t under LINEX by considering Type II censoring for $R(t)$ and $h(t)$ are presented in Table 4 and Table 5.

Table 2. Average ML-II posterior mean and average variance under SELF for $R(t)$.

		$\epsilon \rightarrow$	0	0.2	0.4	0.6	0.8	1
t	$R(t)$	$\mu_o \downarrow$						
0.1	0.90258	0.5	0.8881	0.8955	0.8971	0.8979	0.8984	0.8987
			4.00E-04	4.00E-04	3.00E-04	3.00E-04	3.00E-04	3.00E-04
		1.5	0.8967	0.8974	0.898	0.8986	0.8991	0.8995
			3.00E-04	3.00E-04	3.00E-04	3.00E-04	3.00E-04	3.00E-04
0.5	0.57015	0.5	0.5271	0.5515	0.5568	0.5592	0.5607	0.5616
			0.0038	0.0038	0.0037	0.0036	0.0035	0.0035
		1.5	0.5555	0.5579	0.5601	0.562	0.5638	0.5654
			0.0036	0.0036	0.0036	0.0035	0.0035	0.0035
0.9	0.3342	0.5	0.2973	0.3245	0.3301	0.3326	0.3341	0.335
			0.0042	0.0046	0.0045	0.0044	0.0044	0.0044
		1.5	0.3212	0.324	0.3265	0.3287	0.3307	0.3326
			0.0044	0.0044	0.0044	0.0044	0.0044	0.0044
1.5	0.13312	0.5	0.1153	0.1351	0.1389	0.1405	0.1415	0.1421
			0.0021	0.0026	0.0026	0.0026	0.0026	0.0025
		1.5	0.1334	0.1357	0.1377	0.1395	0.1411	0.1426
			0.0024	0.0025	0.0025	0.0025	0.0026	0.0026

Note: The first row represents the average ML-II posterior mean and second row represents the average variance.

Table 3. Average ML-II posterior mean and average variance under SELF for $h(t)$.

		$\epsilon \rightarrow$	0	0.2	0.4	0.6	0.8	1
t	$h(t)$	$\mu_o \downarrow$						
0.1	1.04996	0.5	1.212	1.1266	1.1075	1.0986	1.0933	1.0899
			0.0525	0.0489	0.046	0.0444	0.0434	0.0426
		1.5	1.1157	1.1078	1.1007	1.0943	1.0884	1.083
			0.0443	0.044	0.0436	0.0431	0.0426	0.0421
0.5	1.24492	0.5	1.4516	1.3507	1.3276	1.3166	1.3102	1.3059
			0.0753	0.0702	0.0662	0.0638	0.0623	0.0612
		1.5	1.3407	1.3317	1.3237	1.3164	1.3097	1.3036
			0.0642	0.0638	0.0632	0.0626	0.0619	0.0611
0.9	1.4219	0.5	1.6348	1.5194	1.4937	1.4817	1.4746	1.47
			0.0957	0.089	0.0839	0.081	0.079	0.0777
		1.5	1.5119	1.5013	1.4918	1.4832	1.4753	1.4681
			0.0815	0.0809	0.0802	0.0793	0.0784	0.0774
1.5	1.63515	0.5	1.8889	1.7563	1.7266	1.7126	1.7043	1.6989
			0.128	0.1192	0.1124	0.1084	0.1058	0.1039
		1.5	1.7574	1.7455	1.735	1.7254	1.7167	1.7087
			0.1103	0.1096	0.1087	0.1076	0.1064	0.1051

Note: The first row represents the average ML-II posterior mean and second row represents the average variance.

Table 4. Average ML-II posterior mean under LINEX loss function for $R(t)$ for $a = 0.01$.

t	0.1		0.5		0.9		1.5	
$R(t)$	0.90258		0.57015		0.3342		0.13312	
$\mu_o \rightarrow$	0.5	1.5	0.5	1.5	0.5	1.5	0.5	1.5
$\epsilon \downarrow$								
0	0.8891	0.8971	0.5306	0.5514	0.2977	0.3233	0.1178	0.1307
0.2	0.8965	0.8978	0.5551	0.5537	0.325	0.3261	0.1381	0.1329
0.4	0.8982	0.8984	0.5603	0.5558	0.3305	0.3286	0.1419	0.1348
0.6	0.8989	0.8989	0.5628	0.5577	0.3331	0.3309	0.1436	0.1366
0.8	0.8994	0.8994	0.5642	0.5593	0.3345	0.3329	0.1445	0.1381
1	0.8997	0.8999	0.5651	0.5609	0.3355	0.3348	0.1451	0.1396

Table 5. Average ML-II posterior mean under LINEX loss function for $h(t)$ for $a = 0.01$.

t	0.1		0.5		0.9		1.5	
h(t)	1.04996		1.24492		1.4219		1.63515	
$\mu_o \rightarrow$	0.5	1.5	0.5	1.5	0.5	1.5	0.5	1.5
$\epsilon \downarrow$								
0	1.2028	1.1158	1.4475	1.3298	1.6413	1.5107	1.8793	1.738
0.2	1.1174	1.1079	1.3468	1.3207	1.526	1.5	1.7467	1.7256
0.4	1.0986	1.1009	1.3239	1.3126	1.5002	1.4905	1.7173	1.7147
0.6	1.0898	1.0945	1.313	1.3052	1.4881	1.4819	1.7035	1.705
0.8	1.0847	1.0887	1.3066	1.2985	1.481	1.474	1.6954	1.6961
1	1.0813	1.0834	1.3024	1.2923	1.4763	1.4668	1.6901	1.6879

Table 6. Average ML-II posterior mean (posterior risk) under SELF and LINEX ($a = 0.01$) of λ.

		SELF		LINEX	
$t_o \downarrow$	$\epsilon \downarrow$ $\mu_o \rightarrow$	0.5	1.5	0.5	1.5
0.3	0.0	2.4076 (0.2456)	2.1356 (0.1947)	2.4059 (1.2e-05)	2.1345 (1e-05)
		0.3372	0.2624		
	0.2	2.1043 (0.3303)	2.1006 (0.2218)	2.1029 (1.6e-05)	2.0995 (1.1e-05)
		0.2932	0.2591		
	0.4	2.0429 (0.3113)	2.0725 (0.2406)	2.0416 (1.6e-05)	2.0714 (1.2e-05)
		0.2641	0.2535		
	0.6	2.0143 (0.2986)	2.0485 (0.2549)	2.013 (1.5e-05)	2.0475 (1.3e-05)
		0.248	0.247		
	0.8	1.9975 (0.2901)	2.0275 (0.2663)	1.9963 (1.4e-05)	2.0265 (1.3e-05)
		0.2377	0.24		
	1	1.9864 (0.284)	2.0086 (0.2756)	1.9853 (1.4e-05)	2.0076 (1.4e-05)
		0.2305	0.2329		
0.5	0.0	2.2658 (0.1712)	2.0847 (0.1502)	2.2649 (9e-06)	2.0839 (8e-06)
		0.1889	0.1604		
	0.2	2.0836 (0.2054)	2.0642 (0.1626)	2.0827 (1e-05)	2.0635 (8e-06)
		0.1718	0.159		
	0.4	2.0479 (0.1974)	2.0471 (0.1717)	2.0471 (1e-05)	2.0464 (9e-06)
		0.1612	0.1569		
	0.6	2.0317 (0.1923)	2.0323 (0.1787)	2.0309 (1e-05)	2.0316 (9e-06)
		0.1556	0.1544		
	0.8	2.0224 (0.1889)	2.0191 (0.1843)	2.0217 (9e-06)	2.0184 (9e-06)
		0.1521	0.1517		
	1	2.0164 (0.1865)	2.0072 (0.1889)	2.0156 (9e-06)	2.0065 (9e-06)
		0.1498	0.1489		

Note: The second row represents the average ML-II posterior variance.

In order to see the performance of estimators under the sampling scheme of Bartholomew, we generated 1000 samples each of size 50 from GHLD with $\lambda = 2$ and $\sigma = 1.2$. For each sample by fixing the time $t_o = 0.3$ and $t_o = 0.5$, we have calculated r and then the estimates under SELF and LINEX loss functions. Since the results were following the same trend, therefore only the ML-II Bayes estimates of λ are presented in Table 6.

6.2. When scale parameter is unknown

Algorithm
This algorithm combines the Metropolis-Hastings with the Gibbs sampling scheme under the gamma proposal distribution.

1. Start with initial guess $\sigma^{(0)}$.
2. Set $i = 1$.
3. Generate $\lambda^{(i)}$ from $\hat{\pi}^*(\lambda \mid \sigma^{(i-1)})$ using (3.5).
4. Generated $\sigma^{(i)}$ from $z^*(\sigma \mid \lambda)$ using the Metropolis-Hasting algorithm with the proposal distribution $q(\sigma) \equiv$ *inverted gamma*(α, β).
   - 4.1. Let $w = \sigma^{(i-1)}$
   - 4.2. Generate *prop* from the proposal distribution q.
   - 4.3. Let $p(w, prop) = \min\left\{1, \frac{z^*(prop|\lambda)q(w)}{z^*(w|\lambda)q(prop)}\right\}$.
   - 4.4. Generate u from *Uniform*$(0, 1)$. If $u \leq p(w, z)$ then accept the proposal and set $\sigma^i = z$; otherwise, set $\sigma^i = w$.
5. Set $i = i + 1$;
6. Repeat Steps 3-5, N times to obtain the posterior sample.

7. Real data study

Now we provide real data analysis to see how the model works in practice. We consider the failure log times to breakdown of an insulating fluid testing experiment (Nelson, 1982 and see Table 8). This data has been utilized by many authors, such as Balakrishnan and Kannan (2001), Balakrishnan et al. (2004), and Kim et al. (2011). Seo, Kim and Kang (2013), applying Kolmogorov test, showed that the data follow GHLD.

Table 7. Failure log times to breakdown of an insulating fluid testing experiment.

0.270027	1.02245	1.15057	1.42311	1.54116	1.57898	1.8718	1.9947
2.08069	2.11263	2.48989	3.45789	3.48186	3.52371	3.60305	4.28895

In order to apply Type II censoring scheme, we obtain first 12 lifetimes from the data and rest 4 observations are considered as censored. We consider the values of the hyper parameters of the base prior to be $(\mu_o = 4, \nu = 4)$ and $(\mu_o = 8, \nu = 4)$. As in this case both the shape and scale parameters are unknown, therefore we use the algorithm given in Section 6.2.

In order to obtain ML-II Bayes estimates under SELF and LINEX loss functions, we use Gibbs sampling technique. We observe from (2.10) and (2.11) that $g_o^*(\lambda \mid \sigma)$ and $g^*(\lambda \mid \sigma)$ both follow gamma distribution, therefore we obtain sample values form conditional posterior of λ using *rgamma*( ) function available in R. However the conditional posterior for σ does not follow any standard distribution. We therefore use Metropolis-Hastings algorithm to generate sample values for σ. We use inverted gamma distribution as proposal density. To obtain sample values from the proposal density, we use inverted gamma function available in MCMCpack in R.

We run MCMC chain with a randomly chosen value of σ and generate 50000 observations. To diminish the effect of the starting distribution, we discard the first 10000 observations and focus on the remaining 40000. The diagnostic plots for λ and σ for different values of ϵ and $\mu_o = 4$ and $\mu_o = 8$ are given in Figure 1 and 2, respectively.

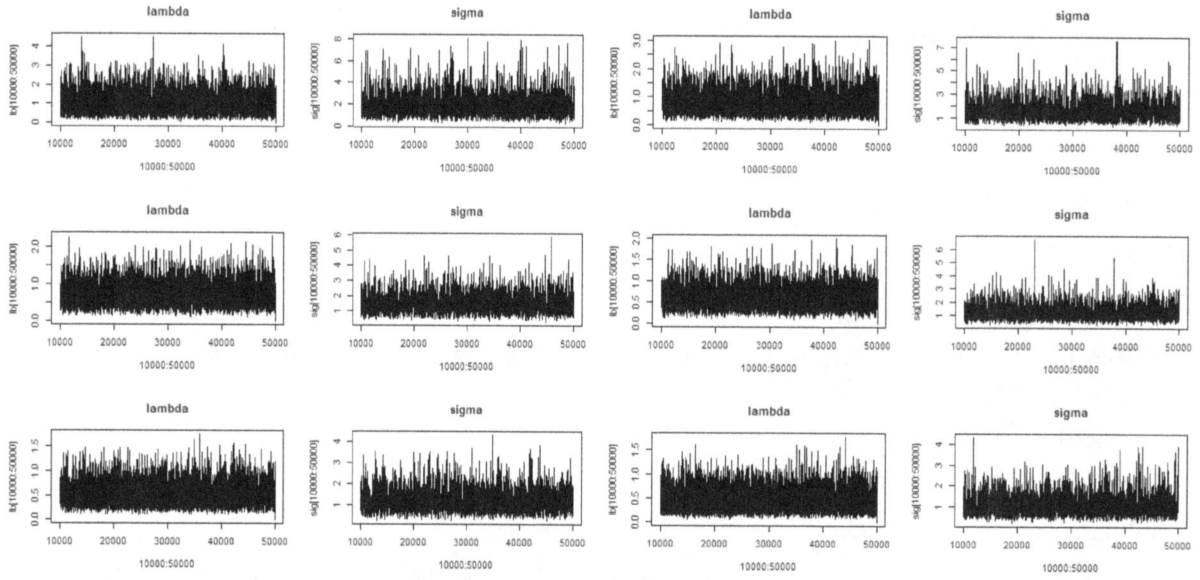

Figure 1. Diagnostic plots for λ and σ for $\mu_o = 4$

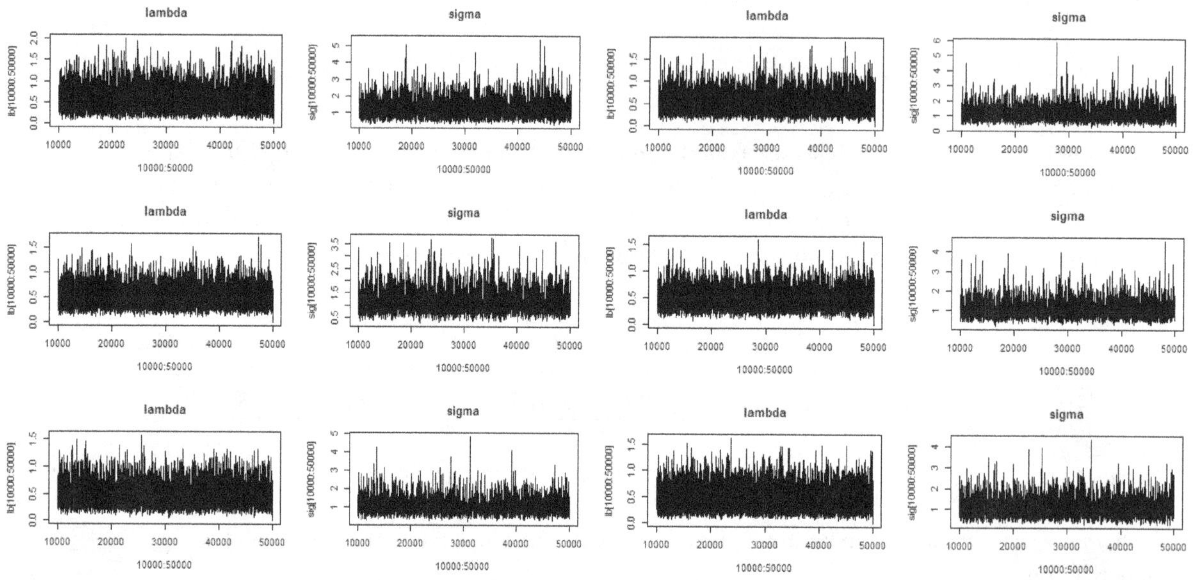

Figure 2. Diagnostic plots for λ and σ for $\mu_o = 8$

The trace plot shows that the mean of the Markov chain is constant over the graph and is stabilized. The chain is able to traverse the support of the target distribution, and the mixing is good. The trace plot shows that the Markov chain appears to have reached a stationary distribution.

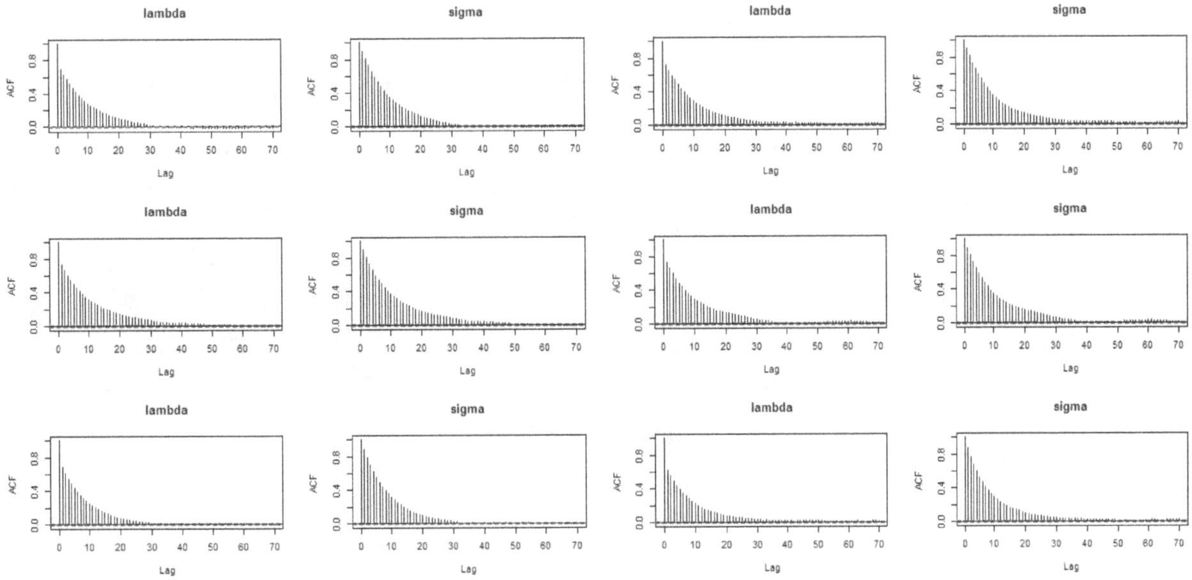

Figure 3. Auto-correlation plots for λ and σ for $\mu_o = 4$

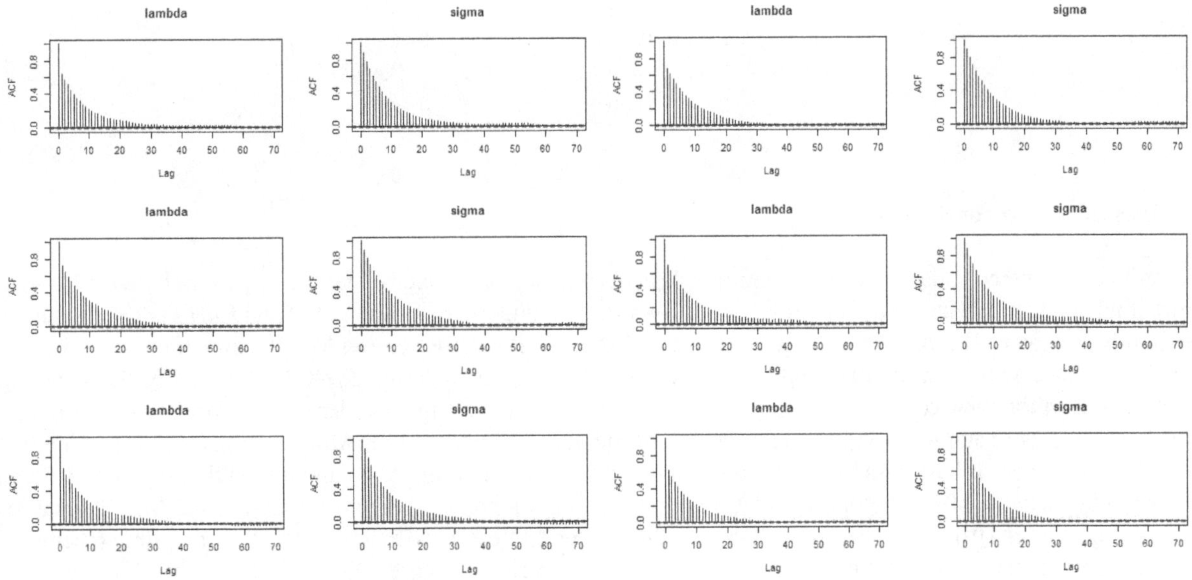

Figure 4. Auto-correlation plots for λ and σ for $\mu_o = 8$

Using auto-correlation plots given in Figure 3 and in Figure 4 for $\mu_o = 4$ and $\mu_o = 8$ we conclude that we should pick every 50th observation. Therefore, to provide relatively independent samples for improvement of prediction accuracy, we calculate the Bayesian MCMC estimates by the means of every sampled values after discarding the first 10000 values from 50000.

In Figure 1, 2, 3 and 4, respectively, graph 1 and 2 are plotted for $\epsilon = 0.0$, graph 3 and 4 are plotted for $\epsilon = 0.2$, graph 5 and 6 are plotted for $\epsilon = 0.4$, graph 7 and 8 are plotted for $\epsilon = 0.6$, graph 9 and 10 are plotted for $\epsilon = 0.8$, graph 11 and 12 are plotted for $\epsilon = 1.0$.

Table 8. ML-II posterior mean under SELF and LINEX for λ and σ.

	SELF				LINEX			
	$\hat{\lambda}$		$\hat{\sigma}$		$\tilde{\lambda}$		$\tilde{\sigma}$	
$\epsilon \downarrow$ $\mu_o \rightarrow$	4	8	4	8	4	8	4	8
0.0	0.91869	0.51275	1.79257	1.14045	0.91566	0.512	1.78243	1.13721
0.2	0.76429	0.50146	1.57316	1.1182	0.76268	0.50087	1.56654	1.11537
0.4	0.65742	0.48062	1.37591	1.08599	0.65632	0.48019	1.37081	1.08364
0.6	0.57724	0.46549	1.25019	1.05408	0.57646	0.46507	1.24636	1.05183
0.8	0.48223	0.44926	1.08714	1.01904	0.48166	0.44884	1.08427	1.01698
1.0	0.43215	0.42207	1.00947	1.00275	0.43166	0.42156	1.00701	1.00008

Table 9. Bayes estimates under SELF for $\epsilon = 0.4$.

	$R(t)$		$Z(t)$	
$\mu_o \rightarrow$	4	8	4	8
$t \downarrow$				
1	0.75538	0.7589	0.32974	0.32732
2	0.52889	0.5329	0.39207	0.38868
4	0.23609	0.24134	0.44094	0.43129
6	0.1027	0.10766	0.45371	0.44066
8	0.04537	0.04902	0.4574	0.44297
10	0.02074	0.0231	0.45858	0.44361
15	0.00398	0.00461	0.45919	0.4439
20	0.00167	0.00182	0.45926	0.44392

8. Discussion and conclusion

From Table 1, we conclude that the posterior risk is decreasing as number of failures i.e., r is increasing. Also from Table 1-3, it can be seen that ML-II posterior variance under SELF for λ, $R(t)$ and $h(t)$ is decreasing as epsilon is increasing. From Table 1-7, it can be seen that as epsilon is increasing ML-II estimators under the two considered base priors under SELF and LINEX loss functions by considering Type II censoring or the sampling scheme of Bartholomew comes near i.e., for, $\epsilon = 0$ we get the Bayes estimate under the base prior. As the value of ϵ increases the estimator tends to include more part of the data dependent prior. All the estimates are showing that ML-II estimator are robust with respect to the change in base. From Table 6, it can be seen that under SELF and the sampling scheme of Bartholomew as termination time t_o is increasing the posterior risk and ML-II posterior variance decreases. It is also seen from Table 6, that under LINEX loss function and the sampling scheme of Bartholomew as termination time t_o is increasing the posterior risk decreases. From Table 1 and Table 6 we also conclude that under the sampling scheme of Bartholomew posterior risk and ML-II posterior variance are higher as compare to Type II censoring scheme. Thus we can say Type II censoring scheme provides better estimates. Also, as the posterior risk under LINEX loss function are lesser than the posterior risk under SELF, therefore LINEX loss function is better than SELF.

Acknowledgment

The authors are thankful to the editor and referees for their valuable suggestions and comments which led to considerable improvement in the original version

REFERENCES

1. S.H. Arora, G.C. Bhimani, and M.N. Patel, *Some results on maximum likelihood estimators of parameters of generalized half logistic distribution under Type-I progressive censoring with changing*, International Journal of Contemporary Mathematical Sciences, vol. 5, pp. 685–698, 2010.
2. A. Asgharzadeh, R. Rezaie, and M. Abdi, *Comparisons of methods of estimation for the half-logistic distribution*, Selçuk J. Appl. Math., Special Issue, pp. 93–108, 2011.
3. R. Azimi, *Bayesian estimation of generalized half logistic type-II doubly censored data*, International Journal of Scientific World, vol. 1, pp. 57–60, 2013.
4. N. Balakrishnan, *Order statistics from the half logistic distribution*, Journal of Statistics and Computer Simulation, vol. 20, pp. 287–309, 1985.
5. N. Balakrishnan, and A. Asgharzadeh, *Inference for the scaled half-logistic distribution based on progressively type II censored samples*, Comunication in Statistics-Theory and Methods, vol. 34, pp. 73–87, 2005.
6. N. Balakrsihnan, and A. Hossain, *Inference for the Type-II generalized logistic distribution under progressive Type-II censoring*, Journal of Statistical Computation and Simulation, vol. 77, pp. 1013–1031, 2007.
7. N. Balakrishnan, and N. Kannan, *Point and interval estimation for the logistic distribution based on progressively Type-II censored samples*, in Handbook of Statistics, N. Balakrishnan and C. R. Rao, Eds., vol. 20, pp. 431–456, 2001.
8. N. Balakrishnan, N. Kannan, C.T. Lin, and S.J.S. Wu, *Inference for the extreme value distribution under progressively Type-II censoring*, Journal of Statistical Computation and Simulation, vol. 74, pp. 25–45, 2004.
9. N. Balakrishnan, and N. Puthenpura, *Best linear unbiased estimators of location and scale parameters of the half logistic distribution*, Journal of Statistics and Computer Simulation, vol. 25, pp. 193–204, 1986.
10. D.J. Bartholomew, *The sampling distribution of an estimate arising in life testing*, Technometrics, vol. 5, pp. 361–374, 1963.
11. J.O. Berger, *Bayesian robustness and Stein effect*, J. Am. Stat. Assoc., vol. 77, pp. 358–368, 1982.
12. J.O. Berger, *The robust Bayesian viewpoint (with discussion)*, In: Kadane, J. (ed.) Robustness in Bayesian Statistics, North Holland, 1983.
13. J.O. Berger, *Statistical Decision Theory and Bayesian Analysis*, New York: Springer-Verlag, 1985.
14. J.O. Berger, and J.M. Berliner, *Robust Bayes and empirical Bayes analysis with ϵ-contaminated priors*, Technical Report No. 83–85, Department of Statistics, Purdue University, West Lafayette, 1983.
15. J.O. Berger, D.R. Insua, and F. Ruggeri, *Bayesian robustness. In: Rios Insua, D., Ruggeri, F. (eds.)*, Robust Bayesian Analysis, Lecture Notes in Statistics, vol. 152, pp. 1–32. Springer, New York, 2000.
16. B. Betro, *Numerical treatment of Bayesian robustness problems*, Int. J. Approx. Reason., vol. 50, pp. 279–288, 2009.
17. G.E.P. Box, and G.C. Tiao, *A further look at robustness in Bayes theorem*, Biometrica, vol. 49, pp. 419–432, 1962.
18. G.E.P. Box, *Sampling and Bayes inferences in scientific modeling and robustness (with discussion)*, J. Roy. Stat. Soc., vol. B41, pp. 113–147, 1980.
19. A. Chaturvedi, *Robust Bayesian analysis of the linear regression model*, J. Stat. Plan. Infer., vol. 50, pp. 175–186, 1993.
20. A. Chaturvedi, *Robust Bayesian analysis of the Exponential failure model*, Pak. J. Stat. vol. 14, no. 24, pp. 115–126, 1998.
21. A. Chaturvedi, S.B. Kang, and A. Pathak, *Estimation and testing procedures for the reliability functions of generalized half logistic distribution*, Journal of The Korean Statistical Society, vol. 45, pp. 314–328, 2016.
22. A. Chaturvedi, M. Pati, and S.K. Tomer, *Robust Bayesian analysis of Weibull failure model*, METRON. DOI 10.1007/s40300-013-0027-7, 2013.
23. A.P. Dempster, *Examples relevant to the robustness of applied inferences*, In: Gupta, S.S., Berger, J.O. (eds.) Stat. Decis. Theory Relat. Topics III. Academic Press, NY, 1976.
24. I.J. Good, *Probability and the Weighting of Evidence*, Griffin: London, 1950.
25. I.J. Good, *The estimation of Probability*, MIT Press, Cambridge, 1965.
26. I.J. Good, *A Bayesian significance test for multinomial distributions*, J. Roy. Stat. Soc., vol. B-29, pp. 399–443, 1967.
27. I.J. Good, *Some history of the hierarchical Bayesian methodology*, In: J.M. Bernardo, M.H. DeGroot, D.V. Lindley, A.F.M. Smith (eds.), Bayesian Statistics. University Press, Valencia, 1980.
28. H. Jeffreys, *Theory of Probability*, 3rd edn., Clarendon Press, Oxford, 1961.
29. Y. Kim, S.B. Kang, and J.I. Seo, *Bayesian estimation in the generalized half logistic distribution under progressively Type II censoring*, Journal of the Korean Data and Information Science Society, vol. 22, pp. 977–987, 2011.
30. D.V. Lindley, *The robustness of interval estimates*, Bull. Int. Stat. Inst., vol. 38, pp. 209–220, 1961.
31. W.B. Nelson, *Applied Life Data Analysis*, John Willey and Sons, New York, 1982.
32. J.I. Seo, and S.B. Kang, *Entropy estimation of generalized half-logistic distribution (GHLD) bassed on Type-II censored samples*, Entropy, vol. 16, pp. 443–454, 2014.
33. J.I. Seo, Y. Kim, and S.B. Kang, *Estimation on the generalized half logistic distribution under Type-II hybrid censoring*, Communications for Statistical Applications and Methods, vol. 20, pp. 63–75, 2013.
34. J.I. Seo, H.J. Lee, and S.B. Kang, *Estimation for generalized half logistic distribution based on records*, Journal of the Korean Data and Information Science Society, vol. 23, pp. 1249–1257, 2012.

11

Proper complex random processes

Yu. V. Kozachenko [1], M.Yu. Petranova [2,*]

[1]*Department of Probability Theory, Statistics and Actuarial Mathematics, Taras Shevchenko National University of Kyiv, Ukraine, Department of Probability Theory and Mathematical Statistics, Vasyl' Stus Donetsk National University, Ukraine, yvk@univ.kiev.ua.*
[2]*Department of Probability Theory and Mathematical Statistics, Vasyl' Stus Donetsk National University, Ukraine, m.petranova@donnu.edu.ua.*

Abstract In this paper we study properties of stationary proper complex random process with stable correlation functions. Estimates are obtained for distribution of supremum of modulus of these processes and normes in spaces L_p on finite and infinite intervals.

Keywords complex random process, stationary Gaussian proper complex random process, stable correlation function, square Gaussian random process.

1. Introduction

This article deals with complex random processes which are one of the most important generalizations of the concept of random process (see [2, 12]). The complex random processes are especially relevant when the narrow-banded processes are investigated. These processes are exploited as models of complex amplitudes of quasi-harmonic oscillations or waves in radiophysics and optics [1]. In this article we presented results of investigation of properties of complex random processes which are useful when solving problems in the listed above areas. Conditions for existence of proper complex random processes are described in [12, 2]. In this article we investigate stationary proper complex random processes, stationary proper complex random processes with stable correlation function. Some results for properties of stable correlation function are presented in paper[11]. In this article some properties of square Gaussian random variables and process are presented (for more results see, for example, [3, 7, 8]). Also, in this paper estimates of distributions of functionals from the module of stationary Gaussian proper complex random processes are obtained (for more results see, for example, [13, 6, 10]). Theorems, which describe behavior of the module of stationary proper complex random process at infinity are developed.

The content of the article is as follows. In Section 2 we introduce the basic definitions related to the complex random processes. Stationary proper complex random processes are introduced and discussed in Section 3. In the next Section 4, we deal with properties of square Gaussian random variables and processes. Section 5 is related to estimates of distributions of some functions from the module of stationary Gaussian proper complex random process. And in the last Section 6 behavior of the module of stationary proper complex random process at infinity is studied.

*Correspondence to: 21, 600-richya str., Vinnytsia, Department of Probability Theory and Mathematical Statistics, Vasyl' Stus Donetsk National University, Ukraine (m.petranova@donnu.edu.ua).

2. Proper Complex Random Process

Definition 2.1. A random process of the form $X(t) = X_c(t) + iX_s(t)$, $t \in \mathbb{R}$, where $X_c(t)$ and $X_s(t)$ are real-valued random processes (c – cosine, s – sine), is called complex random process (see book [2] and paper [12]).

Remark 2.1. In this paper we will consider centered random processes, that is

$$EX(t) = EX_c(t) = EX_s(t) = 0.$$

Definition 2.2. The function

$$r(\tau, t) = EX(t+\tau)\overline{X}(t) = EX_c(t+\tau)X_c(t) + EX_s(t+\tau)X_s(t) + \\ +i\left(EX_s(t+\tau)X_c(t) - EX_c(t+\tau)X_s(t)\right)$$

is called correlation function of the process $X(t)$.

The function

$$\widehat{r}(\tau, t) = EX(t+\tau)X(t) = \\ = EX_c(t+\tau)X_c(t) - EX_s(t+\tau)X_s(t) + i\left(EX_c(t+\tau)X_s(t) + EX_s(t+\tau)X_c(t)\right)$$

is called pseudo correlation function of the process $X(t)$.

Definition 2.3. A complex random process $X(t)$ is called proper complex random process (PCR process), if the pseudo correlation function of this process is equal to zero, $EX(t+\tau)X(t) = 0$, that is when conditions

$$EX_c(t+\tau)X_c(t) = EX_s(t+\tau)X_s(t), \tag{1}$$

$$EX_c(t+\tau)X_s(t) = -EX_s(t+\tau)X_c(t). \tag{2}$$

hold true.

Remark 2.2. Conditions under which PCR processes exist are described in book [2] and paper [12].

Definition 2.4. [12] A proper complex random process is called (wide sense) stationary if for all $\tau, t \in \mathbb{R}$ the following relation holds true

$$r(\tau, t) = EX(t+\tau)\overline{X}(t) = r(\tau).$$

Remark 2.3. In the case where PCR-process $X(t)$ is stationary we can write the following relations

$$EX_c(t+\tau)X_c(t) = EX_s(t+\tau)X_s(t) = \frac{1}{2}\operatorname{Re} r(\tau), \tag{3}$$

$$EX_c(t+\tau)X_s(t) = \frac{1}{2}\operatorname{Im} r(\tau). \tag{4}$$

Definition 2.5. A complex random process $X(t) = X_c(t) + iX_s(t)$ is called Gaussian if the real-valued random processes $X_c(t)$ and $X_s(t)$ are jointly Gaussian processes.

3. Stationary PCR-processes with stable correlation functions

Definition 3.1. The correlation function $r(\tau)$, $\tau \in \mathbb{R}$ of stationary proper complex random process is called stable correlation function if it can be represented in the form

$$r(\tau) = \sigma^2 \exp\left\{-c|\tau|^\alpha \left(1 + i\beta\frac{\tau}{|\tau|}\omega(\tau, \alpha)\right)\right\} \tag{5}$$

where $\sigma^2, c, \beta, \alpha$ are real-valued constants, such that $\sigma^2 > 0, c > 0, |\beta| \le 1, 0 \le \alpha \le 2$,

$$\omega(\tau, \alpha) = \begin{cases} tg\frac{\pi\alpha}{2}, \; 0 \le \alpha \le 2, \alpha \ne 1, \\ \frac{2}{\pi}\log|\tau|, \; \alpha = 1. \end{cases}$$

Remark 3.1. The function $r(\tau)$ is non-negative definite, since $r(\tau)$ is the characteristic function of a stable random variable $\xi, E\xi = 0$, in the case where $\sigma^2 = 1$ (see [11, p.169]).

Definition 3.2. A stationary PCR process is called proper stationary random complex process with stable covariance function (stationary SPCR process) if

$$EX(t+\tau)\overline{X}(t) = r(\tau)$$

where the function $r(\tau)$ is given by formula (5).

Remark 3.2. For the proper stationary random complex process $X(t) = X_c(t) + iX_s(t)$, $EX(t) = 0$, with stable covariance functions the following relations hold true

$$\begin{gathered} EX_c(t+\tau)X_c(t) = \tfrac{1}{2}\operatorname{Re} r(\tau) = \\ = \tfrac{\sigma^2}{2}\exp\left\{-c|\tau|^\alpha\right\}\cos\left(c|\tau|^\alpha\beta\tfrac{\tau}{|\tau|}\omega(\tau,\alpha)\right) = \\ = EX_s(t+\tau)X_s(t), \end{gathered} \tag{6}$$

$$\begin{gathered} EX_c(t+\tau)X_s(t) = \tfrac{1}{2}\operatorname{Im} r(\tau) = \\ = -\tfrac{\sigma^2}{2}\exp\left\{-c|\tau|^\alpha\right\}\sin\left(c|\tau|^\alpha\beta\tfrac{\tau}{|\tau|}\omega(\tau,\alpha)\right), \end{gathered} \tag{7}$$

$$\operatorname{Re} r(-\tau) = \operatorname{Re} r(\tau). \tag{8}$$

4. Square Gaussian random variables and processes

In this section we propose definitions and some properties of square Gaussian random variables and processes.

Definition 4.1. [3, 7] Let (T,ρ) be a metric space and let $\Theta = \{\xi(t), t \in T\}$, $E\xi(t) = 0$, be a family of jointly Gaussian random variables (e.g. $\xi = \{\xi(t), t \in T\}$ is a Gaussian random process). The space of square Gaussian random variable $(SG_\Theta(\Omega))$ is such a space that any element $\eta \in SG_\Theta(\Omega)$ can be presented in the form

$$\eta = \vec{\xi}^\top A\vec{\xi} - E(\vec{\xi}^\top A\vec{\xi}), \tag{9}$$

where $\vec{\xi}^\top = (\xi_1, \xi_2, ..., \xi_n), \xi_k \in \Theta, k = 1, 2, ..., n$ and A is a real-valued matrix, or the element $\eta \in SG_\Theta(\Omega)$ is the mean square limit of a sequence of random variables of the form (9):

$$\eta = l.i.m.\left(\vec{\xi}_n^\top A_n\vec{\xi}_n - E(\vec{\xi}_n^\top A\vec{\xi}_n)\right).$$

Definition 4.2. A random process $\eta = \{\eta(t), t \in T\}$ is called square Gaussian process if the family of random variables $\eta = \{\eta(t), t \in T\}$ forms the space of square Gaussian random variables.

The next theorem is a modification of Theorem 3.2 from the book [7].

Theorem 4.1
Let $X = \{X(t), t \in [a,b]\}$ be a separable square Gaussian random process and let the condition

$$\sup_{|t-s|\le h}\ (Var\,(X(t) - X(s)))^{1/2} \le Ch^\beta \tag{10}$$

holds true for $\beta \in (0,1], C > 0$. Then for all integer $M > 1$ and all

$$x > \frac{\sqrt{2}\gamma_0 M}{\beta}\max\left(1, \left(\frac{b-a}{2}\right)^\beta\frac{C}{\gamma_0}\right)^{\frac{1}{M-1}},$$

where $\gamma_0 = \sup\limits_{a\le t\le b} (VarX(t))^{1/2}$, the tail of distribution of the process $|X(t)|$ can be estimated in the following way

$$P\left\{\sup_{t\in[a,b]} |X(t)| > x\right\} \le 4e^{\frac{M+1}{\beta}} \cdot \exp\left\{-\frac{x}{\sqrt{2}\gamma_0}\right\}\left(\frac{\beta\cdot x}{\sqrt{2}\gamma_0 M}\right)^{\frac{M}{\beta}}\left(1+\frac{\sqrt{2}x}{\gamma_0}\right)^{1/2}. \tag{11}$$

Theorem 4.2
[9] Let $X = \{X(t), t\in[a,b]\}$, where $-\infty \le a < b \le \infty$, be a measurable square-Gaussian random process. Let the Lebesgue integral

$$\int_a^b \left(E(X(t))^2\right)^{p/2} dt$$

be well defined for $p \ge 1$. Then the integral

$$\int_a^b |X(t)|^p dt$$

exists with probability 1 and for all

$$\varepsilon \ge \left(\frac{p}{\sqrt{2}} + \sqrt{\left(\frac{p}{2}+1\right)p}\right)^p C_p,$$

where

$$C_p = \int_a^b \left(E(X(t))^2\right)^{p/2} dt,$$

the following inequality holds true

$$P\left\{\int_a^b |X(t)|^p dt > \varepsilon\right\} \le 2\sqrt{1+\frac{\varepsilon^{1/p}\sqrt{2}}{C_p{}^{1/p}}}\exp\left\{-\frac{\varepsilon^{1/p}}{\sqrt{2}C_p{}^{1/p}}\right\}. \tag{12}$$

Corollary 4.1
Let assumptions of Theorem 4.2 be satisfied. Then for

$$u \ge \left(\frac{p}{\sqrt{2}} + \sqrt{\left(\frac{p}{2}+1\right)p}\right) C_p{}^{1/p}$$

the following inequality holds true

$$P\left\{\|X(t)\|_{L_p(a,b)} > u\right\} \le 2\sqrt{1+\frac{u\sqrt{2}}{C_p{}^{1/p}}}\cdot\exp\left\{-\frac{u}{\sqrt{2}C_p{}^{1/p}}\right\}. \tag{13}$$

5. Estimation of distribution of some functionals from module of stationary Gaussian PCR-processes

Theorem 5.1
Let $X = \{X(t), t\in[a,b]\}$ be a Gaussian SPCR process and let $|X(t)| = \left(X_c^2(t) + X_s^2(t)\right)^{1/2}$. Then for

$$u \ge \left(\frac{p}{\sqrt{2}} + \sqrt{\left(\frac{p}{2}+1\right)p}\right)\sigma^2(b-a)^{1/p}$$

the following inequality holds

$$P\left\{\left\|X(t)^2-\sigma^2\right\|_{L_p([a,b])}>u\right\}\leq 2\sqrt{1+\frac{u\sqrt{2}}{(b-a)^{1/p}\sigma^2}}\cdot\exp\left\{-\frac{u}{\sqrt{2}(b-a)^{1/p}\sigma^2}\right\}. \tag{14}$$

Proof
The proof of this theorem follows from inequality (13). Indeed it follows from (6), that

$$EX_c^2(t)=E(X_s(t))^2=\frac{1}{2}R(r(0))=\frac{\sigma^2}{2}.$$

Therefore $E|X(t)|^2=\sigma^2$ and

$$E\left(|X(t)|^2-\sigma^2\right)^2=E\left(|X(t)|^2-E|X(t)|^2\right)^2=E|X(t)|^4-\left(E|X(t)|^2\right)^2=E|X(t)|^4-\sigma^4. \tag{15}$$

Suppose that (X_1,X_2,X_3,X_4) is a zero-mean Gaussian vector. Then we have:

$$E(X_1X_2X_3X_4)=E(X_1X_2)E(X_3X_4)+E(X_1X_3)E(X_2X_4)+E(X_1X_4)E(X_2X_3).$$

This equality is called Isserlis formula (see, for example [3, p.228]. Making use of this formula and relations (3), (4) we can write

$$E|X(t)|^4=E\left(|X_c(t)|^2+|X_s(t)|^2\right)^2=E|X_c(t)|^4+E|X_s(t)|^4+2E|X_c(t)|^2|X_s(t)|^2,$$

$$E|X_c(t)|^4=3\left(E|X_c(t)|^2\right)^2=3\frac{\sigma^4}{4}=E|X_s(t)|^4,$$

$$E(|X_c(t)|^2|X_s(t)|^2)=E|X_c(t)|^2E|X_s(t)|^2+2(EX_c(t)X_s(t))^2.$$

Since

$$E(X_c(t)X_s(t))=\frac{1}{2}\,\mathrm{Im}(r(0))=0$$

we have

$$E|X(t)|^4=2\sigma^4.$$

It follows from (15) that

$$E\left((X(t))^2-\sigma^2\right)^2=\sigma^4$$

and

$$\int_a^b E\left(|X(t)|^2-\sigma^2\right)^{\frac{p}{2}}dt=\sigma^{2p}(b-a).$$

Now (14) follows from (13). □

Theorem 5.2
Let $X=\{X(t),t\in[a,b]\}$ be a Gaussian SPCR process and let

$$|X(t)|=\left(X_c^2(t)+X_s^2(t)\right)^{1/2}.$$

If $X(t)$ is a separable process, then for all integer $M>1$ and all

$$u>\frac{2\sqrt{2}\sigma^2M}{\alpha}\left(\max\left(1,\left(\frac{b-a}{2}\right)^{\alpha/2}2\sqrt{c}\right)^{\frac{1}{M-1}}\right)$$

we have

$$P\left\{\sup_{a\le t\le b}\left|(X(t))^2-\sigma^2\right|>x\right\}\le 4e^{\frac{2(M+1)}{\alpha}}\exp\left\{-\frac{x}{\sqrt{2}\sigma^2}\right\}\left(\frac{\alpha x}{2\sqrt{2}\sigma^2 M}\right)^{\frac{2M}{\alpha}}\left(1+\frac{\sqrt{2}x}{\sigma^2}\right)^{1/2} \tag{16}$$

Proof
The statement of this Theorem follows from Theorem 4.1. In our case $\gamma_0=\sigma^2$. □

In order to to apply Theorem 4.1 to the process $|X(t)|^2=\left(X_c^2(t)+X_s^2(t)\right)^2$ we have to astimate $E(Y(t)-Y(s))^2$, where $Y(t)=|X(t)|^2-\sigma^2$.
It is easy to see that

$$\begin{gathered}E(Y(t)-Y(s))^2=E\left(X_c^2(t)+X_s^2(t)-X_c^2(s)-X_s^2(s)\right)^2=\\=E\left(X_c^2(t)-X_c^2(s)+X_s^2(t)-X_s^2(s)\right)^2=\\=E\left(X_c^2(t)-X_c^2(s)\right)^2+E\left(X_s^2(t)-X_s^2(s)\right)^2+2E\left(X_c^2(t)-X_c^2(s)\right)E\left(X_s^2(t)-X_s^2(s)\right)=\\w_1+w_2+w_3,\end{gathered}$$

$$\begin{gathered}w_1=E(X_c(t))^4+E(X_c(s))^4-2E(X_c^2(t)X_c^2(s)),\\E(X_c(t))^4=E(X_c(s))^4=\frac{3}{4}\sigma^4,\\E(X_c^2(t)X_c^2(s))=E(X_c(t))^2E(X_c(s))^2+2(EX_c(t)X_c(s))^2=\\=\tfrac{\sigma^4}{4}+2\left(\tfrac{1}{2}\operatorname{Re}(r(t-s)\right)^2\end{gathered}$$

Therefore

$$w_1=\sigma^4-(\operatorname{Re}(r(t-s)))^2.$$

Next¡ we have

$$\begin{gathered}w_1=w_2\\w_1+w_2=2\left(\sigma^4-(\operatorname{Re}(r(t-s)))^2\right),\\\frac{w_3}{2}=E(X_c^2(t)X_s^2(t))+E(X_c^2(s)X_s^2(s))-E(X_c^2(s)X_s^2(t))-E(X_c^2(t)X_s^2(s)).\end{gathered}$$

Since $\operatorname{Im}(r(0))=0$, then

$$E(X_c^2(t)X_s^2(t))=EX_c^2(t)EX_s^2(t)+2(EX_c(t)X_s(t))^2=\frac{\sigma^4}{4}+2\left(\frac{1}{2}\operatorname{Im}(r(0))\right)^2=\frac{\sigma^4}{4}.$$

In the same way we can obtaine that

$$\begin{gathered}E(X_c^2(s)X_s^2(s))=\frac{\sigma^4}{4},\\E(X_c^2(t)X_s^2(s))=\frac{\sigma^4}{4}+\frac{1}{2}(\operatorname{Im}(r(t-s)))^2.\end{gathered}$$

Consequently

$$w_3=-\left((\operatorname{Im}(r(s-t)))^2+(\operatorname{Im}(r(t-s)))^2\right).$$

Since

$$(\operatorname{Im}(r(s-t)))^2=(\operatorname{Im}(r(t-s)))^2,$$

we have

$$w_3=-2(\operatorname{Im}(r(t-s)))^2$$

and

$$E(Y(t)-Y(s))^2 = 2\left(\sigma^4 - \left((\mathrm{Re}\,(r(t-s)))^2 + (\mathrm{Im}\,(r(t-s)))^2\right)\right).$$

Since

$$\begin{gathered}(\mathrm{Re}\,(r(t-s)))^2 + (\mathrm{Im}\,(r(t-s)))^2 = |r(t-s)|^2 = \sigma^4 \exp\{-2c(t-s)\}\cdot \\ = \left|\cos\left(-c\,|\tau^\alpha|\, i\beta \tfrac{(t-s)}{|t-s|} w\,(t,\alpha)\right) + i \sin\left(-c\,|\tau^\alpha|\, i\beta \tfrac{(t-s)}{|t-s|} w\,(t,\alpha)\right)\right|^2 = \\ = \sigma^2 \exp\{-|t-s|^\alpha c\}\end{gathered}$$

we get the following estimate

$$E(Y(t)-Y(s))^2 = 2\sigma^4\left(1-\exp\{-2c|t|^\alpha\}\right) \le 4\sigma^4 c|t|^\alpha.$$

Consequently $\beta = \frac{\alpha}{2}$ and $C = 2\sigma^2\sqrt{c}$. Therefore (16) follows from (11).

6. Behavior of the module of stationary PCR-process at infinity

Theorem 6.1

Let $X = \{X(t), t \in (-\infty,\infty)\}$ be a measurable Gaussian SPCR process, let $|X(t)| = \left(X_c^2(t) + X_s^2(t)\right)^{1/2}$ and let $Y(t) = |X(t)|^2 - E(X(t))^2 = |X(t)|^2 - \sigma^2$. Let $c(t), t \in R$ be a function such that

$$\int_{-\infty}^{\infty} |c(t)|^{-p}\, dt < \infty,\ p \ge 1.$$

Then for

$$u \ge \left(\frac{p}{\sqrt{2}} + \sqrt{\left(\frac{p}{2}+1\right)p}\right)\cdot \sigma^2 \left(\int_{-\infty}^{\infty} |c(t)|^{-p} dt\right)^{1/p}$$

the following inequality holds true

$$P\left\{\left\|\frac{(X(t))^2-\sigma^2}{c(t)}\right\|_{L_p(-\infty,\infty)} > u\right\} \le 2\sqrt{1 + \frac{\sqrt{2}u}{\left(\int_{-\infty}^{\infty} |c(t)|^{-p} dt\right)^{1/p}\sigma^2}} \cdot \exp\left\{-\frac{u}{\sqrt{2}\left(\int_{-\infty}^{\infty} |c(t)|^{-p} dt\right)^{1/p}\sigma^2}\right\}.$$

Proof

The statement of this Theorem follows from Theorem 4.2 (see Corollary 4.1) if we take

$$C_p = 2^{-\frac{p}{2}}\sigma^{2p}\cdot \int_{-\infty}^{\infty} |c(t)|^{-p} dt.$$

□

Corollary 6.1

Let $c(t) > 0$ be an even monotone increasing function for which conditions of Theorem 6.1 are satisfied. Then for all $t \ge 0$ the following inequality holds true with probability one:

$$\left(\int_{-t}^{t} \left|(X(u))^2 - \sigma^2\right|^p du\right)^{1/p} \le c(t)\cdot \xi$$

where $\xi > 0$ is a random variable such that

$$P\{\xi > u\} \leq 2\sqrt{1 + \frac{\sqrt{2}u}{\left(\int\limits_{-\infty}^{\infty} |c(t)|^{-p} dt\right)^{1/p} \cdot \sigma^2}} \cdot \exp\left\{-\frac{u}{\left(\sqrt{2}\int\limits_{-\infty}^{\infty} |c(t)|^{-p} dt\right)^{1/p} \cdot \sigma^2}\right\},$$

for

$$u \geq \left(\frac{p}{\sqrt{2}} + \sqrt{\left(\frac{p}{2}+1\right)p}\right) \cdot \sigma^2 \left(\int\limits_{-\infty}^{\infty} |c(t)|^{-p} dt\right)^{1/p}$$

Proof
The statement of Corollary 6.1 follows from inequalities: $t \geq 0$,

$$\left(\int\limits_{-t}^{t} \left|(X(u))^2 - \sigma^2\right|^p du\right)^{1/p} \leq c(t) \cdot \left(\int\limits_{-t}^{t} \frac{\left|(X(u))^2-\sigma^2\right|^p}{c(u)^p} du\right)^{1/p} \leq$$
$$\leq c(t) \cdot \left\|\frac{\left|(X(t))^2-\sigma^2\right|}{c(t)}\right\|_{L_p(-\infty,\infty)}.$$

□

Remark 6.1. The statement of Corollary 6.1 holds true, for example, for function $c(t) = (1 + |t|^\gamma), \gamma > p$.

Theorem 6.2
Let $X = \{X(t), t \in [a, b]\}$ be a separable Gaussian stationary PCR process. Let there exist a sequence $a_k, k = 0, 1, 2, ...$ such that $a_k < a_{k+1}, a_k \to \infty$, as $k \to \infty$, $a_0 = 0$ and a function $c(t), t \in [0, \infty)$ such that $c(t) \geq 1$, $c(t)$ be an even monotone increasing continuous and $c(t) \to \infty$ if $t \to \infty$, $Y(t) = |X(t)|^2 - E(X(t))^2 = |X(t)|^2 - \sigma^2$. Let the condition

$$\varepsilon^* = \frac{2\sqrt{2}\sigma^2 M}{\alpha} \sup_{0 \leq k \leq \infty} \frac{1}{c(a_k)} \max\left(1, \left(\left(\frac{a_{k+1}-a_k}{2}\right)^{\alpha/2} 2\sqrt{c}\right)^{1/M-1}\right) \leq \infty$$

be satisfied. If for some $\widehat{\varepsilon}, \widehat{\varepsilon} \geq \varepsilon^*$, the following series

$$\sum_{k=0}^{\infty} \exp\left\{-\frac{(c(a_k) - c(a_0))\widehat{\varepsilon}}{\sqrt{2}\sigma^2}\right\}\left(1 + \frac{\sqrt{2}c(a_k)\widehat{\varepsilon}}{\sigma^2}\right)^{1/2} < \infty, \tag{17}$$

then for all $\varepsilon > \widehat{\varepsilon}$

$$P\left\{\sup_{0 \leq t \leq \infty} \frac{\left||X(t)|^2 - \sigma^2\right|}{c(t)} > \varepsilon\right\} \leq \exp\left\{-\frac{\sqrt{2}c(a_0)\varepsilon}{\sigma^2}\right\} \cdot \widehat{z} = Z(\varepsilon), \tag{18}$$

where

$$\widehat{z} = 4e^{\frac{M+1}{\alpha}} \cdot \left(\frac{2}{M}\right)^{\frac{2M}{\alpha}} \sum_{k=0}^{\infty} \exp\left\{-\frac{(c(a_k) - c(a_0))\widehat{\varepsilon}}{\sqrt{2}\sigma^2}\right\}\left(1 + \frac{\sqrt{2}c(a_k)\widehat{\varepsilon}}{\sigma^2}\right)^{1/2}.$$

Proof
For $Y(t) = |X(t)|^2 - \sigma^2$ we have that $Var\,(Y(t) - Y(s)) \leq 2\sigma^2|t-s|^{\frac{\alpha}{2}}$ (see proof of Theorem 5.2). It follows

from Theorem 5.2 that for $M > 1$ and

$$u > \frac{2\sqrt{2}\sigma^2 M}{\alpha}\left(\max\left(1,\left(\left(\frac{a_{k+1}-a_k}{2}\right)^{\frac{\alpha}{2}} 2\sqrt{c}\right)^{\frac{1}{M-1}}\right)\right) \tag{19}$$

we have

$$P\left\{\sup_{a_k \le t \le a_{k+1}} |Y(t)| > u\right\} \le 4e^{\frac{M+1}{\alpha}} \exp\left\{-\frac{u}{\sqrt{2}\sigma^2}\right\}\left(\frac{\alpha u}{2\sqrt{2}\sigma^2 M}\right)^{\frac{2M}{\alpha}}\left(1+\frac{\sqrt{2}u}{\sigma^2}\right)^{1/2} \tag{20}$$

The following inequality is obvious.

$$P\left\{\sup_{t\in[0,\infty)} \frac{|Y(t)|}{c(t)} > \varepsilon\right\} \le \sum_{k=0}^{\infty} P\left\{\sup_{a_k \le t \le a_{k+1}} |Y(t)| > c(a_k)\cdot\varepsilon\right\}. \tag{21}$$

It follows from (19) and (20) that for

$$\varepsilon > \frac{2\sigma M}{\alpha c(a_k)}\cdot\left(\max\left(1,\left(\left(\frac{a_{k+1}-a_k}{2}\right)^{\frac{\alpha}{2}} 2\sqrt{c}\right)^{\frac{1}{M-1}}\right)\right), \tag{22}$$

we have the following estimate

$$P\left\{\sup_{a_k \le t \le a_{k+1}} |Y(t)| > c(a_k)\varepsilon\right\} \le 4e^{\frac{M+1}{\alpha}} \exp\left\{-\frac{c(a_k)\varepsilon}{\sqrt{2}\sigma^2}\right\}\left(\frac{\alpha c(a_k)\varepsilon}{2\sqrt{2}\sigma^2 M}\right)^{\frac{2M}{\alpha}}\left(1+\frac{\sqrt{2}c(a_k)\varepsilon}{\sigma^2}\right)^{1/2}. \tag{23}$$

From this inequality (23) and inequality (21) it follows that under condition (22) we have the following estimate

$$P\left\{\sup_{a_k \le t \le a_{k+1}} \frac{|Y(t)|}{c(t)} > \varepsilon\right\} \le 4e^{\frac{M+1}{\alpha}} \sum_{k=0}^{\infty} \exp\left\{-\frac{c(a_k)\varepsilon}{\sqrt{2}\sigma^2}\right\}\left(\frac{\alpha c(a_k)\varepsilon}{2\sqrt{2}\sigma^2 M}\right)^{\frac{2M}{\alpha}}\left(1+\frac{\sqrt{2}c(a_k)\varepsilon}{\sigma^2}\right)^{1/2} =$$

$$= 4e^{\frac{M+1}{\alpha}} \exp\left\{-\frac{c(a_0)\varepsilon}{\sqrt{2}\sigma^2}\right\}\sum_{k=0}^{\infty} \exp\left\{-\frac{(c(a_k)-c(a_0))\varepsilon}{\sqrt{2}\sigma^2}\right\}\left(\frac{\alpha c(a_k)\varepsilon}{2\sqrt{2}\sigma^2 M}\right)^{\frac{2M}{\alpha}}\left(1+\frac{\sqrt{2}c(a_k)\varepsilon}{\sigma^2}\right)^{1/2} \tag{24}$$

It follows from (22) that $\varepsilon > \frac{2\sigma M}{\alpha c(a_k)}$ and $\frac{\alpha c(a_k)}{2\sigma}\varepsilon > M \ge 2$.
From inequality (24) under condition (22) we have the following estimate

$$P\left\{\sup_{t\in[0,\infty)} \frac{|Y(t)|}{c(t)} > \varepsilon\right\} \le 4e^{\frac{M+1}{\alpha}}\left(\frac{2}{M}\right)^{\frac{2M}{\alpha}} \exp\left\{-\frac{c(a_0)\varepsilon}{\sqrt{2}\sigma^2}\right\}\times$$

$$\times\sum_{k=0}^{\infty} \exp\left\{-\frac{(c(a_k)-c(a_0))\varepsilon}{\sqrt{2}\sigma^2}\right\}\left(1+\frac{\sqrt{2}c(a_k)\varepsilon}{\sigma^2}\right)^{1/2}. \tag{25}$$

The function $f(\varepsilon) = \exp\left\{-\frac{(c(a_k)-c(a_0))\varepsilon}{\sigma}\right\}\left(1+\frac{c(a_k)\varepsilon}{\sigma}\right)^{1/2}$ monotonically decreases for $\varepsilon > 0$. For this reason under the condition

$$\hat{\varepsilon} \ge \varepsilon^* \tag{26}$$

we have that $\forall \varepsilon > \widehat{\varepsilon}$

$$P\left\{\sup_{t\in[0,\infty)} \frac{|Y(t)|}{c(t)} > \varepsilon\right\} \le 4e^{\frac{M+1}{\alpha}}\left(\frac{2}{M}\right)^{\frac{2M}{\alpha}}\times$$

$$\times\exp\left\{-\frac{c(a_0)\varepsilon}{\sqrt{2}\sigma^2}\right\}\sum_{k=0}^{\infty} \exp\left\{-\frac{(c(a_k)-c(a_0))\hat{\varepsilon}}{\sqrt{2}\sigma^2}\right\}\left(1+\frac{\sqrt{2}c(a_k)\hat{\varepsilon}}{\sigma^2}\right)^{1/2}. \tag{27}$$

□

Corollary 6.2
Let a function $c(t)$ satisfies conditions of Theorem 6.2. Then with probability one for all $t \in \mathbb{R}$ the following inequality is satisfied

$$|Y(t)| \leq \eta \cdot c(t),$$

where $\eta > 0$ is a random variable such that for $\varepsilon > \widehat{\varepsilon}$ the inequality

$$P\{\eta > \varepsilon\} \leq Z(\varepsilon)$$

holds true. For definition of the function $Z(\varepsilon)$ see (18).

Remark 6.2. Condition (17) is satisfied if the series

$$\sum_{k=0}^{\infty} \exp\left\{-\frac{c(a_k)\,\hat{\varepsilon}}{\sqrt{2\sigma^2}}\right\} (c(a_k))^{\frac{2M}{\alpha}+\frac{1}{2}}$$

converges. This series converges, for example, in the case where $c(a_k) = \ln\left(k^d\right), k > 1$, where $\frac{d\widehat{\varepsilon}}{\sqrt{2\sigma^2}} > 1$. A special case is $a_k = k$, that is $c(t) = d\ln(t), d > \frac{\sqrt{2\sigma^2}}{\widehat{\varepsilon}}$ and $t > e$.

7. Conclusions

In the article analysis of properties of proper complex random process is presented. Definitions and some properties of proper stationary random complex process with stable covariance function are given. Estimates of distribution of some functionals from module of stationary Gaussian proper complex random processes are obtained. Behaviour of the module of stationary proper complex random processes at infinity is analysed.

8. Acknowledgements

The authors would like to thank professor Mikhail Moklyachuk for valuable discussions which helped to substantially improve the quality of the paper and an anonymous referee for the constructive propositions.

REFERENCES

1. S. A. Akhmanov, J. E. Deacons, A. S. Chirkin, *Introduction to Statistical Radiophysics and Optics*, M.: Science, (1981). (in Russian)
2. J. L. Doob, *Stochastic Processes*, John Wiley & Sons, (1953).
3. Yu. Kozachenko, V.V.Buldygin, *Metric Characterization of Random Variables and Random processes*, American Mathematical Society, Providence, Rhode Island, (2000).
4. Yu. Kozachenko, *Random processes in Orlicz spaces. I*, Theor. Probability and Math. Statist., vol. 30, pp. 92–107, 1984. (in Ukrainian)
5. Yu. Kozachenko, M.M.Perestyuk O.I.Vasylyk *On uniform convergence of wavelet expansions of -sub-Gaussian random processes*, Random Oper. And Stoch. Equations, vol. 14, no. 3, pp. 209–232, 2006.
6. Yu. Kozachenko, I.Rozora, *Simulation of Gaussian stochastic processes*, Random Oper. And Stoch. Equations, vol. 11, no. 3, pp. 275–296, 2003.
7. Yu. Kozachenko, O. Pogorilyak, I. Rozora, A. Tegza, *Simulation of Stochastic Processes with Given Accuracy and Reliability*, ISTE Press - Elsevier, 2016.
8. Yu. Kozachenko, O.V.Stus, *Square-Gaussian Random Processes and Estimators of Covariance Functions*, Math. Communications, vol. 3, no. 1, pp. 83–94, 1998.
9. Yu. Kozachenko, V.Troshki, *A criterion for testing hypotheses about the covariance function of a stationary Gaussian stochastic proses*, Modern Stochastics Theory and Application, vol. 1, no. 2, pp. 139–149, 2014.
10. Yu. Kozachenko, O. Vasylyk R. Jamnenko, *Upper estimate of overrunning by Sub(Ω) random process the level specified by continuous function*, Math. Communications, vol. 3, no. 1, pp. 83–94, 1998.
11. E. Lukacs, *Characteristic Functions*, New York: Hafner Pub. Co., 1970.
12. F. D. Neeser, and J. L. Massey, *Proper complex random processes with applications to information theory*, IEEE Transactions on Information Theory, vol. 39, issue: 4, pp. 1293 - 1302, 1993.
13. M. Petranova, *Simulation of Gaussian stationary quasi Ornstein–Uhlenbeck process with given reliability and accuracy in spaces C([0,T]) and Lp([0,T])*, Journal of Applied Mathematics and Statistics. - vol. 3, no. 1, pp. 144–58, 2016.
14. P. A. Vatnik, *Statistical methods of operational management of prodaction*, M .: Statistics, (1978). (in Russian)

12

New version of the MDR method for stratified samples

Alexander Bulinski, Alexey Kozhevin *

Faculty of Mathematics and Mechanics, Lomonosov Moscow State University, Moscow, Russia 119991

Abstract The new version of the MDR method of performing identification of relevant factors within a given collection $X_1, \ldots, X_n$ is introduced for stratified samples in the case of binary response variable Y. We establish a criterion of strong consistency of estimates (involving K-cross-validation procedure and penalty) for a specified prediction error function. The cost approach is proposed to compare experiments with random and nonrandom number of observations. Analytic results are accompanied by simulations.

Keywords Feature selection, MDR method, Error function estimation, Cross-validation, Stratified sample, Cost approach

1. Introduction

The research direction combining probability, statistics and machine learning for analysis of mathematical problems of feature selection is vastly represented in literature along with various applications of this theory. Quite a number of powerful methods were developed for different models in the course of such investigations. Several new variable selection procedures have emerged during the last 20 years. One can refer, e.g., to the following books [1] – [3], [13] – [15], [19]. Note that many exhaustive, stochastic and heuristic methods to detect epistasis (in genetics) are considered in [6], [16] and [20].

In the paper by M.Ritchie et al. [18] the *multifactor dimensionality reduction* (MDR) method was proposed to identify the relevant (in a sense) factors having influence on a binary response variable. The review [12] demonstrates great popularity of this method. Some 800 papers published between 2001 and 2014 were devoted to extensions, modifications and applications of the general idea suggested in [18]. The goal of the present paper is to extend the dimensionality reduction of factors to stratified samples framework. Here we generalize the approach developed in [4]–[9] for i.i.d. observations.

Recall some notation. Let $X_1, \ldots, X_n$ be a collection of random features (or factors) and Y be a response variable depending on $X := (X_1, \ldots, X_n)$. We assume that all random elements under consideration are defined on a probability space $(\Omega, \mathcal{F}, \mathsf{P})$. Suppose that X_i takes values in a finite set $\mathbb{X}_i$, $i = 1, \ldots, n$. Thus $X : \Omega \to \mathbb{X}$ where $\mathbb{X} := \mathbb{X}_1 \times \ldots \times \mathbb{X}_n$. We consider Y with values in $\{-1, 1\}$ (one uses also the set $\{0, 1\}$). There are important models in medicine and biology where Y characterizes the health state of a patient (for example $Y = 1$ means the *case*, i.e. a patient is sick, and $Y = -1$ corresponds to the *control*, that is a person is healthy) and X comprises both genetic and nongenetic factors. The challenging problem is to predict the risk of certain complex disease on account of the data X and also to identify the collection $(X_{i_1}, \ldots, X_{i_r})$ of relevant factors ($r < n$) which are responsible for the disease provoking.

*Correspondence to: E-mail: kozhevin.alexey@gmail.com (Alexey Kozhevin); bulinski@yandex.ru (Alexander Bulinski)

For any $f: \mathbb{X} \to \{-1,1\}$ the quality of the forecast of Y by means of $f(X)$ can be expressed by the following *Error function* (see [4], [7])

$$Err(f) = \mathsf{E}|Y - f(X)|\psi(Y)$$

where a *penalty function* $\psi\colon \{-1,1\} \to \mathbb{R}_+$ is introduced to weight the importance of incorrect prediction of different values of Y. The trivial cases $\psi \equiv 0$, $Y \equiv 0$ or $Y \equiv 1$ are excluded. Clearly,

$$Err(f) = 2 \sum_{y \in \{-1,1\}} \psi(y)\mathsf{P}(Y = y, f(X) \neq y). \tag{1}$$

All the *optimal functions* f_{opt}, i.e. f rendering minimum to $Err(f)$, were described in [4]. It is convenient to take the optimal function $f^*(x) = \mathbb{I}\{A\}(x) - \mathbb{I}\{\overline{A}\}(x)$, $x \in \mathbb{X}$, where

$$A = \left\{ x \in M\colon \mathsf{P}(Y = 1|X = x) > \frac{\psi(-1)}{\psi(-1) + \psi(1)} \right\}, \tag{2}$$

$M = \{x \in \mathbb{X}\colon \mathsf{P}(X = x) > 0\}$, $\mathbb{I}\{A\}$ stands for indicator of a set A and $\overline{A}$ means the complement of A. Note that any optimal function f_{opt} gives the same value to Error function as $Err(f^*)$. In [4] it is explained why the choice of

$$\psi(y) = \frac{1}{\mathsf{P}(Y = y)}, \quad y \in \{-1,1\}, \tag{3}$$

considered in [22] is natural.

However the joint law of (X,Y) is unknown, and therefore $Err(f)$ is unknown (f^* is unknown as well). So it is reasonable to apply for inference statistical estimates of $Err(f)$ involving independent identically distributed (i.i.d) random vectors $(X^1,Y^1),\dots,(X^N,Y^N)$, having the same law as (X,Y). In [4]–[9] it is shown how one can use such estimates to identify the collection of relevant factors. We call such method MDR-EFE (*multifactor dimensionality reduction - error function estimation*). Moreover, in [5], [8] and [9] the asymptotic normality of introduced statistics was established and in two latter papers a nonbinary response was studied.

Now we concentrate on the following problem. Assume that the probability (of disease) $\mathsf{P}(Y = 1)$ is rather small. Then in the sample $(X^1,Y^1),\dots,(X^N,Y^N)$ for N not too large we could have too small observations amount with positive response variable value (equal to one) for sound conclusions. We discuss several scenarios to overcome this difficulty. Namely, we provide modifications of some previous results using stratification and also consider the random number of random observations.

The paper is organized as follows. After the brief Introduction (Section 1) in Section 2 we prove an auxiliary result concerning the law of observations having a given response value. Section 3 contains the main result. Here we provide a criterion of strong consistency of statistics which are the estimates of prediction error (of approximation of a response Y by means of a function of factors $X_1,\dots,X_n$) for stratified samples. We employ the cross-validation technique or, more precisely, the method of subsampling and averaging. Along with discussion of established result we consider an example showing its application. Then we demonstrate how one can apply the main result for feature selection. Section 4 is devoted to the cost approach to experiments and the XOR-model (see [23]). Here we compare the MDR-EFE method for i.i.d. observations and the version of this method for stratified samples. The simulation results (Section 5) show that even for rather small samples the proposed version of the MDR-EFE method has visible advantages.

2. Auxiliary result

Let $(X,Y),(X^1,Y^1),(X^2,Y^2),\dots$ be a sequence of i.i.d. random vectors defined on a probability space $(\Omega,\mathcal{F},\mathsf{P})$. For each $\omega \in \Omega$ we consider a sequence $Y^1(\omega),Y^2(\omega),\dots$ and pick all the indices $1 \leqslant j^1_{-1}(\omega) < j^2_{-1}(\omega) < \dots$ for which $Y^{j^k_{-1}(\omega)}(\omega) = -1$, $k \in \mathbb{N}$. In a similar way we will write all observations $Y^i(\omega)$ with values 1 as $\{Y^{j^m_1(\omega)}(\omega)\}_{m \in \mathbb{N}}$ where $1 \leqslant j^1_1(\omega) < j^2_1(\omega) < \dots$. Recall that for the Bernoulli trials with probability of success p the Negative binomial random variable $U_{r,p}$ is introduced as the number of successes needed to get r failures where $r \in \mathbb{N}$ (one writes $U_{r,p} \sim NB(r,p)$). Thus

$$\mathsf{P}(U_{r,p} = k) = \binom{k+r-1}{k} p^k (1-p)^r, \quad k = 0, 1, \ldots.$$

If we consider the events $\{Y^i = 1\}$ and $\{Y^i = -1\}$ as a success and failure, respectively (with probability $p = \mathsf{P}(Y = 1)$ of success), then j^r_{-1} has the same law as $U_{r,p} + r$. Therefore

$$\mathsf{P}(j^r_{-1} = m) = \begin{cases} \binom{m-1}{m-r} p^{m-r}(1-p)^r, & m = r, r+1, \ldots, \\ 0, & m = 1, \ldots, r-1. \end{cases} \tag{4}$$

For $r = 1$ we keep only the first line in (4) as in this case $\{1, \ldots, r-1\} = \varnothing$. By similar reasons j^r_1 is distributed as $U_{r,1-p} + r$ where $U_{r,1-p} \sim NB(r, 1-p)$. In other words j^r_1 has the same law as $G^1_p + \ldots + G^r_p$ where $G^1_p, \ldots, G^r_p$ are independent random variables having the Geometric law with parameter p (i.e. $\mathsf{P}(G^1_p = k) = p(1-p)^{k-1}$, $k = 1, 2, \ldots$). Thus $\mathsf{E} j^r_1 = \frac{r}{p}$ and $j^r_1 < \infty$ a.s. for any $r \in \mathbb{N}$ (analogously $\mathsf{E} j^r_{-1} = \frac{r}{1-p}$ for each $r \in \mathbb{N}$). Note also that one can find different definitions of Negative binomial and Geometric laws, that is why we provided the explicit formulae.

Set $Z^k := X^{j^k_1}$ for each $k \in \mathbb{N}$. Introduce a collection $\mathcal{B}$ of all subsets of $\mathbb{X}$. We use the following simple result.

Lemma 1

For each $m \in \mathbb{N}$, the random variables $Z^1, \ldots, Z^m$ are independent and distributed as X given $Y = 1$ (we write $X|Y = 1$), i.e., for any $B \in \mathcal{B}$ and $k = 1, \ldots, m$,

$$\mathsf{P}(Z^k \in B) = \mathsf{P}(X \in B | Y = 1).$$

Proof. First of all we show that, for all $B_i \in \mathcal{B}$, $i = 1, \ldots, m$,

$$\mathsf{P}(Z^1 \in B_1, \ldots, Z^m \in B_m) = \mathsf{P}(Z^1 \in B_1) \ldots \mathsf{P}(Z^m \in B_m). \tag{5}$$

By the total probability formula

$$\mathsf{P}(Z^1 \in B_1, \ldots, Z^m \in B_m) = \mathsf{P}(X^{j^1_1} \in B_1, \ldots, X^{j^m_1} \in B_m)$$
$$= \sum_{\substack{(k_1,\ldots,k_m) \in \mathbb{N}^m: \\ k_1 < \cdots < k_m}} \mathsf{P}(X^{j^1_1} \in B_1, \ldots, X^{j^m_1} \in B_m, j^1_1 = k_1, \ldots, j^m_1 = k_m).$$

Note that for arbitrary positive integers $k_1 < k_2 < \cdots < k_m$

$$\{j^1_1 = k_1, \ldots, j^m_1 = k_m\} = \bigcap_{i=1}^m \{Y^{k_i} = 1\} \cap \bigcap_{r \in T_m} \{Y^r = -1\},$$

where $T_m = T_m(k_1, k_2, \ldots, k_m) := \{1, \ldots, k_m\} \setminus \{k_1, \ldots, k_m\}$. Next, due to independence of the random vectors $(X^1, Y^1), (X^2, Y^2), \ldots$, one has

$$\mathsf{P}(Z^1 \in B_1, \ldots, Z^m \in B_m)$$
$$= \sum_{\substack{(k_1,\ldots,k_m) \in \mathbb{N}^m: \\ k_1 < \cdots < k_m}} \mathsf{P}\left(\{X^{k_1} \in B_1, \ldots, X^{k_m} \in B_m, Y^{k_1} = 1, \ldots, Y^{k_m} = 1\} \cap \left\{\bigcap_{r \in T_m} \{Y^r = -1\}\right\}\right)$$
$$= \sum_{\substack{(k_1,\ldots,k_m) \in \mathbb{N}^m: \\ k_1 < \cdots < k_m}} \mathsf{P}(X^{k_1} \in B_1, Y^{k_1} = 1) \ldots \mathsf{P}(X^{k_m} \in B_m, Y^{k_m} = 1) \mathsf{P}\left(\bigcap_{r \in T_m} \{Y^r = -1\}\right)$$
$$= \sum_{\substack{(k_1,\ldots,k_m) \in \mathbb{N}^m: \\ k_1 < \cdots < k_m}} \prod_{q=1}^m \frac{\mathsf{P}(X^{k_q} \in B_q, Y^{k_q} = 1)}{\mathsf{P}(Y^{k_q} = 1)} \mathsf{P}\left(\bigcap_{i=1}^m \{Y^{k_i} = 1\} \cap \bigcap_{r \in T_m} \{Y^r = -1\}\right)$$

$$= \prod_{q=1}^{m} \mathsf{P}(X \in B_q|Y=1) \sum_{\substack{(k_1,\dots,k_m)\in\mathbb{N}^m:\\ k_1<\dots<k_m}} \mathsf{P}(j_1^1 = k_1,\dots,j_1^m = k_m) = \prod_{q=1}^{m} \mathsf{P}(X \in B_q|Y=1). \quad (6)$$

If we take $B_1 = \dots = B_{k-1} = B_{k+1} = \dots = B_m = \mathbb{X}$ in (6) then

$$\mathsf{P}(Z^k \in B_k) = \mathsf{P}(X \in B_k|Y=1), \quad k=1,\dots,m. \quad (7)$$

Hence, in view of (7) the random variables $Z^1,\dots,Z^m$ are identically distributed and relation (5) holds. □

Remark 1. Lemma 1 is also valid for i.i.d. vectors $(X,Y),(X^1,Y^1),\dots,(X^N,Y^N)$ when X takes values in any space S endowed with some σ-algebra $\mathcal{B}$ (X is measurable w.r.t. σ-algebras $\mathcal{F}$ and $\mathcal{B}$). In the same way one can prove that $X^{j^1_{-1}}, X^{j^2_{-1}},\dots$ are independent and identically distributed as $X|Y=-1$.

3. The main result and discussion

Let $(X^1,Y^1),(X^2,Y^2),\dots$ be i.i.d. observations having the same law as (X,Y). Assume that we have a possibility to form a sample with given nonrandom numbers of cases and controls (N_1 and N_{-1} respectively). More exactly, we take

$$\zeta^1_{N_1} := \{(X^{j_1^1},1),\dots,(X^{j_1^{N_1}},1)\}, \quad \zeta^{-1}_{N_{-1}} := \{(X^{j_{-1}^1},-1),\dots,(X^{j_{-1}^{N_{-1}}},-1)\}$$

where the random indices j_1^k and j_{-1}^k, $k \in \mathbb{N}$, were introduced in Section 2. Let $N := N_1 + N_{-1}$ be the size of our *stratified sample* $\zeta_N := \zeta^1_{N_1} \cup \zeta^{-1}_{N_{-1}}$. In contrast to the i.i.d. sample $\xi_N := \{(X^1,Y^1),\dots,(X^N,Y^N)\}$ taken from the population with $law(X,Y)$ we have (according to Lemma 1) two subsamples $\zeta^1_{N_1}$ and $\zeta^{-1}_{N_{-1}}$ with laws $X|Y=1$ and $X|Y=-1$, respectively. Thus one cannot use the frequency estimates (e.g., of $\mathsf{P}(Y=1)$ or $\mathsf{P}(X\in B, Y=1)$ where $B \subset \mathbb{X}$) constructed by means of ζ_N. Further on we assume that $N_1 = \max\{[aN],1\}$ and $N_{-1} = N - N_1$, here the parameter $a \in (0,1)$, $N \in \mathbb{N}$ and $[\cdot]$ stands for the integer part of a number. Suppose that there are some estimators $\widehat{\mathsf{P}}^y_N$ of $\mathsf{P}(Y=y)$ such that

$$\widehat{\mathsf{P}}^y_N \to \mathsf{P}(Y=y) \text{ a.s.}, N\to\infty, \quad y \in \{-1,1\}. \quad (8)$$

For instance we can assume that $\widehat{\mathsf{P}}^y_N$ involve the data $\{(X^k,Y^k), 1 \leqslant k \leqslant \max\{j_1^{N_1}, j_{-1}^{N_{-1}}\}\}$. In this case the frequency estimates of the probabilities $\mathsf{P}(Y=1)$ or $\mathsf{P}(X \in B, Y=1)$, where $B \subset \mathbb{X}$, mentioned above are strongly consistent since $\max\{j_1^{N_1}, j_{-1}^{N_{-1}}\} \to \infty$ a.s. when $N\to\infty$. In Section 4 of the paper we discuss the advantages and disadvantages of employing the stratified samples. Introduce the vector $\widehat{\mathsf{P}}_N := (\widehat{\mathsf{P}}^{-1}_N, \widehat{\mathsf{P}}^1_N)$. Recall that we exclude the trivial cases $\mathsf{P}(Y=-1)=0$ or $\mathsf{P}(Y=1)=0$.

Let $f_{PA}(x,\zeta_N,\widehat{\mathsf{P}}_N)$ be a function defining a *prediction algorithm* i.e. a function with values in $\{-1,1\}$ constructed by means of $x \in \mathbb{X}$, the sample ζ_N and $\widehat{\mathsf{P}}_N$. In fact we consider a family of functions when instead of ζ_N we use its subsamples. Thus we write $f_{PA}(x,\zeta_N(S),\widehat{\mathsf{P}}_N)$ for $\zeta_N(S) := \{(X^j,Y^j), j \in S\}$, $S \subset (\{j_1^1,\dots,j_1^{N_1}\} \cup \{j_{-1}^1,\dots,j_{-1}^{N_{-1}}\})$. For any $f\colon \mathbb{X} \to \{-1,1\}$ we try to find its estimate f_{PA} which is close in a sense to f and we employ f_{PA} to estimate $Err(f)$. For this purpose we also apply *K-fold cross-validation* procedure or, more precisely, subsampling approach. At first, for some fixed $K \in \mathbb{N}$ and each $y \in \{-1,1\}$, we consider a partition of the set $\{j_y^1,\dots,j_y^{N_y}\}$ into K subsets $S_k^y(N_y,\omega)$, $k=1,\dots,K$, as follows

$$S_k^y(N_y,\omega) := \left\{ j_y^i(\omega)\colon i \in \left\{ (k-1)\left[\frac{N_y}{K}\right]+1,\dots,k\left[\frac{N_y}{K}\right]\mathbb{I}\{k<K\} + N_y\mathbb{I}\{k=K\} \right\} \right\}. \quad (9)$$

Finally, we construct $S_k(N,\omega) = S_k^1(N_1,\omega) \cup S_k^{-1}(N_{-1},\omega)$ and introduce

$$\widehat{Err}_K(f_{PA},\zeta_N,\widehat{\mathsf{P}}_N) := \frac{2}{K} \sum_{y\in\{-1,1\}} \sum_{k=1}^{K} \sum_{j\in S_k^y(N_y)} \frac{\widehat{\psi}(y,\zeta_N(\overline{S_k(N)}),\widehat{\mathsf{P}}_N)\mathbb{I}\{f^j_{PA}(N,k)\neq y\}\widehat{\mathsf{P}}^y_N}{\sharp S_k^y(N_y)} \quad (10)$$

where $f^j_{PA}(N,k) := f_{PA}(X^j, \zeta_N(\overline{S_k(N)}), \widehat{\mathsf{P}}_N)$ and $\widehat{\psi}(y, \zeta_N(\overline{S_k(N)}), \widehat{\mathsf{P}}_N)$ is an estimate of $\psi(y)$, $y \in \{-1,1\}$, constructed by observations $\zeta_N(\overline{S_k(N)})$ and $\widehat{\mathsf{P}}_N$; $\overline{S_k(N)} = \{j^1_1, \dots, j^{N_1}_1\} \cup \{j^1_{-1}, \dots, j^{N_{-1}}_{-1}\} \setminus S_k(N)$ and $\sharp$ stands for the cardinality of a finite set.

We will suppose that, for each $k = 1, \dots, N$,

$$\widehat{\psi}(y, \zeta_N(\overline{S_k(N)}), \widehat{\mathsf{P}}_N) \to \psi(y) \text{ a.s., } N \to \infty, \quad y \in \{-1,1\}. \tag{11}$$

Introduce $L(x) = \psi(1)\mathsf{P}(X = x, Y = 1) - \psi(-1)\mathsf{P}(X = x, Y = -1)$, $x \in \mathbb{X}$. The following result is an analogue of Theorem 1 [4] for stratified samples.

Theorem 1

Let ζ_N be a sample described above, ψ be a penalty function, $f\colon \mathbb{X} \to \{-1,1\}$ be an arbitrary function and f_{PA} define a prediction algorithm. Assume that (11) is valid and there exists a non-empty set $U \subset \mathbb{X}$ such that, for each $x \in U$ and $k = 1, \dots, K$, relation

$$f_{PA}(x, \zeta_N(\overline{S_k(N)}), \widehat{\mathsf{P}}_N) \to f(x) \text{ a.s., } N \to \infty, \tag{12}$$

holds. Then, for each $a \in (0,1)$ (with $N_1 = \max\{[aN], 1\}$, $N_{-1} = N - N_1$),

$$\widehat{Err}_K(f_{PA}, \zeta_N, \widehat{\mathsf{P}}_N) \to Err(f) \text{ a.s., } N \to \infty, \tag{13}$$

if and only if

$$\sum_{k=1}^{K} \sum_{y \in \{-1,1\}} \sum_{x \in \mathbb{X}_y} y\mathbb{I}\{f_{PA}(x, \zeta_N(\overline{S_k(N)}), \widehat{\mathsf{P}}_N) = -y\} L(x) \to 0 \text{ a.s., } \quad N \to \infty, \tag{14}$$

where

$$\mathbb{X}_y = (\mathbb{X} \setminus U) \cap \{x \in \mathbb{X} : f(x) = y\}, \quad y \in \{-1,1\}. \tag{15}$$

Proof. It suffices to consider relation

$$\frac{2}{K} \sum_{k=1}^{K} \sum_{y \in \{-1,1\}} \psi(y) \sum_{j \in S^y_k(N_y)} \frac{\mathbb{I}\{f^j_{PA}(N,k) \neq y\} \mathsf{P}(Y = y)}{\sharp S^y_k(N_y)} \to Err(f) \text{ a.s., } N \to \infty, \tag{16}$$

since the difference between $\widehat{Err}_K(f_{PA}, \zeta_N, \widehat{\mathsf{P}}_N)$ and the left-hand side of (16) tends to zero a.s., $N \to \infty$. Namely, we take into account (8), (11) and the inequality

$$\frac{1}{\sharp S^y_k(N_y)} \sum_{j \in S^y_k(N_y)} \mathbb{I}\{f^j_{PA}(N,k) \neq y\} \leqslant 1$$

which is valid for each $y \in \{-1,1\}$, any $k = 1, \dots, K$ and all N.

Now we note that in view of (1) relation (13) is equivalent to the following one

$$\frac{2}{K} \sum_{k=1}^{K} \sum_{y \in \{-1,1\}} \psi(y) \left(\sum_{j \in S^y_k(N_y)} \frac{\mathbb{I}\{f^j_{PA}(N,k) \neq y\} \mathsf{P}(Y = y)}{\sharp S^y_k(N_y)} - \mathsf{P}(f(X) \neq y, Y = y) \right) \to 0 \ a.s.$$

when $N \to \infty$.

For any $y \in \{-1,1\}$ and $m \in \mathbb{N}$, set $\widetilde{X}^{j^m_y} = \mathbb{I}\{f(X^{j^m_y}) \neq y\} - \mathsf{P}(f(X^{j^m_y}) \neq y)$. Then $|\widetilde{X}^{j^m_y}| \leqslant 2$, $\mathsf{E}\widetilde{X}^{j^m_y} = 0$ for such y and m. Fix any $y \in \{-1,1\}$ and $k \in \{1, \dots, K\}$. Consider an array $\mathcal{A}(y,k) := \{\widetilde{X}^{j^m_y}, j^m_y \in S^y_k(N_y), N_y = N_y(N)\}$ where $N \in \mathbb{N}$. The strong law of large numbers for arrays (SLLNA) given in Theorem 2.1 [21] applies, e.g., when $p = 2$, $\psi(x) = |x|^{5/2}$, $x \in \mathbb{R}$, and $k \in \mathbb{N}$ (we keep here the notation p, ψ and k used in the mentioned paper [21] for other objects). Whence one has

$$\frac{1}{\sharp S_k^y(N_y)} \sum_{j \in S_k^y(N_y)} \mathbb{I}\{f(X^j) \neq y\} \to \mathsf{P}(f(X^{j_y^1}) \neq y) \text{ a.s., } N \to \infty. \tag{17}$$

More precisely, in [21] a triangular array was considered. The study of our array $\mathcal{A}(y,k)$ can be reduced to analysis of a collection of several triangular arrays. Indeed, in view of (9) the contribution to asymptotic behavior of $\frac{1}{\sharp S_K^y(N_y)} \sum_{j \in S_K^y(N_y)} \mathbb{I}\{f(X^j) \neq y\}$ of the last summands appearing in $S_K^y(N_y)$ with indices greater than $[N_y/K]K$ is negligible. Therefore, for each k belonging to $\{1,\dots,K\}$ we have an array $\widetilde{\mathcal{A}}(y,k)$ with rows containing $r_1, r_2, \dots$ elements ($(r_i)_{i \in \mathbb{N}}$ depends on $(N_y(N))_{N \in \mathbb{N}}$) such that $r_1 \leqslant r_2 \leqslant \dots$ ($\widetilde{\mathcal{A}}(y,k) = \mathcal{A}(y,k)$ for $k = 1,\dots,K-1$). Moreover, if $r_m = \dots = r_q$ then $q - m \leqslant I$ where $I = [Ka]$. Now we can consider separately each of I arrays (taking one row among the I rows of equal length) with strictly increasing numbers of elements in rows. Then we introduce auxiliary rows, if necessary, with elements $\mathsf{P}(f(X^{j_y^1}) \neq y)$ to get a triangular array and apply the mentioned SLLNA. Lemma 1 shows that $\mathsf{P}(f(X^{j_y^1}) \neq y) = \mathsf{P}(f(X) \neq y | Y = y)$. Hence

$$\frac{2}{K} \sum_{k=1}^{K} \sum_{y \in \{-1,1\}} \psi(y) \sum_{j \in S_k^y(N_y)} \frac{\mathbb{I}\{f(X^j) \neq y\} \mathsf{P}(Y = y)}{\sharp S_k^y(N_y)} \to Err(f) \text{ a.s., } N \to \infty.$$

Therefore, relation (16) is valid if and only if

$$\sum_{k=1}^{K} \sum_{y \in \{-1,1\}} \psi(y) \frac{1}{\sharp S_k^y(N_y)} \sum_{j \in S_k^y(N_y)} Z_{N,k}^j(y) \to 0 \text{ a.s., } N \to \infty,$$

where

$$Z_{N,k}^j(y) = (\mathbb{I}\{f_{PA}(X^j, \zeta_N(\overline{S_k(N)}), \widehat{\mathsf{P}}_N) \neq y\} - \mathbb{I}\{f(X^j) \neq y\}) \mathsf{P}(Y = y), \quad y \in \{-1,1\}.$$

Set $F_{N,k}(x,y) := \mathbb{I}\{f_{PA}(x, \zeta_N(\overline{S_k(N)}), \widehat{\mathsf{P}}_N) \neq y\} - \mathbb{I}\{f(x) \neq y\}$. For each $y \in \{-1,1\}$, $N \in \mathbb{N}$ and $k = 1,\dots,K$, we introduce the random variables

$$Q_{N,k,y} = \frac{1}{\sharp S_k^y(N_y)} \sum_{j \in S_k^y(N_y)} F_{N,k}(X^j, y).$$

Then (16) is equivalent to the following relation

$$\sum_{k=1}^{K} \sum_{y \in \{-1,1\}} \psi(y) \mathsf{P}(Y = y) Q_{N,k,y} \to 0 \text{ a.s., } N \to \infty.$$

Note that $Q_{N,k,y} = Q_{N,k,y}^{(1)} + Q_{N,k,y}^{(2)}$ where

$$Q_{N,k,y}^{(1)} = \frac{1}{\sharp S_k^y(N_y)} \sum_{j \in S_k^y(N_y)} \mathbb{I}\{X^j \in U\} F_{N,k}(X^j, y),$$

$$Q_{N,k,y}^{(2)} = \frac{1}{\sharp S_k^y(N_y)} \sum_{j \in S_k^y(N_y)} \mathbb{I}\{X^j \notin U\} F_{N,k}(X^j, y).$$

We have

$$|Q_{N,k,y}^{(1)}| \leqslant \sum_{x \in U} \frac{1}{\sharp S_k^y(N_y)} \sum_{j \in S_k^y(N_y)} |\mathbb{I}\{f_{PA}(x, \zeta_N(\overline{S_k(N)}), \widehat{\mathsf{P}}_N) \neq y\} - \mathbb{I}\{f(x) \neq y\}|.$$

Condition (12) entails

$$\sum_{k=1}^{K}\sum_{y\in\{-1,1\}}\psi(y)\mathsf{P}(Y=y)Q^{(1)}_{N,k,y}\to 0 \text{ a.s., } N\to\infty.$$

If $U=\mathbb{X}$, then $Q^{(2)}_{N,k,y}=0$ for all N,k,y and (16) holds. Let $U\neq\mathbb{X}$. Then

$$V_{N,k}:=\sum_{y\in\{-1,1\}}\psi(y)\mathsf{P}(Y=y)Q^{(2)}_{N,k,y}=\sum_{x\in\mathbb{X}_{-1}\cup\mathbb{X}_1}\sum_{y\in\{-1,1\}}G_{N,k,y}(x)$$

where $N\in\mathbb{N}$, $k=1,\dots,K$, the sets $\mathbb{X}_y$ were introduced in (15) and

$$G_{N,k,y}(x):=\frac{\psi(y)\mathsf{P}(Y=y)}{\sharp S^y_k(N_y)}\sum_{j\in S^y_k(N_y)}\mathbb{I}\{X^j=x\}\Big(\mathbb{I}\{f_{PA}(x,\zeta_N(\overline{S_k(N)}),\widehat{\mathsf{P}}_N)\neq y\}-\mathbb{I}\{f(x)\neq y\}\Big).$$

Let $R^y_j(x):=\mathbb{I}\{X^j=x\}\psi(y)\mathsf{P}(Y=y)$, $y\in\{-1,1\}$, $j\in\mathbb{N}$. The following equalities are valid

$$\sum_{x\in\mathbb{X}_1}\sum_{y\in\{-1,1\}}G_{N,k,y}(x)=\sum_{x\in\mathbb{X}_1}\mathbb{I}\{f_{PA}(x,\zeta_N(\overline{S_k(N)}),\widehat{\mathsf{P}}_N)=-1\}\sum_{y\in\{-1,1\}}\frac{y}{\sharp S^y_k(N_y)}\sum_{j\in S^y_k(N_y)}R^y_j(x),$$

$$\sum_{x\in\mathbb{X}_{-1}}\sum_{y\in\{-1,1\}}G_{N,k,y}(x)=-\sum_{x\in\mathbb{X}_{-1}}\mathbb{I}\{f_{PA}(x,\zeta_N(\overline{S_k(N)}),\widehat{\mathsf{P}}_N)=1\}\sum_{y\in\{-1,1\}}\frac{y}{\sharp S^y_k(N_y)}\sum_{j\in S^y_k(N_y)}R^y_j(x).$$

Similarly to (17) it can be shown that, for each $y\in\{-1,1\}$, $k=1,\dots,K$ and $x\in\mathbb{X}$, relation

$$\frac{1}{\sharp S^y_k(N_y)}\sum_{j\in S^y_k(N_y)}R^y_j(x)\to J(x,y) \text{ a.s., } N\to\infty,$$

holds where $J(x,y)=\psi(y)\mathsf{P}(X=x,Y=y)$.

Thus, for any $\varepsilon>0$, $x\in\mathbb{X}$, $y\in\{-1,1\}$ and almost all $\omega\in\Omega$, there exists $N_1(\omega,\varepsilon,k,y)$ such that

$$\left|\frac{1}{\sharp S^y_k(N_y)}\sum_{j\in S^y_k(N_y)}\left(R^y_j(x)-J(x,y)\right)\right|<\varepsilon \tag{18}$$

for $N>N_1(\omega,\varepsilon,k,y)$. We note that $L(x)=\sum_{y\in\{-1,1\}}yJ(x,y)$. Furthermore,

$$V_{N,k}=\sum_{x\in\mathbb{X}_1}\mathbb{I}\{f_{PA}(x,\zeta_N(\overline{S_k(N)}),\widehat{\mathsf{P}}_N)=-1\}L(x)$$

$$+\sum_{x\in\mathbb{X}_1}\mathbb{I}\{f_{PA}(x,\zeta_N(\overline{S_k(N)}),\widehat{\mathsf{P}}_N)=-1\}\sum_{y\in\{-1,1\}}\frac{1}{\sharp S^y_k(N_y)}\sum_{j\in S^y_k(N_y)}y(R^y_j(x)-J(x,y))$$

$$-\sum_{x\in\mathbb{X}_{-1}}\mathbb{I}\{f_{PA}(x,\zeta_N(\overline{S_k(N)}),\widehat{\mathsf{P}}_N)=1\}L(x)$$

$$-\sum_{x\in\mathbb{X}_{-1}}\mathbb{I}\{f_{PA}(x,\zeta_N(\overline{S_k(N)}),\widehat{\mathsf{P}}_N)=1\}\sum_{y\in\{-1,1\}}\frac{1}{\sharp S^y_k(N_y)}\sum_{j\in S^y_k(N_y)}y(R^y_j(x)-J(x,y)).$$

Taking into account (18), for any $\varepsilon > 0$, we obtain the inequality

$$\Bigg|\sum_{k=1}^{K}\Bigg(\sum_{x\in\mathbb{X}_1}\mathbb{I}\{f_{PA}(x,\zeta_N(\overline{S_k(N)}),\widehat{\mathsf{P}}_N)=-1\}\sum_{y\in\{-1,1\}}\frac{1}{\sharp S_k^y(N_y)}\sum_{j\in S_k^y(N_y)}y(R_j^y(x)-J(x,y))$$
$$-\sum_{x\in\mathbb{X}_{-1}}\mathbb{I}\{f_{PA}(x,\zeta_N(\overline{S_k(N)}),\widehat{\mathsf{P}}_N)=1\}\sum_{y\in\{-1,1\}}\frac{1}{\sharp S_k^y(N_y)}\sum_{j\in S_k^y(N_y)}y(R_j^y(x)-J(x,y))\Bigg)\Bigg| < 2K\varepsilon\sharp\mathbb{X}$$

when $N > N_1(\omega,\varepsilon) := \max\limits_{k=1,\dots,K,y\in\{-1,1\}} N_1(\omega,\varepsilon,k,y)$. Thus, $\sum\limits_{k=1}^{K} V_{N,k}\to 0$ a.s., as $N\to\infty$, if and only if condition (14) holds. □

Remark 2. Theorem 1 demonstrates what one has to verify outside the "good set" U (where f_{PA} approximates f point-wise) to guarantee validity of (13). To clarify (14) we can rewrite it (cf. Theorem 1 [9]) in the following way

$$\sum_{k=1}^{K}\sum_{y\in\{-1,1\}}\sum_{x\in\mathbb{X}_y} y\mathbb{I}\{f_{PA}(x,\zeta_N(\overline{S_k(N)}),\widehat{\mathsf{P}}_N)\neq f(x)\}L(x)\to 0 \text{ a.s.},\quad N\to\infty.$$

Note also that condition (14) coincides with the corresponding one proposed for the case of i.i.d. observations without stratification (cf. Theorem 1 [4]).

Remark 3. According to Corollary 1 [4] the choice of

$$U=\left\{x\in M\colon \mathsf{P}(Y=1|X=x)\neq\frac{\psi(-1)}{\psi(-1)+\psi(1)}\right\},\tag{19}$$

where M appears in (2), implies that (14) is satisfied and the problem is to prove that (12) holds on this U for f under consideration and appropriate f_{PA}.

Remark 4. There are various ways to get strongly consistent estimates $\widehat{\mathsf{P}}_N^y$, $y\in\{-1,1\}$. The first one is to estimate $\mathsf{P}(Y=y)$, $y\in\{-1,1\}$, by means of another sample. In this case it is essential to have the samples belonging to the same law. The second approach involves estimates of $\mathsf{P}(Y=y)$ constructed simultaneously with forming of the sample ζ_N. More exactly, $\widehat{\mathsf{P}}_N^y$ is a frequency estimate of the $\mathsf{P}(Y=y)$ by means of a random number of observations $Y^1,Y^2,\dots,Y^{\widetilde{N}}$ where $\widetilde{N}=\max\{j_1^{N_1},j_{-1}^{N_{-1}}\}$. Such estimate is strongly consistent as $N\to\infty$ since $\widetilde{N}\geqslant N$ a.s. We have mentioned in Section 1 that the choice of ψ given by (3) is natural. Moreover, for such ψ we can simplify (10). Namely, now it need not contain $\widehat{\psi}(y,\zeta_N(\overline{S_k(N)}),\widehat{\mathsf{P}}_N)$ and $\widehat{\mathsf{P}}_N^y$.

The estimates of $\mathsf{P}(X\in B,Y=y)$ where $B\subset\mathbb{X}$ and $y\in\{-1,1\}$ are required if $\psi(y)$ depends on the joint distribution of X and Y. For this estimation we cannot employ the sample ζ_N. However the Bayes formula can help to construct the desired estimates. Such approach in the framework of classification problems was employed by J.Park in [17].

Example. We show how for the stratified sample ζ_N and an optimal function f^* one can find f_{PA} to employ the result described by Theorem 1.

For $B\subset\mathbb{X}$ the Bayes theorem yields

$$\mathsf{P}(Y=1|X\in B)=\frac{\mathsf{P}(X\in B|Y=1)\mathsf{P}(Y=1)}{\mathsf{P}(X\in B|Y=1)\mathsf{P}(Y=1)+\mathsf{P}(X\in B|Y=-1)\mathsf{P}(Y=-1)}.$$

Therefore, one can construct the estimates of $\mathsf{P}(Y=1|X\in B)$ by means of ζ_N and $\widehat{\mathsf{P}}_N$. Plug-in principle suggests us how to get the appropriate f_{PA} for f^*. According to this principle we construct estimates of $\mathsf{P}(Y=1|X=x)$ and $\frac{\psi(-1)}{\psi(-1)+\psi(1)}$. Then we substitute them into the definition of A introduced by (2).

For $y \in \{-1, 1\}$, $k = 1, \ldots, K$ and $N \in \mathbb{N}$, set $W_k^y(N_y) := \cup_{m \neq k} S_m^y(N_y)$ and assume that (11) holds. Consider the following estimates for $\mathsf{P}(Y = 1|X = x)$ and $\frac{\psi(-1)}{\psi(-1)+\psi(1)}$, respectively

$$g(x, \zeta_N(\overline{S_k(N)}), \widehat{\mathsf{P}}_N) := \frac{\widehat{\mathsf{P}}_N^1 I_1(x, \zeta_N(\overline{S_k(N)}), \widehat{\mathsf{P}}_N)}{\widehat{\mathsf{P}}_N^{-1} I_{-1}(x, \zeta_N(\overline{S_k(N)}), \widehat{\mathsf{P}}_N) + \widehat{\mathsf{P}}_N^1 I_1(x, \zeta_N(\overline{S_k(N)}), \widehat{\mathsf{P}}_N)},$$

$$h(\zeta_N(\overline{S_k(N)}), \widehat{\mathsf{P}}_N) := \frac{\widehat{\psi}(-1, \zeta_N(\overline{S_k(N)}), \widehat{\mathsf{P}}_N)}{\widehat{\psi}(-1, \zeta_N(\overline{S_k(N)}), \widehat{\mathsf{P}}_N) + \widehat{\psi}(1, \zeta_N(\overline{S_k(N)}), \widehat{\mathsf{P}}_N)} \tag{20}$$

where

$$I_y(x, \zeta_N(\overline{S_k(N)}), \widehat{\mathsf{P}}_N) := \frac{1}{\sharp W_k^y(N_y)} \sum_{j \in W_k^y(N_y)} \mathbb{I}(X^j = x), \quad y \in \{-1, 1\}, \quad x \in \mathbb{X}.$$

As usual we set formally $0/0 := 0$. Let us define

$$A_N = \{x \in \mathbb{X} \colon g(x, \zeta_N(\overline{S_k(N)}), \widehat{\mathsf{P}}_N) > h(\zeta_N(\overline{S_k(N)}), \widehat{\mathsf{P}}_N)\}$$

and

$$f_{PA}(x, \zeta_N(\overline{S_k(N)}), \widehat{\mathsf{P}}_N) := \mathbb{I}\{A_N\}(x) - \mathbb{I}\{\overline{A_N}\}(x). \tag{21}$$

It can be established similarly to the proof of (17) that for $y \in \{-1, 1\}$ and $x \in \mathbb{X}$

$$I_y(x, \zeta_N(\overline{S_k(N)}), \widehat{\mathsf{P}}_N) \to \mathsf{P}(X = x|Y = y) \text{ a.s., } N \to \infty.$$

By virtue of (8) and (11) we come to the following relations

$$g(x, \zeta_N(\overline{S_k(N)}), \widehat{\mathsf{P}}_N) \to \mathsf{P}(Y = 1|X = x) \text{ a.s., } N \to \infty, g$$

$$h(\zeta_N(\overline{S_k(N)}), \widehat{\mathsf{P}}_N) \to \frac{\psi(-1)}{\psi(-1) + \psi(1)} \text{ a.s., } N \to \infty.$$

For any $x \in A$ one has $\mathsf{P}(Y = 1|X = x) > \frac{\psi(-1)}{\psi(-1)+\psi(1)}$. Therefore, for every $\varepsilon > 0$ such that

$$\mathsf{P}(Y = 1|X = x) - \varepsilon > \frac{\psi(-1)}{\psi(-1) + \psi(1)} + \varepsilon,$$

for almost all $\omega \in \Omega$ and any $x \in A$ there exist $N_2(\omega)$ and $N_3(\omega)$ such that

$$g(x, \zeta_N(\overline{S_k(N)}), \widehat{\mathsf{P}}_N) > \mathsf{P}(Y = 1|X = x) - \varepsilon$$

for each $N > N_2(\omega)$ and

$$h(\zeta_N(\overline{S_k(N)}), \widehat{\mathsf{P}}_N) < \frac{\psi(-1)}{\psi(-1) + \psi(1)} + \varepsilon$$

for each $N > N_3(\omega)$. Thus, for almost all $\omega \in \Omega$, any $x \in A$ and $N > \max\{N_2(\omega), N_3(\omega)\}$ the equalities $f(x) = 1$ and $f_{PA}(x, \zeta_N(\overline{S_k(N)}), \widehat{\mathsf{P}}_N) = 1$ are true. Similarly, for

$$x \in U \cap \overline{A} = \left\{ x \in M \colon \mathsf{P}(Y = 1|X = x) < \frac{\psi(-1)}{\psi(-1) + \psi(1)} \right\}$$

it can be shown that for almost all $\omega \in \Omega$ there exists such $N_4(\omega) \in \mathbb{N}$ that, for $N > N_4(\omega)$, the equalities $f(x) = -1$ and $f_{PA}(x, \zeta_N(\overline{S_k(N)}), \widehat{\mathsf{P}_N}) = -1$ are satisfied. Thus condition (12) holds for U introduced by formula (19). Hence, the estimate of $Err(f^*)$ appearing in Theorem 1 is strongly consistent if $f_{PA}(x, \zeta_N(\overline{S_k(N)}), \widehat{\mathsf{P}}_N)$ is defined by (21). The desired f_{PA} is constructed. □

If ψ is introduced by (3) then U and A have the following form

$$U = \{x \in M\colon \mathsf{P}(Y=1|X=x) \neq \mathsf{P}(Y=1)\},$$
$$A = \{x \in M\colon \mathsf{P}(Y=1|X=x) > \mathsf{P}(Y=1)\}.$$

Consequently, instead of (20) we can set $h(\zeta_N(\overline{S_k(N)}), \widehat{\mathsf{P}}_N) := \widehat{\mathsf{P}}_N^1$ and simplify our Example.

Now we turn to the situation when a response variable Y depends only on the part of factors. Let $\{k_1, \dots, k_r\}$ be a subset of $\{1, \dots, n\}$ where $r < n$. We say (cf. [2], Ch. 2) that the factors $X_{k_1}, \dots, X_{k_r}$ are relevant (or collection of indices $k_1, \dots, k_r$ is relevant) whenever for each $x = (x_1, \dots, x_n) \in M$ it turns out that

$$\mathsf{P}(Y=1|X_1=x_1, \dots, X_n=x_n) = \mathsf{P}(Y=1|X_{k_1}=x_{k_1}, \dots, X_{k_r}=x_{k_r}). \tag{22}$$

For an arbitrary collection $\{m_1, \dots, m_r\} \subset \{1, \dots, n\}$ set

$$f^{m_1,\dots,m_r}(x) = \begin{cases} 1, & \mathsf{P}(Y=1|X_{m_1}=x_{m_1}, \dots, X_{m_r}=x_{m_r}) > \frac{\psi(-1)}{\psi(-1)+\psi(1)}, \quad x \in M, \\ -1, & \text{otherwise.} \end{cases}$$

If $\{k_1, \dots, k_r\}$ is a relevant collection then the optimal function f^* has the form $f^{k_1,\dots,k_r}$. Therefore, for any relevant collection $\{k_1, \dots, k_r\} \subset \{1, \dots, n\}$ and an arbitrary subset $\{m_1, \dots, m_r\} \subset \{1, \dots, n\}$, the following inequality holds

$$Err(f^{k_1,\dots,k_r}) \leqslant Err(f^{m_1,\dots,m_r}).$$

For any $\{m_1, \dots, m_r\} \subset \{1, \dots, n\}$, $x \in \mathbb{X}$ and a penalty function ψ, consider a prediction algorithm with $f_{PA}^{m_1,\dots,m_r}$ such that

$$f_{PA}^{m_1,\dots,m_r}(x, \zeta_N(\overline{S_k(N)}), \widehat{\mathsf{P}}_N) = \begin{cases} 1, & g^{m_1,\dots,m_r}(x, \zeta_N(\overline{S_k(N)}), \widehat{\mathsf{P}}_N) > h(\zeta_N(\overline{S_k(N)}), \widehat{\mathsf{P}}_N), \\ -1, & \text{otherwise,} \end{cases}$$

where $k = 1, \dots, K$, h is defined by (20) and

$$g^{m_1,\dots,m_r}(x, \zeta_N(\overline{S_k(N)}), \widehat{\mathsf{P}}_N) = \frac{\widehat{\mathsf{P}}_N^1 I_1^{m_1,\dots,m_r}(x, \zeta_N(\overline{S_k(N)}), \widehat{\mathsf{P}}_N)}{\widehat{\mathsf{P}}_N^{-1} I_{-1}^{m_1,\dots,m_r}(x, \zeta_N(\overline{S_k(N)}), \widehat{\mathsf{P}}_N) + \widehat{\mathsf{P}}_N^1 I_1^{m_1,\dots,m_r}(x, \zeta_N(\overline{S_k(N)}), \widehat{\mathsf{P}}_N)}.$$

Here, for $y \in \{-1, 1\}$ and $x \in \mathbb{X}$,

$$I_y^{m_1,\dots,m_r}(x, \zeta_N(\overline{S_k(N)}), \widehat{\mathsf{P}}_N) = \frac{1}{\sharp W_k^y(N_y)} \sum_{j \in W_k^y(N_y)} \mathbb{I}(X_{m_1}^j = x_{m_1}, \dots, X_{m_r}^j = x_{m_r}).$$

Set

$$U := \left\{ x \in M_{m_1,\dots,m_r}\colon \mathsf{P}(Y=1|X_{m_1}=x_{m_1}, \dots, X_{m_r}=x_{m_r}) \neq \frac{\psi(-1)}{\psi(-1)+\psi(1)} \right\}$$

where $M_{m_1,\dots,m_r} = \{x \in \mathbb{X}\colon \mathsf{P}(X_{m_1}=x_{m_1}, \dots, X_{m_r}=x_{m_r}) > 0\}$. Similarly to [4], it can be shown that for relevant collections $\{k_1, \dots, k_r\} \subset \{1, \dots, n\}$, arbitrary collections $\{m_1, \dots, m_r\} \subset \{1, \dots, n\}$, any $\varepsilon > 0$ and almost all $\omega \in \Omega$,

$$\widehat{Err}_K(f_{PA}^{k_1,\dots,k_r}, \zeta_N, \widehat{\mathsf{P}}_N) \leqslant \widehat{Err}_K(f_{PA}^{m_1,\dots,m_r}, \zeta_N, \widehat{\mathsf{P}}_N) + \varepsilon \ \text{ a.s.}$$

when N is large enough. Thus, it is natural to choose as relevant collection of factors $X_{k_1}, \dots, X_{k_r}$ that one which has the minimal prediction error estimate $\widehat{Err}_K(f_{PA}^{k_1,\dots,k_r}, \zeta_N, \widehat{\mathsf{P}}_N)$.

4. Cost approach to experiments and the XOR-model

We compare two approaches concerning the application of the MDR-EFE method for different sample plans. The first one was described in [4] and consists in the employment of nonrandom number of i.i.d. observations. The second one is considered in this paper and involves the stratified sample. More exactly, stratification means the separation of observations taking into account the values of a response variable under consideration. We will denote these methods as iMDR-EFE and sMDR-EFE, respectively. It seems that the main disadvantage of the second approach is that too large number of independent observations is required to form a stratified sample of N elements with fixed cases to controls ratio when $\mathsf{P}(Y=1)$ is very small. Moreover, we have to skip a lot of observations with $Y^i=-1$. However it is worth to emphasize that no information on predictors X^i is needed at the stage of forming a stratified sample, therefore we need not get X^i for skipped observations. This is essential when the measurement of Y is cheaper than the measurement of X (since X has components $X_1,\ldots,X_n$). Thus it is not interesting to compare iMDR-EFE and sMDR-EFE for the equal sizes of samples and we should take into account additional observations in sMDR-EFE.

We propose to compare two approaches in the sense of the total cost of experiment. Assume that there is a fixed amount of money C ($C\in\mathbb{N}$) for research. Let each observation (X^i,Y^i) cost 1 and let the ratio of the price of measuring Y to that of X be $w\in\mathbb{R}_+$. We consider the maximal sizes $s_{ind}(C,w)$ and $s_{str}(C,w)$ of the samples which are available in experiments organized to apply iMDR-EFE and sMDR-EFE, respectively. Let $C_{ind}(N,w)$ be the total cost of the sample having size N and consisting of independent observations. Let $C_{str}(N,w)$ be the total cost of the stratified sample of size N. Then

$$s_{ind}(C,w)=\max\{N\in\mathbb{N}\colon C_{ind}(N,w)\leqslant C\},$$
$$s_{str}(C,w)=\max\{N\in\mathbb{N}\colon C_{str}(N,w)\leqslant C\}.$$

Clearly, $C_{ind}(N,w)=N$ and therefore $s_{ind}(C,w)=C$. We note that this value is nonrandom and is known before the experiment.

Recall (see Remark 4) that $\widetilde{N}=\widetilde{N}(N,a,\omega)=\max\{j_{-1}^{N_{-1}},j_1^{N_1}\}$. Hence

$$C_{str}(N,w,\omega)=\frac{1}{w+1}N+\frac{w}{w+1}\widetilde{N}(N,a,\omega)$$

because we have to measure X in N observations (which are included into the sample) and to measure Y in $\widetilde{N}$ observations. Thus $C_{str}(N,w)$ is a random variable which is unknown before the experiment. In general case, $s_{str}(C,w)$ is a random variable as well. However we can select such N that $C_{str}(N,w)$ does not exceed C with probability which is not less than $1-\alpha$, $\alpha\in(0,1)$. We assume that if the cost of experiment exceeds C with given small probability then some Reserve Foundation will cover the extra expenses. Now we want to find $s'_{str}(C,w,\alpha)$ so that

$$s'_{str}(C,w,\alpha)=\max\left\{N\in\mathbb{N}\colon \mathsf{P}(C_{str}(N,w)\leqslant C)\geqslant 1-\alpha\right\}.$$

Obviously, $\widetilde{N}(N,a,\omega)\geqslant N$ for each $\omega\in\Omega$. Therefore for any $N\in\mathbb{N}$ and $\omega\in\Omega$

$$C_{str}(N,w,\omega)=N+\frac{w}{w+1}(\tilde{N}(N,a,\omega)-N)\geqslant N.$$

Consequently, for any $N>C$,

$$\mathsf{P}(C_{str}(N,w)\leqslant C)\leqslant \mathsf{P}(C_{str}(N,w)<N)=0.$$

So, for arbitrary $C\in\mathbb{N}$, $\alpha\in(0,1)$ and $w>0$, we observe that

$$s'_{str}(C,w,\alpha)=\max\left\{N\in\{1,2,\ldots,C\}\colon \mathsf{P}(C_{str}(N,w)\leqslant C)\geqslant 1-\alpha\right\}. \tag{23}$$

It is worth mentioning that $s'_{str}(C,w,\alpha)$ is a non-random variable which depends on certain parameters and the distribution of $\widetilde{N}$. We will apply the following result.

Lemma 2
For each $m \in \{0, 1, 2\dots\}$ one has

$$\mathsf{P}(\widetilde{N} = m) = \begin{cases} 0, & m < N, \\ \mathsf{P}(\eta_{-1} = m - N_{-1}) + \mathsf{P}(\eta_1 = m - N_1), & \text{otherwise} \end{cases} \tag{24}$$

where $\eta_{-1} \sim NB(N_{-1}, \mathsf{P}(Y = 1))$, $\eta_1 \sim NB(N_1, \mathsf{P}(Y = -1))$.

Proof. $\mathsf{P}(\widetilde{N} = m) = 0$ for each $m \in \{0, 1, 2, \dots, N-1\}$ since $\widetilde{N}(N, a, \omega) \geqslant N$ for each $\omega \in \Omega$. If $m \in \{N, N+1, \dots\}$ then

$$\mathsf{P}(\widetilde{N} = m) = \mathsf{P}\left(\max\{j_{-1}^{N_{-1}}, j_1^{N_1}\} = m\right)$$
$$= \mathsf{P}\left(j_{-1}^{N_{-1}} = m, j_1^{N_1} < m\right) + \mathsf{P}\left(j_{-1}^{N_{-1}} < m, j_1^{N_1} = m\right)$$
$$+ \mathsf{P}\left(j_{-1}^{N_{-1}} = m, j_1^{N_1} = m\right).$$

Let $\widetilde{S}_y(k) = \sharp\{i \in \{1, \dots, k\} \colon Y^i = y\}$. We note that $\widetilde{S}_y(k) + \widetilde{S}_{-y}(k) = k$ for each $k \in \mathbb{N}$. Thus, for each $m \in \{N, N+1, \dots\}$ and any $y \in \{-1, 1\}$,

$$\left\{j_y^{N_y} = m\right\} = \left\{\widetilde{S}_y(m) = N_y\right\} \cap \left\{\widetilde{S}_y(m-1) = N_y - 1\right\} \subset \left\{\widetilde{S}_y(m-1) = N_y - 1\right\}$$
$$= \left\{\widetilde{S}_{-y}(m-1) = m - N_y\right\} \subset \left\{\widetilde{S}_{-y}(m-1) \geqslant N_{-y}\right\} \subset \left\{j_{-y}^{N_{-y}} < m\right\}.$$

Moreover, $j_{-1}^{N_{-1}}(\omega) \neq j_1^{N_1}(\omega)$ for any $\omega \in \Omega$. Therefore

$$\mathsf{P}(\widetilde{N} = m) = \mathsf{P}(j_{-1}^{N_{-1}} = m) + \mathsf{P}(j_1^{N_1} = m).$$

According to Section 2, $j_y^{N_y}$ has the same distribution as $N_y + \eta_y$. Thus (24) holds. □

Remark 5. Lemma 2 shows that the law of $\widetilde{N}$ depends on the known parameters N_{-1}, N_1 and, in general, unknown $\mathsf{P}(Y = 1)$. However, if $\mathsf{P}(Y = 1)$ is known or we have its estimates constructed by means either of $\xi_{\widetilde{N}}$ or another sample (see Remark 4) then $s'_{str}(C, w, \alpha)$ can be evaluated or estimated.

Using Lemma 2 we can rewrite $\mathsf{P}\left(C_{str}(N, w) \leqslant C\right)$ as

$$\mathsf{P}\left(C_{str}(N, w) \leqslant C\right) = \mathsf{P}\left(\widetilde{N}(N, a, \omega) \leqslant \frac{w+1}{w}C - \frac{N}{w}\right)$$
$$= \mathsf{P}\left(N_1 \leqslant \eta_{-1} \leqslant \frac{w+1}{w}C - \frac{N}{w} - N_{-1}\right) + \mathsf{P}\left(N_{-1} \leqslant \eta_1 \leqslant \frac{w+1}{w}C - \frac{N}{w} - N_1\right) \tag{25}$$

where η_{-1} and η_1 are defined above. Thus, for any N, the probability $\mathsf{P}\left(C_{str}(N, w) \leqslant C\right)$ can be evaluated or estimated (see Remark 5). Moreover,

$$\mathsf{P}\left(C_{str}(N, w) \leqslant C\right) \geqslant \mathsf{P}\left(C_{str}(N+1, \omega) \leqslant C\right) \tag{26}$$

for all $N \in \mathbb{N}$ since, for each $\omega \in \Omega$,

$$\frac{1}{w+1}N + \frac{w}{w+1}\widetilde{N}(N, a, \omega) \leqslant \frac{1}{w+1}(N+1) + \frac{w}{w+1}\widetilde{N}(N+1, a, \omega).$$

Formulae (25) and (26) suggest the following algorithm. If $p := \mathsf{P}(Y = 1)$ is known we evaluate $\mathsf{P}\left(C_{str}(N, w) \leqslant C\right)$ by (25) for each $N = 1, 2, \dots$ (till C) and identify the maximal N belonging to $\{1, \dots, C\}$ such that (23) holds. In this way we determine $s'_{str}(C, w, \alpha)$. If p is unknown but there exists its estimate $\widehat{p}_{N'}$ such that $\widehat{p}_{N'} \to p$ almost surely as $N' \to \infty$ (for instance, it may be an estimate by means of another sample) then for each fixed C, N and w we can estimate $\mathsf{P}\left(C_{str}(N,w) \leqslant C\right)$ by

$$\nu_{N'}(N,w,C) = \sum_{N_1 \leqslant m \leqslant L_1} \binom{N_1+m-1}{m} (1-\widehat{p}_{N'})^{N_{-1}} \widehat{p}_{N'}^{\,m} + \sum_{N_{-1} \leqslant m \leqslant L_{-1}} \binom{N_{-1}+m-1}{m} \widehat{p}_{N'}^{\,N_1} (1-\widehat{p}_{N'})^m$$

where $L_y := \frac{w+1}{w}C - \frac{N}{w} - N_{-y}$, $y \in \{-1, 1\}$. Indeed, $\nu_{N'}(N,w,C)$ tends to the right-hand side of (25) almost surely as $N' \to \infty$. Thus we can introduce

$$\widehat{s'}_{str,N'}(C,w,\alpha) := \max\left\{N \in \{1,2,\dots,C\} \colon \nu_{N'}(N,w,C) \geqslant 1-\alpha\right\}$$

and apply the approach proposed for known p.

To compare iMDR-EFE and sMDR-EFE we turn to the popular XOR-model (see, e.g., [23]). This model is used in genetics to describe epistasis without main effects. Namely, let $\mathbb{X} = \{0,1,2\}^n$ ($0, 1, 2$ correspond to the number of minor alleles of a specified gene). Assume now that the components of a random vector $X = (X_1, \dots, X_n)$ are independent and, for each $i \in \{1, \dots, n\}$, there exists such $p_i \in (0, 0.5]$ that

$$\mathsf{P}(X_i = 0) = (1-p_i)^2, \;\; \mathsf{P}(X_i = 1) = 2p_i(1-p_i), \;\; \mathsf{P}(X_i = 2) = p_i^2. \tag{27}$$

This situation is typical for genome-wide association studies (GWAS) where each X_i corresponds to a single nucleotide polymorphism (SNP) and p_i is a minor allele frequency (MAF). Suppose that collection of relevant factors (describing a binary response variable Y) is $X_{k_1}, \dots, X_{k_r}$ where $\{k_1, \dots, k_r\} \subset \{1, \dots, n\}$. One also says that collection of indices $\overline{k} := \{k_1, \dots, k_r\}$ is relevant.

In our simulations we employ the XOR-model of dependence between predictors and response variable. It is a generalization of the XOR-model described in [23] to the case of more than 2 relevant factors. Namely, for each $x = (x_1, \dots, x_n) \in \mathbb{X}$,

$$\mathsf{P}(Y = 1|X = x) = \begin{cases} \gamma, & (x_{k_1} + \ldots + x_{k_r}) \bmod 2 = 1, \\ 0, & \text{otherwise} \end{cases} \tag{28}$$

where $\gamma \in (0,1)$ and $p_{k_1} = \dots = p_{k_r} = 0.5$. Thus $(X_{k_1}, \dots, X_{k_r})$ is a collection of relevant factors according to (22). XOR-model is interesting as it possess the property described by Lemma 3 below.

For any $\overline{s} := \{s_1, \dots, s_q\} \subset \{1, \dots, n\}$, $q \leqslant n$, set $X_{\overline{s}} := (X_{s_1}, \dots, X_{s_q})$. We write $X_{\overline{s}} = x_{\overline{s}}$ where $x_{\overline{s}} = (x_{s_1}, \dots, x_{s_q}) \in \{0,1,2\}^q$ if $X_{s_i} = x_{s_i}$ for all $i = 1, \dots, q$.

Lemma 3

Let $\mathbb{X}, X, Y$ and $\overline{k}$ be introduced above (the dependence between X and Y is described by (28)). Then, for any collection $\overline{m} = \{m_1, \dots, m_l\} \subset \{1, \dots, n\}$, a response variable Y is dependent with $X_{\overline{m}}$ if and only if $\overline{k} \subset \overline{m}$.

Proof. Formula (28) shows that Y and $X_{\overline{m}}$ are dependent if $\overline{k} \subset \overline{m}$. Let now $\overline{m} = \overline{u} \cup \overline{v}$ where $\overline{u} \subsetneq \overline{k}$ and $\overline{v} \subsetneq \{1, \dots, n\} \setminus \overline{k}$. Introduce $t := \sharp\overline{u}$. Since $\mathsf{P}(X_{\overline{m}} = x_{\overline{m}}) \neq 0$ for each $x_{\overline{m}} \in \{0,1,2\}^l$ it remains to establish that

$$\mathsf{P}(Y = 1|X_{\overline{m}} = x_{\overline{m}}) = \mathsf{P}(Y = 1). \tag{29}$$

Evidently

$$\begin{aligned} &\mathsf{P}(Y = 1|X_{\overline{m}} = x_{\overline{m}}) \\ &= \sum_{x_{\overline{k}\setminus\overline{u}} \in \{0,1,2\}^{r-t}} \mathsf{P}(Y = 1|X_{\overline{u}} = x_{\overline{u}}, X_{\overline{k}\setminus\overline{u}} = x_{\overline{k}\setminus\overline{u}}, X_{\overline{v}} = x_{\overline{v}}) \mathsf{P}(X_{\overline{k}\setminus\overline{u}} = x_{\overline{k}\setminus\overline{u}}) \\ &= \gamma \sum_{x_{\overline{k}\setminus\overline{u}} \in \{0,1,2\}^{r-t}} \mathbb{I}\left\{\left(\sum_{i \in \overline{k}\setminus\overline{u}} x_i + \sum_{i \in \overline{u}} x_i\right) \bmod 2 = 1\right\} \mathsf{P}(X_{\overline{k}\setminus\overline{u}} = x_{\overline{k}\setminus\overline{u}}) \\ &= \gamma \mathsf{P}\left(\sum_{i \in \overline{k}\setminus\overline{u}} X_i \bmod 2 \neq \sum_{i \in \overline{u}} x_i \bmod 2\right) \end{aligned} \tag{30}$$

In view of (27) the laws of $(X_1,\dots,X_n)$ and $(X'_1,\dots,X'_n)$ coincide where $X'_i=\sum_{j=1}^2 B_i^j$ and $B_1^1,B_1^2,\dots,B_n^1,B_n^2$ are independent Bernoulli variables such that $\mathsf{P}(B_i^k=1)=p_i$, $k=1,2$, $i=1,\dots,n$. Hence, $law(\sum_{i\in\overline{s}}X_i)=law(\sum_{i\in\overline{s}}(B_i^1+B_i^2))$ for any $\overline{s}\subset\{1,\dots,n\}$. Thus, for $\overline{s}:=\{s_1,\dots,s_q\}\subset\overline{k}$ ($q\leqslant r$), one has

$$\mathsf{P}\left(\sum_{i\in\overline{s}}X_i \bmod 2=0\right)=\mathsf{P}\left(\sum_{i=1}^{2q}\tilde{B}_i \bmod 2=0\right)=\sum_{j=0}^{q}C_{2q}^{2j}2^{-2q}=\frac{1}{2}$$

where $\widetilde{B}_1,\dots,\widetilde{B}_{2r}$ are i.i.d. Bernoulli random variables with probability of success 0.5. Therefore,

$$\mathsf{P}\left(\sum_{i\in\overline{s}}X_i \bmod 2=1\right)=\frac{1}{2}.$$

Taking into account (30) we observe that $\mathsf{P}(Y=1|X_{\overline{m}}=x_{\overline{m}})=\frac{\gamma}{2}$ for any $x_{\overline{m}}\in\{0,1,2\}^l$. To complete the proof note that

$$\mathsf{P}(Y=1)=\sum_{x\in\mathbb{X}}\mathsf{P}(Y=1|X=x)\mathsf{P}(X=x)$$

$$=\sum_{x_{\overline{k}}\in\{0,1,2\}^r}\mathsf{P}(Y=1|X_{\overline{k}}=x_{\overline{k}})\mathsf{P}(X_{\overline{k}}=x_{\overline{k}})=\gamma\mathsf{P}\left(\sum_{i\in\overline{k}}X_i \bmod 2=1\right)=\frac{\gamma}{2}.$$

Thus (29) holds. □

Remark 6. Lemma 3 shows that only the whole collection of relevant factors $X_{k_1},\dots,X_{k_r}$ and not some of its subcollections determines the response Y in XOR-model.

To complete this Section we discuss some asymptotical properties of a random sample as $C\to\infty$. For a given C the iMDR-EFE method involves C observations whereas its sMDR-EFE counterpart (using the same amount of money C) leads to N observations and

$$\frac{1}{w+1}N+\frac{w}{w+1}\max\{j_{-1}^{N-([aN]\vee 1)},j_1^{[aN]\vee 1}\}\leqslant C, \tag{31}$$

here $a\in(0,1)$. Clearly, (31) implies that $N\leqslant C$. Now we consider the problem whether one can take $N=\lambda C$ for some $\lambda\in(0,1)$ such that with probability close to one inequality (31) is satisfied.

Lemma 4

For an arbitrary (fixed) $a\in(0,1)$ and w inequality (31) is valid with probability tending to one as $C\to\infty$ if and only if

$$\lambda<\lambda_0:=(1+w)\left(1+w\max\left\{\frac{a}{p},\frac{1-a}{1-p}\right\}\right)^{-1}. \tag{32}$$

Proof. We can rewrite (31) in the following way

$$\max\{j_{-1}^{[\lambda C]-[a[\lambda C]]},j_1^{[a[\lambda C]]}\}\leqslant\frac{C(w+1)-[\lambda C]}{w}.$$

Note that

$$\mathsf{P}\left(j_1^{[a[\lambda C]]}\leqslant\frac{C(w+1)-[\lambda C]}{w}\right)$$

$$=\mathsf{P}\left(\frac{j_1^{[a[\lambda C]]}-\frac{[a[\lambda C]]}{p}}{\sigma\sqrt{[a[\lambda C]]}}\leqslant\frac{1}{\sigma\sqrt{[a[\lambda C]]}}\left(\frac{C(w+1)-[\lambda C]}{w}-\frac{[a[\lambda C]]}{p}\right)\right)$$

where $\sigma^2 := \frac{1-p}{p^2}$ (the variance of a random variable following the Geometric law with parameter p). The central limit theorem for i.i.d. random variables having finite variance yields that $j_1^{[a[\lambda C]]} \leqslant \frac{C(w+1)-[\lambda C]}{w}$ with probability tending to one as $C \to \infty$ if and only if $(w+1-\lambda)/w - a\lambda/p > 0$, that is

$$\lambda < (1+w)\left(1+\frac{aw}{p}\right)^{-1}. \tag{33}$$

In a similar way we can claim that

$$\mathsf{P}\left(j_{-1}^{[\lambda C]-[a[\lambda C]]} \leqslant \frac{C(w+1)-[\lambda C]}{w}\right) \to 1, \quad C \to \infty,$$

if and only if

$$\lambda < (1+w)\left(1+\frac{(1-a)w}{1-p}\right)^{-1}. \tag{34}$$

Thus the statement of Lemma 1 is valid if and only if (33) and (34) are true simultaneously, i.e. relation (32) holds. □

Thus the optimal boundary λ_0 is found.

5. Simulations

We will consider different levels of the total cost C and parameter γ of XOR-model. Note that now $\mathsf{P}(Y=1) = \frac{\gamma}{2}$, i.e. $p = \frac{\gamma}{2}$. We employ two levels of parameter w. The value $w = 0$ corresponds to the extreme case when values of a response variable are obtained free of charge. This situation can be viewed as the limit one when the price of the measurement of Y is very low w.r.t. the price of X measurement. According to [22] the application of MDR method is reasonable for balanced datasets, i.e. with equal number of cases and controls. Therefore, in the case of the stratified samples we consider $a = 0.5$ and only even N in (23). For each combination of these parameters we generate D independent datasets for iMDR-EFE and for sMDR-EFE applications in order to evaluate performance of these methods by Monte Carlo experiments.

Parameter	**Value**
$\psi(y)$	$(\mathsf{P}(Y=y))^{-1}$
D	100
n	100
K	5
r	3
$(k_1,\ldots,k_r)$	(2,3,5)
a	0.5
C	5 levels: $\{100, 200, 300, 400, 500\}$
w	2 levels: $\{0, 0.1\}$
γ	3 levels: $\{0.05, 0.1, 0.2\}$
α	0.05

Table 1. Simulation scheme parameters.

The values of parameters used in our simulations are given in Table 1. For relevant factors $X_{k_1},\ldots,X_{k_r}$ we set $p_{k_i} = 0.5$, $i = 1,\ldots,r$ (according to XOR-model) whereas for other (non-relevant) factors X_i the corresponding

p_i are drawn independently from the uniform distribution $U(0.05, 0.5)$ to generate each dataset. For our datasets collections of p_i are drawn independently. Such simulation setting for p_i was proposed in [10]. The interval $[0.05, 0.5]$ for uniform law was taken since p_i is considered as MAF.

We take a Bernoulli variable $B(\gamma)$ having the success probability γ. Let $B(\gamma)$ and X be independent. Introduce $Y := 2B(\gamma)\mathbb{I}\{\sum_{v=1}^{r} X_{k_v} \bmod 2 = 1\} - 1$ where $X_{k_1}, \ldots, X_{k_r}$ are the relevant factors. Then one can verify that (28) holds. In such a way we generate independent vectors (X^j, Y^j), $j = 1, 2, \ldots$.

Taking $w = 0.1$ we will assume that p is known in order to use it for determining $s'_{str}(C, w, \alpha)$ introduced in (23). For $w = 0$ the size of the sample $s'_{str}(C, 0, \alpha)$ is equal to C as in the case of independent observations. Then it is non-random and fixed before the experiment and we estimate $\mathsf{P}(Y = 1)$ by means of $Y^1, \ldots, Y^{\widetilde{C}}$. Here $\widetilde{C} = \widetilde{C}(C, a, \omega) = \max\{j_{-1}^{C_{-1}}, j_1^{C_1}\}$, $C_1 = [aC]$ and $C_{-1} = C - [aC]$. In order to measure the method performance power we use TMR (true model rate) which is defined as

$$TMR := \frac{1}{D} \sum_{d=1}^{D} T_d$$

where

$$T_d = \begin{cases} 1, & \text{if all relevant factors are identified correctly in } d\text{-th dataset,} \\ 0, & \text{otherwise.} \end{cases}$$

Below the phrase "implementation of iMDR-EFE (sMDR-EFE) for sample" means that we select such collection of r factors which has the minimal $\widehat{Err}_K$.

The simulation procedure can be described in the following way.

I. Fix some values γ and C from Table 1, other parameters except w are fixed as well.

II. Calculate $s'_{str}(C, w, \alpha)$ where $w = 0.1$ and $\alpha = 0.5$ assuming that $p = \mathsf{P}(Y = 1)$ is known (and equals $\gamma/2$).

III. Perform D times

1. generation of independent $p_i \sim U(0.05, 0.5)$ for non-relevant factors;
2. generation of independent X^j, having law introduced by (27), and corresponding Y^j until we have 3 samples:
   (a) sample ξ_C which consists of C independent observations;
   (b) stratified sample ζ_N which consists of $N = s'_{str}(C, w, \alpha)$ observations (for $w = 0.1$);
   (c) stratified sample ζ_C which consists of C observations (for $w = 0$). While generating observations (X^j, Y^j) in order to form ζ_C we estimate p (and write "p is estimated by means of observations $Y^1, \ldots, Y^{\widetilde{C}}$");
3. implementation of iMDR-EFE for ξ_C, p is unknown ($\psi(y)$ as well);
4. implementation of iMDR-EFE for ξ_C, p is known therefore $\psi(y)$ is known and
$$\widehat{\psi}(y, \xi_C(\overline{S_k(C)})) = 1/\mathsf{P}(Y = y),$$
here we use the notation $\widehat{\psi}(y, \xi_C(\overline{S_k(C)}))$ of [4] for i.i.d. observations;
5. implementation of sMDR-EFE for ζ_N where $N = s'_{str}(C, w, \alpha)$, p is known therefore $\widehat{\mathsf{P}}_N = (1 - p, p)$ and
$$\widehat{\psi}(y, \zeta_N(\overline{S_k(N)}), \widehat{\mathsf{P}}_N) = 1/\mathsf{P}(Y = y);$$
6. implementation of sMDR-EFE for ζ_C, p is unknown and estimated by means of observations $Y^1, \ldots, Y^{\widetilde{C}}$;
7. implementation of sMDR-EFE for ζ_C, p is known thus $\widehat{\mathsf{P}}_C = (1 - p, p)$ and
$$\widehat{\psi}(y, \zeta_C(\overline{S_k(C)}), \widehat{\mathsf{P}}_C) = 1/\mathsf{P}(Y = y);$$
8. assignment of T_d to every implementation of the method under consideration.

IV. Evaluate TMR for each approach.

The results of our simulations are shown in Figures 1 and 2. Thus for the fixed total cost C a stratified sampling gives better results than independent one. Despite of the fact that $s_{ind}(C, \alpha) > s'_{str}(C, w, \alpha)$ (when $w > 0$, see Figure 2), sMDR-EFE demonstrates better performance than iMDR-EFE in all 3 models for each value of C and different scenarios concerning p. Moreover, small w permits to include more observations into the stratified sample. Figure 1 suggests that it leads to better performance of the sMDR-EFE method.

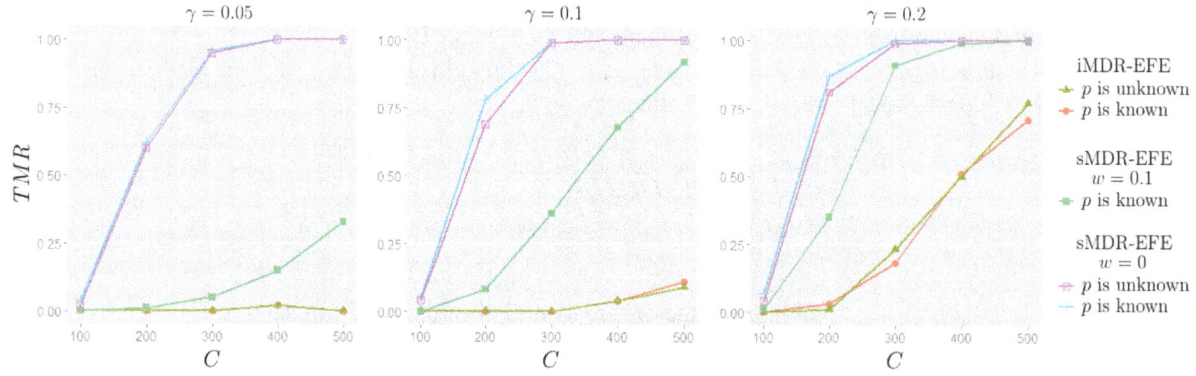

Figure 1. Performance (TMR) of iMDR-EFE and sMDR-EFE for different levels of $\gamma = 2p$.

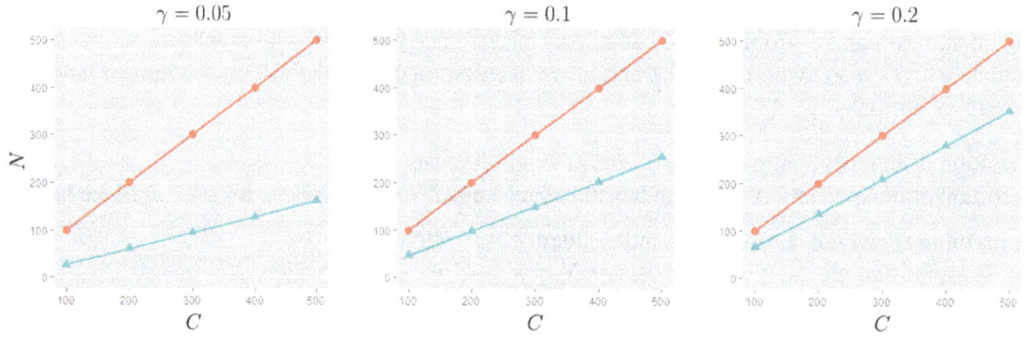

Figure 2. Sizes of the samples depending on the cost C for different levels of $\gamma = 2p$. The red line corresponds to iMDR-EFE and sMDR-EFE with $w = 0$, the blue line corresponds to sMDR-EFE with $w = 0.1$.

Let $a = 0.5$ and $w = 0.1$. Then, for γ taking values 0.05, 0.1 and 0.2 ($p = \gamma/2$) Lemma 4 yields that the corresponding values of λ_0 are equal to $0.366(6)$, 0.55 and $0.733(3)$, respectively. For any $a \in (0, 1)$, all $p \in (0, 1)$ and $w = 0$ we get $\lambda_0 = 1$. Clearly if w is close to 0 then λ_0 is close to 1.

We note that there are various possibilities to take into account the cost of experiments with random and nonrandom number of observations. This interesting problem will be considered separately.

6. Acknowledgements

The work of A.V.Bulinski is supported by the Russian Science Foundation under grant 14-21-00162 and performed at the Steklov Mathematical Institute of Russian Academy of Sciences. The problems setting belongs to A.V.Bulinski, all theoretical results are established jointly with A.A.Kozhevin. Computer simulations are carried out by A.A.Kozhevin.

REFERENCES

1. S.E.Ahmed. Penalty, Shrinkage and Pretest Strategies. Variable Selection and Estimation. Springer, Cham, 2014.
2. V.Bolón-Canedo, N.Sanchez-Maroño and A.Alonso-Betanzos. Feature Selection for High-Dimensional Data. Springer, Cham, 2015.
3. P.Bühlmann, S.van de Geer. Statistics for High-Dimensional Data. Methods, Theory and Applications. Springer, Heidelberg, 2011.
4. A. Bulinski. On foundation of the dimensionality reduction method for explanatory variables. Journal of Mathematical Sciences, v. 199, No. 2, 113-122 (2014).
5. A.Bulinski. Central limit theorem related to MDR-method. In: Asymptotic Laws and Methods in Stochastics. A volume in Hounor of Miklos Csorgo. Fields Institute Communications, v. 76, 113-128. Springer, New York, 2015.
6. A.Bulinski. Some statistical methods in genetics. In: V.Schmidt (Ed.). Stochastic Geometry, Spatial Statistics and Random Fields. Lecture Notes in Mathematics, v. 2120, 293-320. Springer-Verlag, Berlin, 2014.
7. A.Bulinski, O.Butkovsky, V.Sadovnichy, A.Shashkin, P.Yaskov, A.Balatskiy, L.Samokhodskaya and V.Tkachuk. Statistical methods of SNP data analysis and applications. Open Journal of Statistics, v. 2, No 1, 73-87 (2012).
8. A.Bulinski, A.Rakitko. Simulation and analytical approach to the identification of significant factors. Commun. in Statistics. Part B: Simulation and Computation, v. 44, 1-23 (2015).
9. A.Bulinski, A.Rakitko. MDR method for nonbinary response variable. J. of Multivariate Analysis, v. 135, 25-42 (2015).
10. A.Dehman, C.Ambroise and P.Neuvial. Performance of a blockwise approach in variable selection using linkage disequilibrium information. BMC Bioinformatics 16:148 (2015).
11. K-A.Do, Z.S.Qin and M.Vannucci (Eds.). Advances in Statistical Bioinformatics. Models and Integrative Inference for High-Throughput Data. Cambridge University Press, Cambridge, 2013.
12. D.Gola, J.M.M.John, K. van Steen and R.König. A roadmap to multifactor dimensionality reduction methods. Briefings in Bioinformatics, June 24, 1-16 (2015).
13. G. James, D. Witten, T. Hastie and R. Tibshirani. An Introduction to Statistical Learning with Applications in R, Springer Science + Business Media, New York, 2013.
14. I.Koch. Analysis of Multivariate and High-Dimensional Data. Cambridge University Press, Cambridge, 2014.
15. J-M.Marin, C.Robert. Bayessian Essentials with R. Springer Science + Business Media, New York, 2014.
16. J.H.Moore, S.M.Williams (Eds.). Epistasis: Methods and Protocols. Methods in Molecular Biology. v. 1253. Springer Science + Business Media, New York, 2015.
17. J. Park, Independent rule in classification of multi- variate Binary Data, J. of Multivariate Analysis, v. 100, No. 10, 2270-2286 (2009).
18. M. D. Ritchie, L. W. Hahn, N. Roodi, R. Bailey, W. D. Dupont, F. F. Parl and J.H. Moore, Multifactor dimensionality reduction reveals high-order interactions among estrogen-metabolism genes in sporadic breast cancer, Amer. J. Human Genetics, v. 69, 139-147 (2001).
19. G.Ritter. Robust Claster Analysis and Variable Selection. CRC Press, Boca Raton, 2015.
20. J. Shang, J. Zhang, Y. Sun, D. Liu, D. Ye and Y. Yin, Performance analysis of novel methods for detecting epistasis BMC Bioinformatics 12:475 (2011).
21. R. L. Taylor and T.-C. Hu. Strong laws of large numbers for arrays of row-wise independent random elements, Int. J. Math. Math. Sci., v. 10, 805-814 (1987).
22. D. Velez, B. White, A. Motsinger, W. Bush, M. Ritchie, S. Williams, and J. Moore. Balanced accuracy function for epistasis modeling in imbalanced datasets using multifactor dimensionality reduction, Genetic Epidemiology, v. 31, 306-315 (2007).
23. S.J.Winham, A.J.Slater and A.A.Motsinger-Reif. A comparison of internal validation techniques for multifactor dimensionality reduction. BMC Bioinformatics, 11:394 (2010).

13

Developing industries in cooperative interaction: equilibrium and stability in processes with lag

A.I.Kirjanen, O.A.Malafeyev, N.D.Redinskikh*

Faculty of Applied Mathematics and Control Processes, Saint-Petersburg State University, Russia.

Abstract A mathematical model of dynamic interaction between mining and processing industries is formalized and studied in the paper. The process of interaction is described by a system of two delay differential equations. The criterion for asymptotic stability of nontrivial equilibrium point is obtained when both industries co-work steadily. The problem is reduced to finding stability criterion for quasi-polynomial of second order. Time intervals between deliveries of raw materials which make it possible to preserve stable interaction between the two industries are found.

Keywords Dynamic cooperative interaction, Mining and processing industries, Delay, Differential equations, Coefficient criteria for asymptotic stability

1. Introduction

A mathematical model of dynamic interaction between mining and processing industries is formalized and studied in the paper. It is supposed, that the resource is mined by the first industry and then it is transformed into some product by the second industry. Let us denote the amount of the resources mined by P and the number of companies, producing the final product from this resource. Similarly, it is possible to consider Q as the total output in the processing industry.

Let us assume, that the volume of the mined resources is growing with a coefficient $a > 0$ due to ongoing mineral exploration and the amount of raw materials for processing industry is unlimited. We assume that the increase in the number of processing companies leads to the reduction of the volume of extracted raw materials with coefficient $b > 0$ and, conversely, an increase in the amount of raw materials implies an increase in the number of processing companies with coefficient $d > 0$. Although the extraction of raw materials occurs continuously, it is shipped to producers in portions with some positive time lag h. So the amount of raw materials mined depends on its volume obtained earlier at time moment $(t - h)$ when the last supply was made. If this quantity of "old" mined materials is large, the rate of its extraction must decrease with coefficient $e > 0$.

In the absence of raw materials the number of processing companies is reduced with the coefficient $c > 0$. In the model described by system (1) the number of processing companies at time moment t, also depends on the number of already operating companies at time moment $(t - h)$.

So we get that the process of mutual interaction between industries can be described by the differential equation

*Correspondence to: N.D.Redinskikh (Email: redinskich@yandex.ru). Faculty of Applied Mathematics and Control Processes, Saint-Petersburg State University, Universitetskii prospekt 35, Petergof, Saint Petersburg, Russia 198504.

system as follows

$$\frac{dP(t)}{dt} = (a - eP(t-h) - bQ(t))P(t)$$
$$\frac{dQ(t)}{dt} = (-c + dP(t) - fQ(t-h))Q(t). \tag{1}$$

This system has a nontrivial equilibrium

$$P^* = \frac{af + bc}{bd + ef}; \; Q^* = \frac{ad - ce}{bd + ef}, \tag{2}$$

provided that

$$ad > ce. \tag{3}$$

Our aim is to find conditions on the coefficients of system (1) for stabilizing the equilibrium point (2). This would mean that the first industry produces such a quantity of raw materials that they will be processed by the second industry. In this case, it will not be overstocking in the warehouses and there will be sufficient volumes of raw materials.

Necessary and sufficient criterion for stable coexistence of two competitors was obtained in [6]. Graphically the stability areas were described in the form of multidimensional cones in [16]. In this article, our aim is to express conditions for asymptotic stability in the form of inequalities. In this case these conditions help to solve problems of control and stabilization. A delay effect on the stability of the equilibrium point was studied in [5],[2],[3].

A number of mathematical models describing the interaction between agents based on the game theory was considered in [1, 11, 12, 13, 14, 15, 17, 18]. The results on the business security, the impact of external factors on the growth of the business are given in [10, 9]. Mathematical models for delay-dependent linear systems with multiple time delays, for growing tumor, on optimal properties of special nonlinear and semi-infinite problems arising in parametric optimization, criterion for testing hypothesis about impulse response function, filtering problem for functionals of stationary sequences, stability criteria for high even order delay differential equations were considered in [7, 8][20]–[36].

2. Stability criterion for quasi-polynomial

By changing variables $x = P - P^*$, $y = Q - Q^*$ in the system (1) and writing down a linear approximation system, we get the characteristic quasi-polynomial and the characteristic equation as follows

$$H(z) = z^2 e^{2z} + c_1 e^{2z} + c_2 z e^z + c_3 = 0. \tag{4}$$

Here $c_1 = bdh^2 P^* Q^*$, $c_2 = (eP^* + fQ^*)h$, $c_3 = efh^2 P^* Q^*$. To get the conditions under which the roots of the quasi-multinomial (4) lie in the left half-plane, we use Pontryagin and Hermite - Biehler criteria [6, 5, 2, 4, 19].

Theorem 1

The roots of the quasi-polynomial (4) with positive coefficients lie in the left half-plane if one of two following assertions A or B is fulfilled:

Assertion A:

I.1. $0 < c_1 < \pi^2$;

I.2. $0 < c_2 < \frac{z(\dot{y})}{\dot{y}} = \frac{2(\dot{y}^2 - c_1)\sin\dot{y}}{\dot{y}}$, here $\dot{y} \in (\sqrt{c_1}, \pi)$ is a unique root of the equation $\tan y = \frac{(c_1 - y^2)y}{c_1 + y^2}$; notice, that $\dot{y} \in (\frac{\pi}{2}; \pi)$;

I.3.1. If $y_1 \in (0; \frac{\pi}{2})$ is a unique root of the equation $2(y^2 - c_1)\sin y = c_2 y$ then the following conditions are fulfilled:

I.3.1.1. $c_1 + c_3 < y_1^2$ and I.3.1.2. $c_1 + c_2\frac{\pi}{2} < c_3 + \frac{\pi^2}{4}$.

I.3.2. If $y_1, y_2 \in (\frac{\pi}{2}; \pi)$, $y_1 < y_2$ are the roots of the equation $2(y^2 - c_1)\sin y = c_2 y$ then the following conditions are fulfilled:

I.3.2.1. $y_1^2 < c_1 + c_3 < y_2^2$ and I.3.2.2. $c_1 + c_2\frac{\pi}{2} > c_3 + \frac{\pi^2}{4}$

Assertion B:

II.1. $\pi^2 < c_1 < 4\pi^2$;

II.2. $0 < c_2 < \frac{z(\dot{y})}{\dot{y}} = \frac{2(\dot{y}^2 - c_1)\sin\dot{y}}{\dot{y}}$, here $\dot{y} \in (\pi; \sqrt{c_1})$ is the unique root of the equation $\tan y = \frac{(c_1 - y^2)y}{c_1 + y^2}$; notice, that $\dot{y} \in (\pi; \frac{3\pi}{2})$;

II.3.1. If $y_1 < y_2$ are the roots of the equation $2(y^2 - c_1)\sin y = c_2 y$ and$y_1, C_2 \in (\pi; \sqrt{c_1})$, then the following conditions are fulfilled:

II.3.1.1.$y_1^2 < c_1 + c_3 < y_2^2$ and II.3.1.2.$c_1 + c_2\frac{\pi}{2} > c_3 + \frac{\pi^2}{4}$.

II.3.2. If $y_2 \in (\frac{3\pi}{2}; \sqrt{c_1})$ is the unique root of the equation $2(y^2 - c_1)\sin y = c_2 y$ then the following conditions are fulfilled:

II.3.2.1. $c_1 + c_3 > y_2^2$ and II.3.2.2. $\frac{9\pi^2}{4} + c_2\frac{3\pi}{2} + c_3 < c_1$.

Proof

Let us substitute $z = iy$ into the quasi-polynomial (4) and write down its real $F(y)$ and imaginary $G(y)$ parts:

$$F(y) = (-y^2 + c_1)\cos 2y - c_2 y \sin y + c_3, \tag{5}$$

$$G(y) = (-2y^2 \sin y + 2c_1 \sin y + c_2 y)\cos y = (c_2 y - z(y))\cos y = 0. \tag{6}$$

Then derivative $G'(y)$ may be written as follows

$$G'(y) = (c_2 - z'(y))\cos y - (c_2 y - z(y))\sin y. \tag{7}$$

Here $z(y) = 2(y^2 - c_1)\sin y$.

From Pontryagin and Hermite - Biehler criteria it is known that the roots of quasi-polinomial (4) have negative real parts if the vector of gain-phase cha-racteristic (amplitude-phase characteristic) $w = H(iy)$ monotonically rotates counterclockwise round the origin with positive rate [6, 5, 2, 4, 19]. It means that the gain-phase characteristic turning around the origin crosses every line passing through the point (0,0) at the positive angle without touching it [6, 5]. In this case all roots of the functions $F(y)$ and $G(y)$ are real, simple, alternate and the inequality

$$F(y)G'(y) - F'(y)G(y) > 0 \tag{8}$$

holds for all y. For the stability of quasi-polinomial (4) it is sufficient for the inequality (8) to be valid only at the roots of function $G(y)$. In the future, we will consider the inequality

$$F(y)G'(y) > 0 \tag{9}$$

at the roots of the function $G(y)$.

Due to the Pontryagin criterion [19] inequalities (8) and (9) are valid iff the rotation angle of gain-phase characteristic $(F(y); G(y))$ around the origin asymptotically tends to $\varphi(-2\pi k + \varepsilon \le y \le 2\pi k + \varepsilon) \approx (4ks + r)\pi$ as k tends to infinity. Here s is the degree of quasi-polynomial (4) with respect to $e^z$and r is the degree of quasi-polynomial (4) with respect to z. In our case $s = r = 2$ so, polynomials $F(y)$ and $G(y)$ have each 10 roots on the segment $[-2\pi + \varepsilon; 2\pi + \varepsilon]$. As the function $G(y)$ is odd it can not have more than 9 roots over the segment $[-2\pi; 2\pi]$. One root is $y_0 = 0$, the other ones should lie symmetrically with 4 on each side. The tenth root will be discussed later.

Proof of Assertion A. Note that $\frac{\pi}{2}$ and $\frac{3\pi}{2}$ are the roots of the function $G(y)$. Other two roots $y_1 < y_2 \in (0; 2\pi)$ are the solutions of the equation

$$z(y) = 2(y^2 - c_1)\sin y = c_2 y. \tag{10}$$

Since $z'(0) = -2c_1 < 0$, the equation (10) has no roots for small positive y. On the other hand, the function $z(y)$ crosses the x-axis at the points $y = \sqrt{c_1}$ and $y = \pi$, so two cases can occur.

Let $0 < c_1 < \pi^2$. Then the function $z(y) = 2(y^2 - c_1)\sin y$ is positive over $(\sqrt{c_1}, \pi)$ and equation (10) has 2 solutions on this interval if the coefficient A_2 is less than the slope of tangent line to the graph of the function

$z = z(y)$, drawn from the origin. Then the touch point $\dot{y}$ is the solution of the equation $z(y) = z'(y)y$ or $\tan y = \frac{(c_1 - y^2)y}{c_1 + y^2}$. As we consider the solution of this equation on $(\sqrt{c_1}, \pi)$, so $\tan y < 0$ and the touch point $\dot{y} \in (\frac{\pi}{2}; \pi)$. In this case equation (10) has two roots $y_1 < y_2 \in (0; \pi)$ if the condition I.2 of the Theorem 1 is satisfied. To satisfy the Hermite - Biehler conditions it is necessary to have alternation of signs both of derivative $G'(y)$ and function $F(y)$ in the roots of $G(y)$ and their multiplication should satisfy condition (9).

For the root $y_0 = 0$ we get $G'(0) = 2c_1 + c_2 > 0, F(0) = c_1 + c_3 > 0$. Let's assume that other roots of the function $G(y)$ are ordered as follows: $y_1 < \frac{\pi}{2} < y_2 < \frac{3\pi}{2}$. Then $c_2 < z'(y_1)$ and the inequality (7) is equivalent to the condition $G'(y_1) < 0$. For the root y_2 the inverse inequality is valid. But $\cos y_2 < 0$ so $G'(y_2) < 0$. Between these two roots a straight line $z = c_2 y$ lies under the graph of the function $z = z(y)$ so the inequality $c_2 y < z(y)$ is valid at the point $y = \frac{\pi}{2}$. From (7) we get $G'(\frac{\pi}{2}) > 0$. Taking into account the condition I.1, we obtain $G'(\frac{3\pi}{2}) > 0$. So the signs of the derivative $G'(y)$ in roots of function $G(y)$ alternate. The function $F(y)$ has the form $F(y) = c_1 + c_3 - y_j^2$, j $= 1, 2$ at the roots y_1 and y_2 of the function $G(y)$. If $c_1 + c_3 < y_1^2$ then the both inequalities $F(y_j) < 0,\ \ j = 1, 2$ are valid. Inequality $F(\frac{\pi}{2}) > 0$ corresponds to the condition I.3.1.2 of the theorem 1 and then inequality $F(\frac{3\pi}{2}) > 0$ is valid as well. So the signs of the function $F(y)$ alternate in roots of $G(y)$.

We can get the tenth root of $G(y)$ by shifting the segment $[-2\pi; 2\pi]$ to the right so that the root of the function $G(y)$ $y_3 \in (2\pi; 2\pi + \frac{\pi}{2})$ will be in this segment. As there are no roots of the function $G(y)$ on interval $(-2\pi; -\frac{3\pi}{2})$ we don't lose any of the root of the function $G(y)$ on the left side. So we have constructed a segment of 4π-length, and there are 10 roots of the function $G(y)$ on this segment.

Let the roots of the function $G(y)$ be ordered as follows: $\frac{\pi}{2} < y_1 < y_2 < \frac{3\pi}{2}$. In this case a straight line$z = c_2 y$ is located over graph of the function $z = z(y)$ on the interval $(0; y_1)$, so at point $y = \frac{\pi}{2}$ inequality $c_2 \frac{\pi}{2} > z(\frac{\pi}{2})$ is valid. From (7) we get $G'(\frac{\pi}{2}) < 0$. Similarly to the previous case, $c_2 < z'(y_1)$ and taking into account the inequality $\cos y_1 < 0$ from (7) we obtain $G'(y_1) > 0$. The inverse inequality $c_2 > z'(y_2)$ with $\cos y_2 < 0$ yields $G'(y_2) < 0$. Inequality $G'(\frac{3\pi}{2}) > 0$ is also valid, and we get the alternation of signs of the derivative $G'(y)$ in the roots of the function $G(y)$. From the condition I.3.2.2 of the Theorem 1 we obtain $F(\frac{\pi}{2}) < 0$ and from inequalities $y_1^2 < c_1 + c_3 < y_2^2$ we obtain $F(y_1) > 0$ and $F(y_2) < 0$. Inequality $F(\frac{3\pi}{2}) > 0$ is valid as well. So the signs of the function $F(y)$ in roots of $G(y)$ alternate. Further, all 2π-long segments will include four roots of the function $G(y)$ and there will be a similar alternation of the signs of the derivative $G'(y)$ and the signs of the function $F(y)$ in these roots and inequality (9) is valid as well. Assertion A is proved.

Proof of Assertion B. Let $\pi^2 < c_1 < 4\pi^2$. Then function $z(y) = 2(y^2 - c_1)\sin y$ is positive on $(\pi; \sqrt{c_1})$ and equation (10) has 2 solutions on this interval if coefficient c_2 is less than the slope of tangent to function graph $z = z(y)$, drawn from the origin. It was noted at the proof of the assertion that the touch point $\dot{y}$ is the solution of the equation $\tan y = \frac{(c_1 - y^2)y}{c_1 + y^2}$. We consider the solution of this equation on $(\pi; \sqrt{c_1})$, so the touch point $\dot{y} \in (\pi; \frac{3\pi}{2})$. Let the roots of $G(y)$ are ordered as follows $\frac{\pi}{2} < y_1 < y_2 < \frac{3\pi}{2}$. We have considered such sequence of roots in the previous part of the proof but now $\pi < y_1$. As before we have $c_2 \frac{\pi}{2} > z(\frac{\pi}{2})$ and from (7) we get $G'(\frac{\pi}{2}) < 0$. At the point y_1 we have $c_2 < z'(y_1)$ and $\cos y_1 < 0$ so the inequality (7) is equivalent to the inequality $G'(y_1) > 0$. At the point y_2 we have $c_2 > z'(y_2)$ and $\cos y_2 < 0$, so from (7) we obtain $G'(y_2) < 0$. Inequality $G'(\frac{3\pi}{2}) > 0$ is fulfilled as well. Inequalities $F(\frac{\pi}{2}) < 0, F(y_1) > 0, F(y_2) < 0, F(\frac{3\pi}{2}) > 0$ correspond to the inequalities $c_1 + c_2 \frac{\pi}{2} < c_3 + \frac{\pi^2}{4}$, $y_1^2 < c_1 + c_3 < y_2^2$, so the signs of the function $F(y)$ in roots of $G(y)$ alternate. If $\frac{\pi}{2} < y_1 < \frac{3\pi}{2} < y_2$, and $\pi < y_1$, then $G'(\frac{\pi}{2}) < 0$. The root $y = \frac{3\pi}{2}$ lies between two roots $y_1 < y_2$ so the straight line $z = c_2 y$ is located under the graph of the function $z = z(y)$ or $c_2 y < z(y)$. From this inequality we obtain $G'(y) = -\sin y\ (c_2 y - z(y) < 0$ for $y = \frac{3\pi}{2}$. So the signs of the derivative $G'(y)$ in roots of function $G(y)$ alternate. The corresponding alternation of signs of function $F(y)$ takes place iff the inequalities $\frac{9\pi^2}{4} + c_2 \frac{3\pi}{2} + c_3 < c_1$ and $c_1 + c_3 > y_2$ are satisfied. If $\frac{\pi}{2} < \frac{3\pi}{2} < y_1 < y_2$, then the both inequalities $c_1 + c_3 < y_1$ and $c_1 + c_3 > y_2$ are fulfilled together. It contradicts to the inequality $y_1 < y_2$. So the assertion B of the Theorem 1 is proved.

If $c_1 > 4\pi^2$, the segment $[0; 2\pi]$ contains only two roots of the function $G(y)$, whereas Pontryagin criterion requires four roots. □

3. Criterion for stable co-functioning of the two industries

In this paragraph we obtain conditions under which the goods quantity meets the demand. Under these conditions there is as overproducing of raw materials and its shortage as well. Proving the theorem 1 we have considered only linear system associated with the equilibrium point $(P^*;Q^*)$ without square members. From the theory of the differential equations it is known that in this case the equilibrium point of system (1) is asymptotically stable if the approximating linear system is asymptotically stable. The last statement is valid if the roots of the quasi-polynomial (4) lie in the left half-plane. Thus, we obtain the following theorem.

Theorem 2

A non-trivial equilibrium $(P^*;Q^*)$ of the system (1) is asymptotically stable iff the following conditions are fulfilled:

Assertion A:

I.1. $0 < bdh^2P^*Q^* < \pi^2$

I.2. $0 < (eP^* + fQ^*)h < \frac{z(\dot{y})}{\dot{y}} = \frac{2(\dot{y}^2 - bdh^2P^*Q^*)\sin\dot{y}}{\dot{y}}$, here $\dot{y} \in (\sqrt{bdh^2P^*Q^*}, \pi)$ is the unique root of the equation $\tan y = \frac{(bdh^2P^*Q^* - y^2)y}{bdh^2P^*Q^* + y^2}$;

I.3.1. If $y_1 \in (0; \frac{\pi}{2})$ is a root of equation $2(y^2 - bdh^2P^*Q^*)\sin y = (eP^* + fQ^*)hy$ then the following conditions are fulfilled:

I.3.1.1. $bdh^2P^*Q^* + efh^2P^*Q^* < y_1^2$ and

I.3.1.2. $bdh^2P^*Q^* + (eP^* + fQ^*)h\frac{\pi}{2} < efh^2P^*Q^* + \frac{\pi^2}{4}$.

I.3.2. If $y_1, y_2 \in (\frac{\pi}{2}; \pi)$, $y_1 < y_2$ are the roots of the equation $2(y^2 - bdh^2P^*Q^*)\sin y = (eP^* + fQ^*)hy$ then the following conditions are fulfilled:

I.3.2.1. $y_1^2 < bdh^2P^*Q^* + efh^2P^*Q^* < y_2^2$ and

I.3.2.2. $bdh^2P^*Q^* + (eP^* + fQ^*)h\frac{\pi}{2} > efh^2P^*Q^* + \frac{\pi^2}{4}$

Assertion B:

A non-trivial equilibrium $(P^*;Q^*)$ of the system (1) is asymptotically stable if the following conditions are fulfilled:

II.1. $\pi^2 < bdh^2P^*Q^* < 4\pi^2$;

II.2. $0 < (eP^* + fQ^*)h < \frac{z(\dot{y})}{\dot{y}} = \frac{2(\dot{y}^2 - bdh^2P^*Q^*)\sin\dot{y}}{\dot{y}}$,here $\dot{y} \in (\pi; \sqrt{bdh^2P^*Q^*})$ is the unique root of the equation $\tan y = \frac{(bdh^2P^*Q^* - y^2)y}{bdh^2P^*Q^* + y^2}$;

II.3.1. If $y_1, y_2 \in (\pi; \frac{3\pi}{2})$ $y_1 < y_2$ are the roots of the equation $2(y^2 - bdh^2P^*Q^*)\sin y = (eP^* + fQ^*)hy$ then the following conditions are valid:

II.3.1.1. $y_1^2 < bdh^2P^*Q^* + efh^2P^*Q^* < y_2^2$ and

II.3.1.2. $bdh^2P^*Q^* + (eP^* + fQ^*)h\frac{\pi}{2} > efh^2P^*Q^* + \frac{\pi^2}{4}$.

II.3.2. If $y_2 \in (\frac{3\pi}{2}; \sqrt{bdh^2P^*Q^*})$ is a root of the equation$2(y^2 - bdh^2P^*Q^*)\sin y = (eP^* + fQ^*)hy$ then the following conditions are valid:

II.3.2.1. $bdh^2P^*Q^* + efh^2P^*Q^* > y_2^2$,

II.3.2.2. $\frac{9\pi^2}{4} + (eP^* + fQ^*)h\frac{3\pi}{2} + efh^2P^*Q^* < bdh^2P^*Q^*$.

4. Conclusion and Discussion

1. A mathematical model of dynamic interaction between mining and processing industries is described by the system of two nonlinear delay differential equations. At the proposed model we take into account the volume of raw materials mined and shipped in the preceding time $(t - h)$. In the Theorem 2 we find conditions on the coefficients of system (1) for stabilizing the equilibrium point (2). It means that the first industry produces such a quantity of raw materials that they will immediately be processed by the second industry. It means that there is a balance between the amount of extracted raw materials and the number of processing enterprises.

2. The conditions under which the solutions of system (1) tend to equilibrium point $(Q^*; P^*)$ are given in Theorem 2. The conditions I.1, I.2, I.3.1.1 and I.3.1.2 taken together, give a size of admissible interval between deliveries of raw material at which the balance between industries is kept. The size of this $h^2 < \frac{y_1^2}{(bd+ef)P^*Q^*}$, where $y_1 \in (0; \frac{\pi}{2})$ is a root of equation $2(y^2 - bdh^2P^*Q^*)\sin y = (eP^* + fQ^*)hy$.
3. If the interval between deliveries of raw material is not small but conditions I.1, I.2, I.3.2.1 and I.3.2.2 of Theorem 2 taken together are fulfilled then the balance between industries is kept also. In this case size of this admissible interval is $\frac{y_1^2}{(bd+ef)P^*Q^*} < h^2 < \frac{y_2^2}{(bd+ef)P^*Q^*}$, where $y_1, y_2 \in (\frac{\pi}{2}; \pi)$, $y_1 < y_2$ are the roots of the equation $2(y^2 - bdh^2P^*Q^*)\sin y = (eP^* + fQ^*)hy$.
4. Moreover if delay h is rather large $\pi^2 < bdh^2P^*Q^* < 4\pi^2$ (condition II.I of Theorem 2) the stability of the equilibrium point $(Q^*; P^*)$ can be restored if conditions of Assertion B of Theorem 2 are fulfilled.
5. If this time interval between supplies of raw materials increases, the asymptotic stability of equilibrium point stabilization. Restrictions obtained in the Theorem 2 allow to solve the problem of stabilization of equilibrium point.
6. Assume that a priori it is known the quantity of final products required to meet demand in the region, and to manifacture these final products we need a certain amount of raw materials. It means that initially the equilibrium point is known, but the point $(Q^*; P^*)$ determined by the formula (2), does not meet the demand for these products. In this case, it is necessary to change the technological process of extraction and processing (change coefficients a, b, c, d, e, f) so that point $(Q^*; P^*)$ is consistent with the economically justified demand. Then it is possible to solve the above mentioned problem of finding time intervals between deliveries of raw materials which make it possible to preserve stable interaction between the two industries.

REFERENCES

1. X. Grigorieva and O.A. Malafeev, *Competitive many-period postman problem with varying parameters*, Applied Mathematical Sciences, vol. 8, no. 145-148, pp. 7249–7258, 2014.
2. A. Kirjanen and V. Veselov, *The influence of capital turnover time and coefficient criteria of asymptotic stability of two competing firms*, Proceedings of V International Conference State and Business, pp. 50-52, 2013.
3. A. Kirjanen and O. Shalyapina, *Pointwise synthes controllability of systems with delay*, Contemporary Engineering Sciences, vol. 2, pp. 14–18, 1989.
4. J. K. Hale, *Theory of functional differential difference equations*, Springer Verlag, New York, 1977.
5. A. I. Kirjanen, *Stability of systems with aftereffect and there applications*, St.Petersburg State University, Russia, Saint-Petersburg, 1994.
6. A.Kirjanen, *Coefficient criteria of stable coexistence of two competitors*, Stability and Control Processes 2015 - Proceedings, pp.463–466, 2015.
7. L.A. Bondarenko and A.V. Zubov and V.B. Orlov and V.A. Petrova and N.S. Ugegov, *Application in practice and optimization of industrial information systems*, Journal of Theoretical and Applied Information Technology, vol. 85, no.3, pp. 305–308, 2016.
8. A.V.Zubov and A.Y. Murashko and L.G. Kolyada and E.A. Volkova and O.A. Zubova, *Fidelity issue of engineering analysis and computer aided calculations in sign models of dynamic systems*, Global Journal of Pure and Applied Mathematics, vol. 12, no.5, pp. 4203-4217, 2016.
9. A.I. Kirjanen and A.A. Samodurov, *Influence of external factors on business process of company in dynamics*, Proc. of the 7th International Research and Practical Conference, pp. 88–92, 2015.
10. A.I.Kirjanen and A.V.Labudin and A.A.Samodurov, *Business-Process Approach to Economic Security Management of the Enterprise with Attraction of the Equation of Profitability*, Russian Public Policy Journal (RuPP) Upravlenceskoe konsultirovanie, vol. 3, pp. 96–105, 2016.
11. V.N.Kolokoltsov and O.A.Malafeyev, *Mean-Field-Game Model of Corruptions*, Dynamic Games and Applications, vol. 7, no.1, pp. 34–47, 2017.
12. O.A. Malafeyev and N.D. Redinskikh and G.V. Alferov, *Electric circuits analogies in economics modeling: Corruption networks*, BDO 2014 - Proceedings, http://dx.doi.org/10.1109/Emission.2014.6893965, 2014.
13. O.A. Malafeyev and L.A. Petrosyan, *Differential search games - Dynamic-games with complete information* Vestnik Leningradskogo Universiteta, Seriya Matematika, Mekhanika, Astronomiya, vol. 2, pp. 26-30, 1983.
14. O.A. Malafeev, *Existence of equilibrium points in differential noncooperative many-person games*, Vestnik Leningradskogo Universiteta, Seriya Matematika, Mekhanika, Astronomiya, vol.3, pp.40-46, 1982.
15. O.A. Malafeev, *On the existence of equilibrium points in 2-person differential games with separated dynamics*, Vestnik Leningradskogo Universiteta, Seriya Matematika, Mekhanika, Astronomiya, pp.12-16, 1980.
16. V.V. Malygina and M.M. Kipnis, *The stability cone for a matrix delay difference equation*, International Journal of Mathematics and Mathematical Sciences, http://dx.doi.org/10.1155/2011/860326, 2011.
17. E.G. Neverova and O.A. Malafeyef, *A model of interaction between anticorruption authority and corruption groups*, AIP Conference Proceedings, http://dx.doi.org/10.1063/1.4912671, 2015.

18. Yu.A. Pichugin and O.A. Malafeev, *Statistical estimation of corruption indicators in the firm*, Applied Mathematical Sciences, vol.10, no.42, pp.2065-2073, 2016.
19. L.S. Pontryagin, *On the zeros of some elementary transcendental functions*, American Mathematical Society, vol.2, pp. 95–110, 1955.
20. Y.V.Kozachenko and I.V.Rozora, *A criterion for testing hypothesis about impulse response function*, Statistics, Optimization and Information Computing, vol. 4, pp. 214–232, 2016.
21. O.Kostyukova and T.Tchemisova and M.Kurdina, *On optimal properties of special nonlinear and semi-infinite problems arising in parametric optimization*, Statistics, Optimization and Information Computing, vol. 5, pp. 99-108, 2017.
22. N.N. Subbotina and E.A. Kolpakova, *Method of Characteristics for Optimal Control Problems and Conservation Laws*, Journal of Mathematical Sciences, vol. 199, no.5, pp. 588–595, 2014.
23. M.Luz and M.Moklyachuk, *Filtering problem for functionals of stationary sequences*, Statistics, Optimization and Information Computing, vol.4, pp.68–83, 2016.
24. O.A.Malafeev and S.A.Nemnyugin, *Generalized dynamic model of a system moving in an external field with stochastic components*, Theoretical and Mathematical Physics, vol.107, pp.770–774, 1996.
25. E.G.Neverova and O.A.Malafeyev and G.V.Alferov and T.E.Smirnova, *Model of interaction between anticorruption authorities and corruption groups*, Stability and Control Processes 2015 - Proceedings, pp.488–490, 2015.
26. G.V.Alferov and O.A.Malafeyev, *The robot control strategy in a domain with dynamical obstacles*, Lecture Notes in Computer Science (including subseries Lecture Notes in Artificial Intelligence and Lecture Notes in Bioinformatics), vol.1093, pp.211–217, 1996.
27. O.A. Malafeev, *Stationary strategies in differential games*, USSR Computational Mathematics and Mathematical Physics, vol.17, pp.37–46, 1977.
28. O.A.Malafeev, *Equilibrium situations in dynamic games*, Cybernetics, vol.10, pp.504–513, 1974.
29. Z.Taushanov and A.Berchtold, *A direct local search method and its application to a markovian model*, Statistics, Optimization and Information Computing, vol. 5, no. 1, pp. 19-34, 2017.
30. Y.V.Kozachenko and M.Y.Petranova, *Proper complex random processes*, Statistics, Optimization and Information Computing, vol. 5, no. 2, pp. 137-146, 2017.
31. M.Moklyachuk and M.Sidei, *Filtering problem for stationary sequences with missing observations*, Statistics, Optimization and Information Computing, vol. 4, no. 4, pp. 308-325, 2016.
32. X.Naihua, *A SQP method for general nonlinear complementarity problems*, Applied Mathematics, vol.15, no.4, pp. 433-442, 2000.
33. X.Naihua, *A class of trust region methods for linear inequality constrained optimization and its theory analysis: I. Algorithm and global convergence*, Applied Mathematics, vol.10, no.3, pp. 287-296, 1995.
34. O.A.Malafeyev and S.A.Nemnyugin and G.A.Ivaniukovich, *Stochastic models of social-economic dynamics*, Stability and Control Processes 2015 - Proceedings, pp.483–485, 2015.
35. G.D.Drozdov and O.A.Malafeyev and S.A.Nemnyugin, *Multicomponent dynamics of competitive single-sector economy development*, Stability and Control Processes 2015 - Proceedings, pp.457–459, 2015.
36. O.A.Malafeev, *The existence of situations of ε-equilibrium in dynamic games with dependent movements*, USSR Computational Mathematics and Mathematical Physics, vol.14, pp.88–99, 1974.

14

Existence theorem and optimality conditions for a class of convex semi-infinite problems with noncompact index sets

Olga Kostyukova [1], Tatiana Tchemisova [2,*], Maryia Kurdina [1]

[1]*Institute of Mathematics, National Academy of Sciences of Belarus, Minsk, Belarus*
[2]*Center for Research and Development in Mathematics and Applications, University of Aveiro, Aveiro, Portugal*

Abstract The paper is devoted to study of a special class of semi-infinite problems arising in nonlinear parametric optimization. These semi-infinite problems are convex and possess noncompact index sets. In the paper, we present conditions, which guarantee the existence of optimal solutions, and prove new optimality criterion. An example illustrating the obtained results is presented.

Keywords Semi-Infinite Programming (SIP), Linear Programming (LP), Quadratic Programming (QP), Constraint Qualifications (CQ).

1. Introduction

Semi-infinite Programming (SIP) deals with optimization problems, which have an infinite number of constraints. SIP has always been a topic of a special interest due to the numerous theoretical and practical applications such as robotic, classical engineering, optimal design, the Chebyshev approximations, etc. (see [6, 7, 8], and the references therein). Nowadays, SIP models are efficiently used in dynamic processes, biomedical and chemical engineering, biology, tissue engineering, polymer reaction engineering, etc. (see [1, 16], and others). A general SIP problem can be formulated as

$$\min_{\varkappa\in\mathbb{R}^n} c(\varkappa) \quad \text{s.t.} \quad f(\varkappa,\tau)\leq 0 \ \ \forall \tau\in T,$$

where $\varkappa\in\mathbb{R}^n$ is a decision variable, τ is a constraint' index, $T\subset\mathbb{R}^p$ is an infinite index set. When, additionally, the index set T depends on the decision variable $\varkappa$, one gets a problem of the *generalized* SIP (see [9]). We say that a SIP problem is *continuous* whenever the index set T is a compact Hausdorff topological space and the functions $c(\varkappa)$ and $f(\varkappa,\tau)$ are continuous w.r.t. their variables. The compactness of the index set T ensures the existence of global maximizers of the so-called *lower level problem*: $\max\limits_{t\in T} f(\varkappa,\tau)$. The continuity of the functions defining the SIP problem is a natural condition, which permits to apply the methods of continuous optimization. Usually, it is assumed that a SIP problem is continuous. However, it should be noted that there are some classes of problems, for which noncompact index set is commonplace. Without the compactness of T and/or the continuity of the inequality

*Correspondence to: Tatiana Tchemisova (Email: tatiana@ua.pt). Center for Research and Development in Mathematics and Applications, Department of Mathematics, University of Aveiro, Campus Universitário Santiago, 3810-193, Aveiro, Portugal.

constraint function with respect to the index variable, the properties of SIP problems change dramatically and the known SIP theory and methods may fail.

In this paper, we explore a class of SIP problems, in which the corresponding index sets have a conic structure and hence are not compact. Such problems arise in nonlinear parametric SIP when the differential properties of the solutions are being studied, and therefore, it is important to study these problems, which is the main purpose of this paper.

The rest of the paper is organized as follows. In Section 2, we state an optimization problem arising in nonlinear parametric SIP, and formulate the main aims of the paper. In Section 3, we show that the optimization problem formulated in Section 2 can be reduced to a convex SIP problem of special form with noncompact index set. In Section 4, for the special class of convex SIP problems with noncompact index sets, we present the existence theorem and discuss a possibility of its strengthening, and in Section 5, we formulate and prove the optimality conditions. Section 6 contains an example, which illustrates that the results obtained in the paper can be applied in situations, where the classical optimality results are not efficient due to noncompactness of the index set. The conclusions and final remarks are done in the final Section 7. In Appendix, we prove Proposition 1, which is need for the proof of the main results of the paper.

2. Problem statement

Let finite index sets $J \subset \mathbb{N},\ S_* \subset \mathbb{N}$, and $S \subset \mathbb{N}$, such that $S_* \cap S = \emptyset$, as well as matrices, vectors and numbers

$$\bar{W}_j \in \mathbb{R}^{n\times n},\ \bar{d}_j \in \mathbb{R}^n,\ \bar{r}_j \in \mathbb{R},\ g_j \in \mathbb{R}^n,\ \ j \in J,\ \ c \in \mathbb{R}^n,$$

$$W_0 \in \mathbb{R}^{n\times n},\ d_0 \in \mathbb{R}^n,\ r_0 \in \mathbb{R};\ \ q_j \in \mathbb{R}^n,\ \omega_j \in \mathbb{R},\ j \in S_* \cup S,$$

$$D_j \in \mathbb{R}^{p\times p},\ A_j \in \mathbb{R}^{n\times p},\ B_j \in \mathbb{R}^{m_j\times p},\ c_j \in \mathbb{R}^p, j \in J,$$

be given. Define the following sets:

$$Y = \{y = (y_j, j \in J) : \sum_{j\in J} g_j y_j = c, y_j \geq 0, j \in J\},$$
$$\mathcal{X} = \{x \in \mathbb{R}^n : q_j^T x + \omega_j = 0, j \in S_*,\ \ q_j^T x + \omega_j \leq 0, j \in S\},$$
$$K(j) = \{t \in \mathbb{R}^p : B_j t \leq 0\},\ \ j \in J,$$

and suppose that

1) the (polyhedral) set Y is nonempty and bounded;
2) the set $\mathcal{X}$ is nonempty;
3) the following inequalities are satisfied:

$$x^T W_0 x \geq 0,\ x^T \bar{W}_j x \geq 0,\ \forall x \in \mathbb{R}^n,\ \ t^T D_j t \geq 0\ \forall t \in K(j),\ j \in J.$$

Consider the following optimization problem:

$$\min_{x\in\mathcal{X}} \left[\Omega_0(x) + \max_{y\in Y} \sum_{j\in J} y_j \left(\Omega_j(x) - \min_{t\in K(j)} \Psi_j(x,t) \right) \right], \tag{1}$$

where

$$\Omega_0(x) := \frac{1}{2} x^T W_0 x + d_0^T x + r_0,\ \ \Omega_j(x) := \frac{1}{2} x^T \bar{W}_j x + \bar{d}_j^T x + \bar{r}_j,\ j \in J;$$

$$\Psi_j(x,t) := \frac{1}{2} t^T D_j t + (c_j^T - x^T A_j) t,\ j \in J.$$

Note that in (1), functions $\Omega_j(x), j \in J \cup \{0\}$, are convex w.r.t. $x \in \mathbb{R}^n$, and functions $\Psi_j(x,t),\ j \in J$, are linear w.r.t. $x \in \mathbb{R}^n$, and, in general, are non-convex w.r.t. $t \in K(j), j \in J$.

The problems in the form (1) arise in nonlinear parametric SIP when the differential properties of solutions are being studied [10, 12]. Therefore, it is important to study the properties of these optimization problems.

The main aims of this paper are as follows:

- to show that the optimization problem (1) is equivalent to some special SIP problem;
- to study the issues connected with conditions, which guarantee the existence of optimal solutions of the special SIP problem (it should be mentioned that, at the moment, such conditions are not very well studied in SIP);
- to formulate the optimality conditions for the special SIP problem.

3. Semi-infinite formulation

Let $y^{(i)} = (y_j^{(i)}, j \in J),\ i \in I$, be the vertices (the extremal points) of the polyhedral set Y defined in Section 2. In what follows, we suppose that these vertices are known.

Let us show that

$$\max_{y\in Y} y^T f = \max_{i\in I} {y^{(i)}}^T f \tag{2}$$

for any $f \in \mathbb{R}^{|J|}$.

In fact, since $y^{(i)} \in Y, i \in I$, then $\max\limits_{y\in Y} y^T f \geq \max\limits_{i\in I} {y^{(i)}}^T f$. Suppose that

$$\max_{y\in Y} y^T f = {y^0}^T f > \max_{i\in I} {y^{(i)}}^T f, \tag{3}$$

where $y^0 \in Y$. Consequently,

$${y^0}^T f > {y^{(i)}}^T f, \quad i \in I. \tag{4}$$

Since $y^0 \in Y$, then there exist numbers $\lambda_i \geq 0,\ i \in I,\ \sum\limits_{i\in I} \lambda_i = 1$, such that $y^0 = \sum\limits_{i\in I} \lambda_i y^{(i)}$.

Hence, ${y^0}^T f = \sum\limits_{i\in I} \lambda_i {y^{(i)}}^T f$. From the last equality and (4), it follows that

$${y^0}^T f = \sum_{i\in I} \lambda_i {y^{(i)}}^T f < \sum_{i\in I} \lambda_i {y^0}^T f = {y^0}^T f.$$

The resulting contradiction proves that (3) is false, and hence (2) holds.

Taking into account equality (2), we conclude that the optimization problem (1) is equivalent to the following one:

$$\min_{x\in\mathcal{X}} \left[\Omega_0(x) + \max_{i\in I} \sum_{j\in J} y_j^{(i)} \left(\Omega_j(x) - \min_{t\in K(j)} \Psi_j(x,t)\right)\right]. \tag{5}$$

Denote $W_i := \sum\limits_{j\in J} y_j^{(i)} \bar{W}_j,\quad d_i := \sum\limits_{j\in J} y_j^{(i)} \bar{d}_j,\quad r_i := \sum\limits_{j\in J} y_j^{(i)} \bar{r}_j,\quad i \in I$, and

$$\rho_j(x) := \min_{t\in K(j)} \left(\frac{1}{2} t^T D_j t + (c_j^T - x^T A_j)t\right),\ j \in J. \tag{6}$$

Then, problem (5) can be rewritten in the form

$$\min_{x,\beta} \quad \frac{1}{2} x^T W_0 x + d_0^T x + \beta$$

$$\text{s.t.} \quad \frac{1}{2} x^T W_i x + d_i^T x + r_i - \sum_{j\in J} y_j^{(i)} \rho_j(x) \leq \beta,\ i \in I;\ \ x \in \mathcal{X}.$$

Taking into account that $y_j^{(i)} \geq 0, i \in I, j \in J$, one can show that the last problem is equivalent to the following SIP problem:

$$\begin{aligned} \min_{x,\ \rho_j, j\in J,\ \beta} \quad & \frac{1}{2}x^T W_0 x + d_0^T x + \beta \\ \text{s.t.} \quad & \frac{1}{2}x^T W_i x + d_i^T x + r_i - \sum_{j\in J} y_j^{(i)}\rho_j \leq \beta,\ i \in I;\ x \in \mathcal{X}, \\ & \rho_j \leq \frac{1}{2}t_j^T D_j t_j + (c_j^T - x^T A_j)t_j,\ \ \forall\, t_j \in K(j),\ \ j \in J. \end{aligned} \tag{7}$$

In problem (7), the decision variables form the vector $\phi = (x \in \mathbb{R}^n, \rho_j \in \mathbb{R}, j \in J, \beta \in \mathbb{R})$. Without loss of generality, we may suppose that

$$\sum_{i\in I} y_j^{(i)} > 0,\ j \in J;\ \sum_{j\in J} y_j^{(i)} > 0,\ i \in I. \tag{8}$$

Let us make some observations concerning problem (7).

- The constraints and the cost function of this problem are linear-quadratic w.r.t. decision variables $x \in \mathbb{R}^n$, $\rho_j \in \mathbb{R}, j \in J$, and β. Hence, it is evident that this problem is convex, and each its local optimal solution is a global one.
- If problem (7) is consistent, then its linear constraints (infinite number)

$$-\frac{1}{2}t_j^T D_j t_j - (c_j - A_j^T x)^T t_j + \rho_j \leq 0,\ \ \forall t_j \in K(j),\ j \in J,$$

  and linear-quadratic constraints (finite number)

$$x^T W_i x + d_i^T x + r_i - \sum_{j\in J} y_j^{(i)}\rho_j - \beta \leq 0,\ i \in I,$$

  satisfy the Slater condition (i.e. these constraints are strictly satisfied for some $\bar{x} \in \mathcal{X}, \bar{\rho} \in \mathbb{R}^{|J|}, \bar{\beta} \in \mathbb{R}$).
- Since $t_j = 0 \in K(j)$, it is evident that

$$\rho_j \leq 0,\ j \in J. \tag{9}$$

- In problem (7), the index sets $K(j), j \in J$, are not compacts.
- For any feasible solution $(x, \rho_j, j \in J, \beta)$, the following relations take place:

$$(c_j^T - x^T A_j)\tau_j \geq 0\ \ \forall \tau_j \in \Delta K(j) := \{\tau \in K(j) : \tau^T D_j \tau = 0\}, j \in J. \tag{10}$$

Let us define the function

$$\beta(x) := \max_{i\in I}\Big(\frac{1}{2}x^T W_j x + d_j^T x + r_j - \sum_{j\in J} y_j^{(i)}\rho_j(x)\Big). \tag{11}$$

Note that for any feasible solution $(x, \rho_j, j \in J, \beta)$ of problem (7), there exists a feasible solution

$$(x, \rho_j(x), j \in J, \beta(x)), \tag{12}$$

such that $\beta(x)$ satisfies (11) and $\beta(x) \leq \beta$. Hence, without loss of generality, in what follows, we can consider only feasible solutions in the form (12).

As it was revealed above, our interest to the problems in the form (7) arose from the study of the solutions' properties in parametric SIP problems w.r.t. perturbations of parameters. Though, it is worth mentioning that problem (7) is an interesting subject itself. Similar SIP problems were considered, for example, in [3, 5], et al. The references that one can find in these papers, indicate also on other areas, where such problems appear.

4. On existence of optimal solutions of problem (7)

4.1. Sufficient conditions guaranteeing existence of optimal solutions

The main result of this section is Theorem 1, which gives a sufficient conditions of solvability of the convex SIP problem (7).

Given problem (7), consider the set

$$\Delta X = \{\Delta x \in \mathbb{R}^n : q_j^T \Delta x = 0,\ j \in S_*;\ q_j^T \Delta x \le 0,\ j \in S;\ W_i \Delta x = 0, i \in I \cup \{0\}; \\ \exists \mu^*(j) = \mu^*(\Delta x, j) \ge 0 \text{ such that } \Delta x^T A_j = \mu^{*T}(j) B_j, j \in J\}. \tag{13}$$

Theorem 1

Suppose that

(A) there exists $\bar{x} \in \mathcal{X}$ such that

$$(c_j^T - \bar{x}^T A_j)\tau_j \ge 0 \ \ \forall \tau_j \in \Delta K(j),\ j \in J, \tag{14}$$

(B) either the set $\Delta X \setminus \{0\}$ is empty or the following implication takes place:

$$\Delta x \in \Delta X \setminus \{0\} \ \Rightarrow \ d_0^T \Delta x + \max_{i \in I} d_i^T \Delta x > 0.$$

Then problem (7) has an optimal solution.

Proof. Let us prove, first, that under condition (A), the feasible set X in problem (7) is nonempty. Note that it follows from assumption 2) (see Section 2) and condition (A), that $\mathcal{X} \neq \emptyset$ and there exists $\bar{x} \in \mathcal{X}$ satisfying (14).

Given $\bar{x} \in \mathcal{X}$ and $j \in J$, consider a Quadratic Programming (QP) problem

$$\rho_j(\bar{x}) = \min_{t \in K(j)} (\frac{1}{2} t^T D_j t + c_j^T t - \bar{x}^T A_j t).$$

Taking into account condition (A) and the results from [4], we conclude that the problem above has an optimal solution. Consequently, the vector $(\bar{x}, \rho_j(\bar{x}), j \in J, \beta(\bar{x}))$ (with $\beta(x)$ defined in (11)) is a feasible solution of problem (7) and the set of feasible solutions X in this problem is nonempty. As it was noted in Section 3, for any $(x, \rho_j, j \in J, \beta) \in X$, there exists a feasible solution $(x, \rho_j(x), j \in J, \beta(x))$, such that $\beta(x) \le \beta$. Hence, without loss of generality, we can consider only such feasible solutions.

Now, let us prove that, if, additionally, the condition (B) is satisfied, then the convex problem (7) has an optimal solution. Consider any sequence of feasible solutions $x^k \in X$, $k = 1, 2, ...$, such that the corresponding sequence of the cost function values of problem (7) decreases. Hence the following inequalities take place:

$$x^{k^T} W_0 x^k + d_0^T x^k + \beta(x^k) \le x^{1^T} W_0 x^1 + d_0^T x^1 + \beta(x^1) = const, \ \ k = 1, 2... \tag{15}$$

Let us show that there exists a number $M_0 > 0$ such that $||x^k|| \le M_0$, $k = 1, 2, ...$ Having supposed that on the contrary, such M_0 does not exist, without loss of generality, we can consider that $||x^k|| \to \infty$ as $k \to \infty$. It follows from the constraints of (7) that

$$\frac{1}{2} x^{kT} W_i x^k + d_i^T x^k + r_i - \sum_{j \in J} y_j^{(i)} \rho_j(x^k) \le \beta(x^k),\ i \in I. \tag{16}$$

Divide inequalities (15) and (16) by $||x^k||^2$, and pass to the limit as $k \to \infty$. As a result, we obtain

$$\Delta x^{*T} W_0 \Delta x^* + \lim_{k \to \infty} \frac{\beta(x^k)}{||x^k||^2} \le 0,\ \frac{1}{2} \Delta x^{*T} W_i \Delta x^* - \sum_{j \in J} y_j^{(i)} \Delta \rho_j^* \le \lim_{k \to \infty} \frac{\beta(x^k)}{||x^k||^2},\ i \in I, \tag{17}$$

where $\Delta x^* = \lim\limits_{k\to\infty} \frac{x^k}{||x^k||}$, $\Delta \rho_j^* = \lim\limits_{k\to\infty} \frac{\rho_j(x^k)}{||x^k||^2}, j \in J$.

Taking into account inequalities (8), (9), and the positive semi-definitiveness of the matrices $W_i,\ i \in I \cup \{0\}$, we conclude from (17) that

$$W_i \Delta x^* = 0,\ i \in I \cup \{0\};\ \Delta \rho_j^* = 0,\ j \in J. \quad (18)$$

Moreover, it is easy to show that

$$q_j^T \Delta x^* = 0,\ j \in S_*;\ q_j^T \Delta x^* \leq 0,\ j \in S. \quad (19)$$

For the fixed $j \in J$ and $k \in \mathbb{N}$, consider the problem

$$\min_{t \in K(j)} (\frac{1}{2} t^T D_j t + c_j^T t - {x^k}^T A_j t). \quad (20)$$

Since $x^k \in X$, this problem has an optimal solution, which we denote here by $t_j^k \in K(j)$, $\rho_j(x^k)$ being the optimal value of the cost function. Without loss of generality, let us consider that t_j^k is an optimal solution, which has the minimal norm. For t_j^k, the following first order (necessary) optimality conditions take place:

$$D_j t_j^k + c_j - A_j^T x^k + B_j^T \mu(k,j) = 0,\ \mu(k,j) \geq 0,\ {t_j^k}^T B_j^T \mu(k,j) = 0. \quad (21)$$

Below, with no loss of generality, we will suppose that in (21) it holds $\mu(k,j) = \mu^*(k,j)$ with

$$\mu^*(k,j) := \arg \min_{\mu \in \mathcal{M}(k,j)} ||\mu||, \quad (22)$$

where $\mathcal{M}(k,j) \subset \mathbb{R}^{m_j}$ is the set of all $\mu(k,j)$ satisfying (21). It follows from (21), that

$$\rho_j(x^k) = -\frac{1}{2} {t_j^k}^T D_j t_j^k.$$

Having divided both sides of this equality by $||x^k||^2$ and passing to the limit as $k \to \infty$, we get

$$0 = \Delta \rho_j^* = -\frac{1}{2} \Delta t_j^T D_j \Delta t_j \sigma^2 \text{ with } \Delta t_j = \lim_{k\to\infty} \frac{t_j^k}{||t_j^k||},\ \sigma = \lim_{k\to\infty} \frac{||t_j^k||}{||x^k||}.$$

There are two possible situations here: I) $\sigma > 0,\ \ \Delta t_j^T D_j \Delta t_j = 0,$ and II) $\sigma = 0.$

Suppose, first, that the situation I) takes place. Hence, there exists Δt_j such that

$$\Delta t_j \neq 0,\ \Delta t_j^T D_j \Delta t_j = 0,\ B_j \Delta t_j \leq 0.$$

Denote

$$M_a^k(j) := \{m \in \{1, ..., m_j\} : b_{mj}^T t_j^k = 0\},\ M_a(j) := \{m \in \{1, ..., m_j\} : b_{mj}^T \Delta t_j = 0\}.$$

Here and in what follows, b_{mj}^T denotes the m-th row of matrix B_j.

For sufficiently large k, it is evident that $M_a^k(j) \subset M_a(j)$ and from (21) we conclude that $\mu_m(k,j) = 0$, $m \in \{1, ..., m_j\} \setminus M_a^k(j)$. Here $\mu_m(k,j)$ denotes the m-th element of vector $\mu(k,j) \in \mathbb{R}^{m_j}$. Based on these observations, it is easy to show that $(\mu(k,j))^T B_j \Delta t_j = 0$ for sufficiently large k. Taking into account this equality, let us multiply the first equality in (21) by Δt_j^T:

$$\Delta t_j^T D_j t_j^k + (c_j^T - {x^k}^T A_j) \Delta t_j = 0. \quad (23)$$

If suppose that $(c_j^T - {x^k}^T A_j) \Delta t_j < 0$, then we should conclude that the cost function of problem (20) in not bounded from below on the feasible set. But this is impossible, since this problem admits an optimal solution.

Therefore, the following inequality holds true:

$$(c_j^T - x^{k^T} A_j)\Delta t_j \geq 0. \tag{24}$$

Let us show that for any $j \in J$,

$$t^T D_j \tau \geq 0 \text{ for all } t \in K(j),\ \tau \in \Delta K(j). \tag{25}$$

In fact, for any $t \in K(j)$, $\tau \in \Delta K(j)$ and $\theta \geq 0$, we have $(\theta t + \tau) \in K(j)$. Hence for all $\theta \geq 0$ it holds $0 \leq (\theta t + \tau)^T D_j(\theta t + \tau) = \theta^2 t^T D_j t + \theta t^T D_j \tau$. For a sufficiently small $\theta \geq 0$, it follows from the last inequality that $t^T D_j \tau \geq 0$ and the relations (25) are proved.

Since $t_j^k \in K(j), \Delta t_j^T D_j \Delta t_j = 0, \Delta t_j \in \Delta K(j)$, it follows from (25) that $\Delta t_j^T D_j t_j^k \geq 0$. Then the last inequality together with (23) and (24) imply

$$(c_j^T - x^{k^T} A_j)\Delta t_j = 0,\ \ \Delta t_j^T D_j t_j^k = 0. \tag{26}$$

By construction, the vector t_j^k can be written in the form

$$t_j^k = (\Delta t_j + \varepsilon_k)\tilde{\theta}_k, \text{ where } \tilde{\theta}_k = ||t_j^k||,\ \varepsilon_k \to 0 \text{ as } k \to \infty. \tag{27}$$

Taking into account this representation, we conclude from (26) that

$$0 = \Delta t_j^T D_j(\Delta t_j + \varepsilon_k)\tilde{\theta}_k \ \ \Rightarrow \ \ \Delta t_j^T D_j \varepsilon_k = 0,\ k = 1, 2, ..., \tag{28}$$

and from the inequalities $b_{mj}^T t_j^k \leq 0, m = 1, ..., m_j$, we get $b_{mj}^T(\Delta t_j + \varepsilon_k) \leq 0, m = 1, ..., m_j$. Consequently, for $m = 1, ..., m_j$, the following implication is valid: if $b_{mj}^T \varepsilon_k > 0$, then $b_{mj}^T \Delta t_j < 0$.

Let us set $\tilde{\alpha}_m(k) = 0$ if $b_{mj}^T \varepsilon_k \leq 0$, $\tilde{\alpha}_m(k) = -\frac{b_{mj}^T \varepsilon_k}{b_{mj}^T \Delta t_j}$ if $b_{mj}^T \varepsilon_k > 0, m = 1, ..., m_j$, and calculate

$$\tilde{\alpha}(k) := \max\{\tilde{\alpha}_m(k), m = 1, ..., m_j\}.$$

It is easy to check that $\tilde{\alpha}(k) \to 0$ as $k \to \infty$. Consider vector $\tilde{t}_j^k := (\tilde{\alpha}(k)\Delta t_j + \varepsilon_k)\tilde{\theta}_k$. By construction, it holds

$$B_j \tilde{t}_j^k \leq 0 \ \ \Leftrightarrow \ \ \tilde{t}_j^k \in K(j).$$

Taking into account (26)-(28), we get

$$\tilde{t}_j^{kT} D_j \tilde{t}_j^k = \tilde{\theta}_k^2 \varepsilon_k^T D_j \varepsilon_k,\ \ t_j^{kT} D_j t_j^k = \tilde{\theta}_k^2 \varepsilon_k^T D_j \varepsilon_k,$$

$$(c_j^T - x^{kT} A_j)\tilde{t}_j^k = \tilde{\theta}_k(c_j^T - x^{kT} A_j)\varepsilon_k,\ \ (c_j^T - x^{kT} A_j)t_j^k = \tilde{\theta}_k(c_j^T - x^{kT} A_j)\varepsilon_k.$$

From the last equalities, it follows that for $k = 1, 2, ...$, both vectors, $\tilde{t}_j^k$ and t_j^k are optimal solutions of problem (20). Note that for the large numbers k, the inequality $||\tilde{t}_j^k|| < ||t_j^k||$ takes place, that is impossible since t_j^k is the minimal norm optimal solution of problem (20). The obtained contradiction permits to conclude that the situation I) is not possible.

Now, suppose that the situation II) takes a place:

$$\sigma = \lim_{k \to \infty} \frac{||t_j^k||}{||x^k||} = 0. \tag{29}$$

Let us show that $\lim\limits_{k \to \infty} \frac{||\mu(k,j)||}{||x^k||} < \infty$ for $\mu(k, j)$ defined in (21). Having supposed that, on the contrary, $\lim\limits_{k \to \infty} \frac{||\mu(k,j)||}{||x^k||} = \infty$, we get $\lim\limits_{k \to \infty} \frac{||x^k||}{||\mu(k,j)||} = 0$. Taking into account the last equality and (29), let us divide all relations in (21) by $||\mu(k,j)||$ and pass to the limit as $k \to \infty$. As a result, we obtain

$$B^T\Delta\mu(j) = 0,\ \Delta\mu(j) \geq 0,\ ||\Delta\mu(j)|| = 1 \text{ with } \Delta\mu(j) = \lim_{k\to\infty} \frac{\mu(k,j)}{||\mu(k,j)||}.$$

Hence, vector $\mu(k,j)$ admits representation $\mu(k,j) = (\Delta\mu(j) + w(k,j))\theta_k$, where $\theta_k = ||\mu(k,j)||$, $w(k,j) \to 0$ as $k \to \infty$. The relations (21) can be rewritten in the form

$$D_j t_j^k + c_j - A_j^T x^k + B_j^T w(k,j)\theta_k = 0,\ \Delta\mu(j) + w(k,j) \geq 0,\ t_j^{kT} B_j^T w(k,j) = 0, k = 1,2,... \quad (30)$$

Let $\Delta\mu_i(j), w_i(k,j)$ be the i-th components of the vectors $\Delta\mu(j), w(k,j)$, $i = 1,...,m_j$. From (30), we conclude that given $i \in \{1,...,m_j\}$, the inequality $w_i(k,j) < 0$ implies $\Delta\mu_i(j) > 0$.

Let us calculate $\alpha(k) = \max\{\alpha_i(k), i = 1,...m_j\}$, where

$$\alpha_i(k) = 0 \text{ if } w_i(k,j) \geq 0, \text{ and } \alpha_i(k) = -w_i(k,j)/\Delta\mu_i(j) \text{ if } w_i(k,j) < 0, i = 1,...,m_j.$$

It is evident that $\alpha(k) \geq 0$, $\lim_{k\to\infty} \alpha(k) = 0$, and, by construction, for $\bar\mu(k,j) := \theta_k(\alpha(k)\Delta\mu(j) + w(k,j))$ we have

$$0 = D_j t_j^k + c_j - A_j^T x^k + B_j^T \bar\mu(k,j),\ \bar\mu(k,j) \geq 0,\ 0 = t_j^{kT} B_j^T \bar\mu(k,j), k = 1,2,...$$

Hence, $\bar\mu(k,j) \in \mathcal{M}(k,j)$ and $||\bar\mu(k,j)|| = ||\mu(k,j)|| \cdot ||(\alpha(k)\Delta\mu(j) + w(k,j))||$.

Note that $||(\alpha(k)\Delta\mu(j) + w(k,j))|| \to 0$ as $k \to \infty$. Then, for sufficiently large k, we have $||\bar\mu(k,j)|| < ||\mu(k,j)||$, that contradicts (22). Therefore, $\lim_{k\to\infty} \frac{||\mu(k,j)||}{||x^k||} < \infty$.

Now divide both sides of (21) by $||x^k||$ and pass to the limit as $k \to \infty$, taking into account (29). As a result, we obtain

$$-A_j^T \Delta x^* + B_j^T \Delta\mu_j^* = 0,\ \Delta\mu_j^* \geq 0 \text{ with } \Delta\mu_j^* = \lim_{k\to\infty} \frac{\mu(k,j)}{||x^k||}.$$

The last relations together with (18), (19) permit to conclude that $\Delta x^* \in \Delta X$.

It follows from (15) and (16) that $d_0^T x^k + \beta(x^k) \leq const$, $d_i^T x^k + r_i \leq \beta(x^k)$, $i \in I$. Hence

$$d_0^T x^k + \max_{i\in I}(d_i^T x^k + r_i) \leq const.$$

Divide both sides of the last inequality by $||x^k||$ and pass to the limit as $k \to \infty$. As a result, we get inequality $d_0^T \Delta x^* + d_i^T \Delta x^* \leq 0,\ i \in I$, which contradicts the assumption (B) of the theorem. Hence situation II) is impossible as well as situation I).

The obtained contradictions lead us to conclude that for any sequence of feasible solutions $x^k \in X$, $k = 1,2,...$, of problem (7), where x^k satisfy inequalities (15), there exists $M_0 > 0$ such that $||x^k|| \leq M_0$, $k = 1,2,...$ This fact permits to conclude that problem (7) admits an optimal solution. The theorem is proved. □

At the end of this section, we would like to make the following remarks.

1. It can be shown that the feasible set X of problem (7) is not empty if and only if the condition (A) of Theorem 1 is satisfied, that, in turn, always happens when $t_j^T D_j t_j > 0$ for all $t_j \in K(j) \setminus \{0\}$, $j \in J$.
2. Note that the condition (B) is considered to be satisfied if $\Delta X \setminus \{0\} = \emptyset$. Hence, from (13), one can see that the condition (B) holds true, if the matrix $W := \sum_{i\in I\cup\{0\}} W_i$ is (strictly) positive definite on the set

$$\Delta\tilde{X} := \{\Delta x \in \mathbb{R}^n : q_j^T \Delta x = 0,\ j \in S_*;\ \ q_j^T \Delta x \leq 0,\ j \in S;$$
$$\exists\mu^*(j) = \mu^*(\Delta x, j) \geq 0,\ \text{ such that } \Delta x^T A_j = \mu(j)^{*T} B_j, j \in J\}.$$

This condition will be satisfied if at least one of the matrices $W_i, i \in I\cup\{0\}$, is positive definite on the set $\Delta\tilde{X}$.

3. The implication

$$\Delta x \in \Delta X \ \Rightarrow \ d_0^T \Delta x + \max_{i \in I} d_i^T \Delta x \geq 0, \tag{31}$$

which is a slight modification of the condition (B), is a necessary condition for boundedness from below of the cost function in problem (7).

4.2. Relaxation of the sufficient condition

From the remarks done at the end of the previous subsection, one can conclude that the sufficient conditions formulated in Theorem 1, are also "almost" necessary for existence of an optimal solution in the convex SIP problem (7). The differences between the necessary and sufficient conditions appear only in the case, when there exists a vector $\Delta x \neq 0$ such that $\Delta x \in \Delta X$, $d_0^T \Delta x + \max_{i \in I} d_i^T \Delta x = 0$.

The following conjecture of more strong statement naturally arises from our considerations.

Conjecture. *Problem (7) admits an optimal solution if conditions (A) and (31) (a relaxed condition (B)) are satisfied.*

At the moment, we have no proof of this statement and also have certain doubts about its truthfulness. The following example indicates that, possibly, the proposed conjecture is not true.

Example. Consider a linear SIP problem

$$\min_z c^T z \quad \text{s.t. } a(t)^T z + b(t) \leq 0, \forall t \in T. \tag{32}$$

Problem (7) has the form (32), if we set:

$$z^T = (x^T, \rho) \in \mathbb{R}^{n+1}, \ J = \{1\}, \ I = \{1\}, \ d_0 = 0, \ c^T = (d_0^T, -y_1^{(1)}),$$

$$W_0 = \mathbb{O}, W_1 = \mathbb{O}, \ d_1 = 0, r_1 = 0, \ \mathcal{X} = \mathbb{R}^n, K(1) = T,$$

$$a^T(t) = (t^T A_1^T, 1), \ b(t) = -\frac{1}{2} t^T D_1 t - c_1^T t,$$

where $\mathbb{O}$ denotes the $n \times n$ null matrix.

For problem (32), the conditions (A) and (31) take the form of the following two conditions:

(A1) the constraints of problem (32) are consistent;
(B1) for any $\Delta z \in \{\Delta z \in \mathbb{R}^{n+1} : a(t)^T \Delta z \leq 0, \ \forall \, t \in T\}$, it holds $c^T \Delta z \geq 0$.

The conditions (A1) and (B1), in general, do not guarantee the existence of optimal solutions in problem (32) even if the index set T is compact and the cost function of this problem is bounded in the feasible set. Indeed, let us consider example 5.101. from [2]:

$$\min_z \bar{c}^T z = z_2 \ \text{ s.t. } \bar{a}(t)^T z + \bar{b}(t) = -t^2 z_1 - z_2 + 2t \leq 0, \ \forall t \in T = [0, 1]. \tag{33}$$

The condition (A1) is satisfied here, since for any $k \geq 1$ vector $z^k = (k, 1/k)^T$ is a feasible solution of this problem. The condition (B1) is fulfilled as well. Nevertheless, problem (33) does not have an optimal solution. Notice that there exists vector $\Delta \bar{z} = (1, 0)^T$, such that

$$\Delta \bar{z} \in \{\Delta z \in \mathbb{R}^2 : \bar{a}(t)^T \Delta z \leq 0, \ \forall t \in T\} \text{ and } \bar{c}^T \Delta \bar{z} = 0,$$

and the vector z^k admits representation $z^k = (k, 1/k)^T = k(\Delta \bar{z} + \omega_k)^T$, where $\omega_k = (0, 1/k^2)^T \to (0, 0)^T$ as $k \to \infty$.

5. Optimality conditions for problem (7)

The aim of this section is to formulate and prove optimality conditions for a given feasible solution of problem (7). It was mentioned in the Introduction, that in the majority of papers on optimality conditions for SIP problems, the compactness of the (infinite) index set of the (continual) constraints is being assumed. In our study of problem (7), this assumption is not satisfied: it was already noticed above that the sets $K(j), j \in J$, of the indices, which correspond to the continual constraints, are not compact.

In few papers dedicated to study of optimality for SIP problems without the assumption of the compactness of the index set (see, for example, [3, 5, 13], and references therein), it is supposed that the Farkas-Minkowski Constraint Qualification (CQ) is satisfied. In this paper, we do not have such an assumption for our problem (7) (see Theorem below).

In the literature, there is an approach, a so called technique of homogenization, [15], which also permits to bypass the difficulties caused by non-compactness of the index set. Having applied this approach to our problem, we obtain an equivalent SIP problem, which has a compact index set but does not satisfy the Slater condition: if $\exists j_0 \in J, \Delta K(j_0) \neq \emptyset$, this new problem has immobile indices with infinite immobility orders [11]. Therefore, the known from the SIP literature necessary optimality conditions (see [2, 11, 14]) cannot be applied here too. Despite the above, in this section, for problem (7) with noncompact index set, we will prove the optimality criterion without any additional condition (CQ) on the constraints.

Let $(x, \rho_j, j \in J, \beta)$ be a feasible solution in problem (7). As it was mentioned above, without loss of generality, we can consider that $(x, \rho_j, j \in J, \beta) = (x, \rho_j(x), j \in J, \beta(x))$, where $\rho_j(x)$, $j \in J$, and $\beta(x)$ are defined in (6) and (11). Denote

$$K_a(j, x) := \{t_j \in K(j) : \frac{1}{2} t_j^T D_j t_j + (c_j - A_j^T x)^T t_j = \rho_j(x)\}, \ j \in J,$$
$$\Delta K(j, x) := \{\tau \in \Delta K(j) : (c_j^T - x^T A_j)\tau = 0, \ ||\tau|| = 1\}, j \in J.$$

Theorem 2

A feasible solution $(x^0 \in \mathbb{R}^n, \ \rho_j^0, j \in J, \ \beta^0)$ of the convex SIP problem (7) is optimal in this problem iff there exist vectors

$$t_{kj}^* \in K_a(j, x^0), k = 1, ..., p_j, \ \ \tau_{kj}^* \in \Delta K(j, x^0), k = 1, ..., l_j, \ \ j \in J; \tag{34}$$

and numbers

$$\lambda_i^*, i \in I; \ \eta_j^*, j \in S_* \cup S; \ y_{kj}^*, \ k = 1, ..., p_j, \ \mu_{kj}^*, k = 1, ..., l_j, \ j \in J, \quad \text{with } p_j \geq 0, \ l_j \geq 0, j \in J, \ \ \sum_{j \in J} (p_j + l_j) \leq 1 + n, \tag{35}$$

such that

$$\lambda_i^* \geq 0, \ \lambda_i^* (\frac{1}{2} x^{0T} W_i x^0 + d_i^T x^0 + r_i - \sum_{j \in J} y_j^{(i)} \rho_j^0 - \beta^0) = 0, i \in I; \sum_{i \in I} \lambda_i^* = 1;$$

$$\eta_j^* \geq 0, \ \eta_j^* (q_j^T x^0 + \omega_j) = 0, \ j \in S; \tag{36}$$

$$y_{kj}^* \geq 0, k = 1, ..., p_j, \ \sum_{k=1}^{p_j} y_{kj}^* = \sum_{i \in I} \lambda_i^* y_j^{(i)}, \ \mu_{kj}^* \geq 0, k = 1, ..., l_j, \ \ j \in J,$$

$$W_0 x^0 + d_0 + \sum_{i \in I} \lambda_i^* (W_i x^0 + d_i) + \sum_{j \in J} A_j (\sum_{k=1}^{p_j} y_{kj}^* t_{kj}^* + \sum_{k=1}^{l_j} \mu_{kj}^* \tau_{kj}^*) + \sum_{j \in S \cup S_*} q_j \eta_j^* = 0.$$

Note that here and in what follows, we suppose that, if $\kappa = 0$, then the set $\{1, ..., \kappa\}$ is empty and $\sum_{i=1}^{\kappa} ... = 0$.

Since the sets S, S_*, I and J consist of finite numbers of elements, then to simplify calculations, without loss of generality, we can present a proof of Theorem 2, considering problem (7) with $|I| = 1$, $|J| = 1$, and $\mathcal{X} = \mathbb{R}^n$. In other words, we will prove here the theorem for problem (7) in the following form:

$$\begin{aligned} &\min_{x\in\mathbb{R}^n,\rho\in\mathbb{R}} \quad \tfrac{1}{2}x^TWx + d^Tx - y^*\rho \\ &\text{s.t.} \quad \rho \leq f(t,x) := \tfrac{1}{2}t^TDt + (c^T - x^TA)t, \;\; \forall t \in K = \{t \in \mathbb{R}^p : Bt \leq 0\}, \end{aligned} \tag{37}$$

where $W = W_0 + W_1$, $d = d_0 + d_1$, $y^* = y_1^{(1)}$, $K = K(1)$, $D = D_1$, $A = A_1$, $B = B_1$.

Theorem 3 (A particular case of Theorem 2)
A feasible solution $(x^0,\ \rho^0) \in \mathbb{R}^{n+1}$ of problem (37) is optimal in this problem iff there exist vectors

$$t_k^* \in K_a(1,x^0) =: K_a(x^0), k = 1,...,p_1, \;\; \tau_k^* \in \Delta K(1,x^0) =: \Delta K(x^0), k = 1,...,l_1, \tag{38}$$

and numbers

$$y_k^*,\; k = 1,...,p_1,\; \mu_k^*, k = 1,...,l_1, \;\; \text{with } p_1 \geq 0,\; l_1 \geq 0,\;\; p_1 + l_1 \leq 1 + n, \tag{39}$$

such that

$$y_k^* \geq 0, k = 1,...,p_1, \; \sum_{k=1}^{p_1} y_k^* = y^*, \; \mu_{kj}^* \geq 0, k = 1,...,l_1, \tag{40}$$

$$Wx^0 + d + A\Big(\sum_{k=1}^{p_1} y_k^* t_k^* + \sum_{k=1}^{l_1} \mu_k^* \tau_k^*\Big) = 0. \tag{41}$$

Before proceeding with the proof, let us introduce some notation and fulfill necessary preparatory calculations.

Let $(x^0,\ \rho^0 = \rho(x^0))$ be a feasible solution of problem (37). Then the set $K_a(x^0)$ is the set of optimal solutions in the QP problem

$$\min\ f(t,x^0) \quad \text{s.t. } t \in K.$$

Note that, in general, this problem is nonconvex.

It follows from [2], that the set $K_a(x^0)$ can be represented in the form $K_a(x^0) = \bigcup_{s\in S} K_s(x^0)$, where for any $s \in S$, the set $K_s(x^0)$ is a convex polyhedron and $|S| < \infty$. Hence, for any $s \in S$, there exist finite sets of (extremal) vectors $\bar{t}(s,i) \in K_s(x^0), i \in J(s)$, and rays $\bar{\tau}(s,i) \in \Delta K(x^0), i \in I(s)$, such that

$$K_s(x^0) = \{t \in \mathbb{R}^p : t = \sum_{i\in J(s)} \alpha_i \bar{t}(s,i) + \sum_{i\in I(s)} \beta_i \bar{\tau}(s,i),\ \sum_{i\in J(s)} \alpha_i = 1, \alpha_i \geq 0, i \in J(s);\ \beta_i \geq 0, i \in I(s)\}.$$

Moreover, one can show that

$$\begin{aligned} &\bar{t}^T(s,i)D\bar{t}(s,j) = const,\ (c^T - {x^0}^TA)\bar{t}(s,j) = const_*,\ \ i \in J(s),\ j \in J(s), \\ &\bar{t}^T(s,i)D\bar{\tau}(s,j) = 0,\ i \in J(s),\ j \in I(s), \\ &\bar{\tau}^T(s,i)D\bar{\tau}(s,j) = 0,\ \ \bar{\tau}(s,i) \in \Delta K(x^0),\ \ i \in I(s),\ j \in I(s). \end{aligned} \tag{42}$$

Now, consider the set $\Delta K(x^0)$ defined in (38). One can prove that this set is a union of a finite number of bounded convex polyhedra $\Delta K_s(x^0), s \in \Delta S : \Delta K(x^0) = \bigcup_{s\in\Delta S} \Delta K_s(x^0)$. Then, for any $s \in \Delta S$, there exist finite sets of (extremal) vectors $\hat{\tau}(s,i) \in \Delta K(x^0), i \in \Delta I(s)$, such that

$$\Delta K_s(x^0) = \{\tau \in \mathbb{R}^p : \tau = \sum_{i\in\Delta I(s)} \alpha_i \hat{\tau}(s,i),\ \sum_{i\in\Delta I(s)} \alpha_i = 1, \alpha_i \geq 0, i \in \Delta I(s)\}.$$

From the above considerations, it follows that relations

$$-\Delta x^T At \geq \Delta\rho,\ \forall t \in K_a(x^0),\quad -\Delta x^T A\tau \geq 0,\ \forall \tau \in \Delta K(x^0), \tag{43}$$

are equivalent to the following ones:

$$-\Delta x^T A\bar{t}(s,i) \geq \Delta\rho,\ i \in J(s), s \in \mathcal{S},\quad -\Delta x^T A\hat{\tau}(s,i) \geq 0,\ i \in \Delta I(s), s \in \Delta\mathcal{S}. \tag{44}$$

Here we have taken into account that $\bar{\tau}(s,i) \in \Delta K(x^0)$ (see (42)). Now we can prove the theorem formulated above.

Proof of Theorem 3.

Necessity. Suppose that $(x^0,\ \rho^0 = \rho(x^0))$ is an optimal solution of problem (37). Consider the following problem of Linear Programming (LP):

$$\min_{\Delta x \in \mathbb{R}^n, \Delta\rho \in \mathbb{R}} \quad (Wx^0 + d)^T \Delta x - y^* \Delta\rho, \text{ s.t. conditions (44).} \tag{45}$$

Suppose that the vector $(\Delta x^0, \Delta\rho^0) = 0 \in \mathbb{R}^{n+1}$ is an optimal solution of this problem. Then, according to the theory of LP, there exist subsets

$$\{t_k^*, k = 1, ..., p_1\} \subset \{\bar{t}(s,i), i \in J(s), s \in \mathcal{S}\} \subset K_a(x^0),$$
$$\{\tau_k^*, k = 1, ..., l_1\} \subset \{\hat{\tau}(s,i), i \in \Delta I(s), s \in \Delta\mathcal{S}\} \subset \Delta K(x^0),$$

and numbers (39) such that relations (40) and (41) are satisfied, and the necessity is proved.

Now, suppose that the vector $0 \in \mathbb{R}^{n+1}$ is not an optimal solution of problem (45). Then there exists a vector $(\Delta x,\ \Delta\rho)$, for which conditions (44) and the inequality

$$(Wx^0 + d)^T \Delta x - y^* \Delta\rho < 0$$

are satisfied. Vector $(\Delta x,\ \Delta\rho)$ satisfies conditions (44), and therefore, it satisfies relations (43) as well. Consequently, according to Proposition 1 (see Appendix), for any $\delta > 0$, there exists $\varepsilon_0 = \varepsilon_0(\delta) > 0$ such that for all $\varepsilon \in [0, \varepsilon_0]$, the vector $(x(\varepsilon) := x^0 + \varepsilon\Delta x, \rho(\varepsilon) := \rho^0 + \varepsilon(\Delta\rho - \delta))$ is a feasible solution of problem (37).

Let us choose $0 < \delta < -[(Wx^0 + d)^T \Delta x - y^* \Delta\rho]/y^*$ and calculate

$$\frac{1}{2}x^T(\varepsilon)Wx(\varepsilon) + d^T x(\varepsilon) - y^*\rho(\varepsilon) = \frac{1}{2}x^{0T}Wx^0 + d^T x^0 - y^*\rho^0 + \varepsilon^2 \frac{1}{2}\Delta x^T W \Delta x + \varepsilon\alpha_*,$$

where $\alpha_* = (Wx^0 + d)^T \Delta x - y^*(\Delta\rho - \delta) < 0$. Hence, for a sufficiently small $\varepsilon > 0$ we get

$$\frac{1}{2}x^T(\varepsilon)Wx(\varepsilon) + d^T x(\varepsilon) - y^*\rho(\varepsilon) < \frac{1}{2}x^{0T}Wx^0 + d^T x^0 - y^*\rho^0.$$

But this contradicts the assumption that $(x^0,\ \rho^0 = \rho(x^0))$ is an optimal solution of problem (37). The necessity is proved.

Sufficiency. Suppose that there exist vectors (38) and numbers (39) such that relations (40) and (41) are fulfilled. Consider the convex QP problem

$$\min_{x \in \mathbb{R}^n, \rho \in \mathbb{R}} \quad \frac{1}{2}x^T Wx + d^T x - y^*\rho$$
$$\text{s.t. } \rho \leq f(t_k^*, x), k = 1, 2, ..., p_1;\ (c^T - x^T A)\tau_k^* \geq 0,\ k = 1, 2, ..., l_1. \tag{46}$$

Since $\{t_k^*, k = 1, ..., p_1\} \subset K_a(x^0)$, $\{\tau_k^*, k = 1, ..., l_1\} \subset \Delta K(x^0) \subset \Delta K$, it follows from (10) and (37) that the set of feasible solutions of problem (37) belongs to the set of feasible solutions of problem (46). Hence, vector (x^0, ρ^0) is a feasible solution of problem (46) and the fulfillment of relations (40) and (41) implies that the vector is optimal in problem (46). Consequently, it is optimal in problem (37) as well. The sufficiency is proved. □

The approach that was used in the proof of Theorem 3, can be applied to the proof of Theorem 2 as well, but the calculations in this case are much more bulky. Therefore, we can consider that Theorem 2 is valid.

Observation. Let $(x^0, \rho_j^0,\ j \in J, \beta^0)$ be a feasible solution of problem (7). From (10), it follows that

$$(c_j^T - x^{0T} A_j)\tau_j \geq 0 \ \ \forall \tau_j \in \Delta K(j),\ j \in J.$$

Suppose that the last inequalities are strongly satisfied:

$$(c_j^T - {x^0}^T A_j)\tau_j > 0 \ \ \forall \tau_j \in \Delta K(j) \backslash \{0\}, j \in J. \tag{47}$$

Then $\Delta K(j, x^0) = \emptyset,\ j \in J$, and, hence, in (34)-(36) we have $l_j = 0,\ \sum_{k=1}^{l_j} \mu_{kj}^* \tau_{kj}^* = 0,\ j \in J$, (or $l_1 = 0$, $\sum_{k=1}^{l_1} \mu_k^* \tau_k^* = 0$ in (38)-(41), respectively). In this case the optimality conditions of Theorem 2 are the same as the optimality conditions formulated in [3, 5, 13] under the assumption that problem (7) satisfies the Farkas-Minkowski CQ.

Finally, we mention two cases where the terms $\sum_{k=1}^{l_j} \mu_{kj}^* \tau_{kj}^*,\ j \in J$, can be omitted in (34)-(36):

1. $t^T D_j t > 0, t \in K(j) \setminus \{0\},\ j \in J$,
2. $t^T D_j t \geq 0,\ \forall t \in \mathbb{R}^p,\ j \in J$.

Remind that, by assumption, the inequalities $t^T D_j t \geq 0, t \in K(j),\ j \in J$, take place.

6. Example

In this section, we present an example which shows that in a general case (when condition (47) is not satisfied) of problem (7), the presence of the terms $\sum_{k=1}^{l_j} \mu_{kj}^* \tau_{kj}^*,\ j \in J$, in the optimality conditions given by relations (34)-(36) is essential.

We consider here the SIP problem in the form (37), where

$$n = p = 2,\ y^* = 1,\ W = \mathbb{O}_{2\times 2},\ B = A = -\mathbb{E}_{2\times 2}, \tag{48}$$

$$c = \begin{pmatrix} 0 \\ 0 \end{pmatrix},\ d = \begin{pmatrix} d_1 \\ d_2 \end{pmatrix},\ D_1 = \begin{pmatrix} 0 & b \\ b & 1 \end{pmatrix},\ \ d_1 > 0,\ d_2 > 0,\ b > 0.$$

Here $\mathbb{E}_{n\times n}$ denotes a $n \times n$ identity matrix.

Let us check the optimality conditions (38)-(41) of Theorem 3 (a particular case of conditions (34)-(36) of Theorem 2). Rewrite the problem in the equivalent form:

$$\min_{x\in\mathbb{R}^2,\rho} \ d_1 x_1 + d_2 x_2 - \rho \tag{49}$$

$$\text{s.t.}\ \ f(t,x) := \tfrac{1}{2} t_2^2 + t_1 t_2 b + x_1 t_1 + x_2 t_2 \geq \rho,\ \forall t \in K := \{t \in \mathbb{R}^2 : t \geq 0\},$$

with $d_1 > 0,\ d_2 > 0,\ b > 0$, and consider the vector

$$(x^0, \rho^0) = \left(x_1^0 := 0, x_2^0 := -d_2,\ \rho^0 := -\frac{1}{2} d_2^2 \right). \tag{50}$$

To make sure that (x^0, ρ^0) is a feasible point in SIP problem (49), it is enough to notice the following:

$$\min_{t\geq 0} f(t, x^0) = \min_{t\geq 0} \left(t_1 t_2 b + \frac{1}{2} t_2^2 - d_2 t_2 \right) = \min_{t_1=0, t_2\geq 0} \left(\frac{1}{2} t_2^2 - d_2 t_2 \right) = -\frac{1}{2} d_2^2 = \rho^0.$$

Having applied the results obtained in the previous section, one can confirm that vector (x^0, ρ^0) defined in (50) is a unique optimal solution of problem (49). Indeed, in our example, we have $K_a(x^0) = \{t^0\}$, $\Delta K(x^0) = \{\tau^0\}$ with $t^0 = (t_1^0 = 0, t_2^0 = d_2)$, $\tau^0 = (1, 0)$ and hence, according to Theorem 3, the optimality conditions (see (38)-(41)) take the form:

$$\begin{pmatrix} d_1 \\ d_2 \end{pmatrix} - y_1^* t_1^* - \mu_1^* \tau_1^* = 0,\ y_1^* \geq 0,\ y_1^* = 1,\ \mu_1^* \geq 0, \tag{51}$$

with some $t_1^* \in K_a(x^0)$ and $\tau_1^* \in \Delta K(x^0)$. If set $t_1^* := t^0$, $p_1 := 1$, $\tau_1^* := \tau^0$, $l_1 := 1$, $y_1^* := 1$, $\mu_1^* := d_1 > 0$, and substitute in (51), we can confirm that the optimality conditions of Theorem 3 are satisfied.

Now, let us formulate for the same problem the known optimality conditions from [5, 13, 3] without checking the fulfillment of the Farkas-Minkowski CQ . Then the optimality conditions have the form:

$$\begin{pmatrix} d_1 \\ d_2 \end{pmatrix} - y_1^* t_1^* = 0,\ y_1^* \geq 0,\ y_1^* = 1,$$

with some $t_1^* \in K_a(x^0)$. It is evident that these conditions are not fulfilled.

Note, additionally, that for problem (37) with data (48), the Farkas-Minkowski CQ is not satisfied. According to [5], the Farkas-Minkowski CQ is satisfied if the set

$$\mathcal{K} := cone \left\{ \begin{pmatrix} At \\ 1 \\ 0.5t^T Dt + c^T t \end{pmatrix},\ t \in K; \begin{pmatrix} \mathbb{O} \\ 0 \\ 1 \end{pmatrix} \right\}$$

is closed. Here for $\Omega \subset \mathbb{R}^n$, $cone\,\Omega := \{v = \sum_{i=1}^{n+1} \alpha_i \omega_i,\ \alpha_i \geq 0,\ \omega_i \in \Omega, i = 1, 2, ..., n+1\}$.

Consider the sequence of vectors and numbers

$$t(k) = t^0 + k\tau^0 \in K,\ \alpha(k) = \frac{1}{k};\quad w(k) = \alpha(k) \begin{pmatrix} At(k) \\ 1 \\ 0.5t^T(k) Dt(k) + c^T t(k) \end{pmatrix} \in \mathcal{K},\quad k = 1, 2, ...$$

It is evident that there exists the limit

$$\lim_{k\to\infty} w(k) =: w = \begin{pmatrix} A\tau^0 \\ 0 \\ t^{0T} D\tau^0 \end{pmatrix} = \begin{pmatrix} -\tau^0 \\ 0 \\ bd_2 \end{pmatrix}.$$

Let us show that $w \notin \mathcal{K}$. Suppose the contrary. Then there exist $t_i \in K, \alpha_i, i \in I$, and α_* such that

$$\tau^0 = \sum_{i\in I} t_i \alpha_i,\quad 0 = \sum_{i\in I} \alpha_i,\quad bd_2 = \sum_{i\in I} \alpha_i (0.5 t_i^T D t_i) + \alpha_*;\quad \alpha_i > 0, i \in I,\ \alpha_* \geq 0.$$

Evidently, the last relations are irreconcilable, that permits to conclude that the set $\mathcal{K}$ is not closed.

7. Conclusion

In the paper, we considered a special class of convex SIP problems with noncompact index sets, which arise in nonlinear parametric optimization. For the problems of this class, we have formulated and proved the existence theorem and new optimality conditions.

The results of the paper will be used in study of differential properties of parametric SIP problems. Moreover, these results may be used as the basis of new approach to study of special classes of SIP problems, such as that of Copositive Programming, Semi-Infinite Polynomial Programming, and others, for which the noncompactness of index sets is commonplace.

Acknowledgement

This work was partially supported by Belarusian State Scientific Program "Convergence" and Portuguese funds through CIDMA - Center for Research and Development in Mathematics and Applications, and FCT - Portuguese Foundation for Science and Technology, within the project UID/MAT/04106/2013.

REFERENCES

1. Asprey S.P., Maccietto S. Eds. *Dynamic model development: methods, theory and applications,* In: Proceedings of the Workshop on the Life af a Process Model-From Conception to Action, Imperial Colledge, London, UK, October 25-26, 2000.
2. Bonnans J.F., Shapiro A. *Perturbation analysis of optimization problems,* Springer-Verlag, New-York, 2000.
3. Cánovas M. J., López M. A., Mordukhovich B. S., and Parra J. *Variational analysis in semi-infinite and infinite programming, II: necessary optimality conditions,* SIAM Journal on Optimization, vol. 20, no. 6, pp. 2788-2806, 2010.
4. Eaves B.C. *On Quadratic Programming,* Management Science, Theory Series, vol. 17, no. 11, pp. 698–711, 1971.
5. Goberna M.A., and López M. A. *Linear Semi-Infinite Optimization,* Wiley, Chichester, 1998.
6. Goberna M.A., Lópes M.A. (eds.) *Semi-Infinite Programming: recent advances,* Kluwer, Dordrecht, 2001.
7. Hettich R., Jongen H.Th. *Semi-Infinite Programming: conditions of optimality and applications,* in: *J. Stoer, ed., Optimization Techniques,* Part 2, Lecture Notes in Control and Information Sciences, no. 7, pp. 1–11, 1978.
8. Hettich R., Kortanek K.O. *Semi-Infinite Programming: theory, methods and applications,* SIAM Rev., vol. 35, pp. 380–429, 1993.
9. Hettich R., Still G. *Second order optimality conditions for generalized semi-infinite programming problems,* Optimization, vol. 34, pp. 195–211, 1995.
10. Kostyukova O.I., Tchemisova T.V., Kurdina M.A. *A study of one class of NLP problems arising in parametric Semi-Infinite Programming,* Optimization Methods and Software, Vol. 32 , Iss. 6, pp.1218-1243, 2017.
11. Kostyukova O.I., Tchemisova T.V. *Implicit optimality criterion for convex SIP problem with box constrained index set,* TOP, vol. 20, no. 2, pp. 475–502, 2012.
12. Kostyukova O.I., Tchemisova T.V., and Kurdina M.A. *On Optimal Properties of Special Nonlinear and Semi-infinite Problems Arising in Parametric Optimization,* Stat., Optim. Inf. Comput., vol. 5, pp. 99-108, 2017.
13. Mordukhovich B., Nghia T. T. A. *Constraint qualifications and optimality conditions for nonconvex semi-infinite and infinite programs,* Math. Program., Ser. B, vol. 139, pp. 271-300, 2013. DOI 10.1007/s10107-013-0672-x.
14. Stein O., Still G. *On optimality conditions for generalized semi-infinite programming problems,* J. Optim. Theory Appl., vol. 104, no. 2, pp. 443–458, 2000.
15. Wang L., Guo F. *Semidefinite relaxations for semi-infinite polynomial programming,* Computational Optimization and Applications, vol. 58, no. 1, pp. 133–159, 2013.
16. Weber G.-W., Kropat E., Alparslan Gök S.Z. *Semi-Infinite and Conic Optimization in Modern Human Life and Financial Sciences under Uncertainty,* In: ISI Proceedings of 20th Mini-EURO conference, pp. 180–185, 2008.

Appendix

Proposition 1

Assume that conditions (43) are satisfied for $(\Delta x, \Delta\rho)$. Then for any $\delta > 0$, there exists $\varepsilon_0 = \varepsilon_0(\delta) > 0$, such that for all $\varepsilon \in [0, \varepsilon_0]$, vector $(x(\varepsilon) := x^0 + \varepsilon\Delta x, \rho(\varepsilon) := \rho^0 + \varepsilon(\Delta\rho - \delta))$ is a feasible solution of problem (37), i.e.

$$f(t, x(\varepsilon)) \geq \rho(\varepsilon) \;\; \forall t \in K. \tag{52}$$

Proof. Rewrite relations (52) as follows:

$$f(t, x^0) - \rho^0 + \varepsilon g(t) \geq 0, \quad \forall t \in K, \;\; \varepsilon \in [0, \varepsilon_0], \tag{53}$$

where $g(t) := -\Delta x^T At - \Delta\rho + \delta$.

Consider the optimization problem

$$\varepsilon^* := \inf_t \frac{f(t,x^0)-\rho^0}{-g(t)} \quad \text{s.t. } t \in T := \{t \in K : g(t) \le 0\}. \tag{54}$$

By construction, we have

$$f(t,x^0)-\rho^0 > 0 \quad \text{for all } t \in T. \tag{55}$$

Hence, in (54) we have $0 \le \varepsilon^* \le \infty$. It is evident that, if $\varepsilon^* > 0$, then relations (53) take place with $\varepsilon_0 = \varepsilon^*$ and the proposition is proved in this case.

Let us show that $\varepsilon^* \neq 0$. Suppose that, on the contrary, $\varepsilon^* = 0$. Let $\bar{t}^k \in T, k = 1,2,...$ be a minimizing sequence in problem (54):

$$M_k := \frac{f(\bar{t}^k,x^0)-\rho^0}{-g(\bar{t}^k)} \to 0, \quad \text{as } k \to \infty.$$

Let us consider another sequence

$$t^k = \arg\left\{\min ||t||, \quad \text{s.t. } \frac{f(t,x^0)-\rho^0}{-g(t)} \le M_k,\ t \in T\right\}. \tag{56}$$

If suppose that the sequence $t^k, k = 1,2,...$ has a converging subsequence $t^{k_i}, i = 1,2,...$:

$$\lim_{i\to\infty} t^{k_i} = t^*, \ \lim_{i\to\infty} k_i = \infty,$$

then, evidently, $t^* \in T$ and $0 = \varepsilon^* = \frac{f(t^*,x^0)-\rho^0}{-g(t^*)}$. But this contradicts relations (55). Thus, we can conclude that no one subsequence of the sequence $t^k, k = 1,2,...$ converges and, therefore, $||t^k|| \to \infty$ as $k \to \infty$.

By assumption,

$$0 = \varepsilon^* = \lim_{k\to\infty} \frac{f(t^k,x^0)-\rho^0}{-g(t^k)} = \lim_{k\to\infty} \frac{[f(t^k,x^0)-\rho^0]/||t^k||^2}{-g(t^k)/||t^k||^2} \ge \frac{1}{2}\Delta\bar{\tau}^T D\Delta\bar{\tau},$$

where $\Delta\bar{\tau} = \lim_{k\to\infty} t^k/||t^k||$. It is evident that $\Delta\bar{\tau} \in K$. Consequently, $\Delta\bar{\tau}^T D\Delta\bar{\tau} = 0$ and $\Delta\bar{\tau} \in \Delta K$. Moreover, the inequalities $g(t^k) = -\Delta x^T At^k - \Delta\rho + \delta \le 0, k = 1,2,...$ imply that $-\Delta x^T A\Delta\bar{\tau} \le 0$.

By assumption,

$$0 = \varepsilon^* = \lim_{k\to\infty} \frac{f(t^k,x^0)-\rho^0}{-g(t^k)} \ge \lim_{k\to\infty} \frac{[(c^T - x^{0^T}A)t^k - \rho^0]/||t^k||}{-g(t^k)/||t^k||} = \frac{(c^T - x^{0^T}A)\Delta\bar{\tau}}{\Delta x^T A\Delta\bar{\tau}},$$

where $(c^T - x^{0^T}A)\Delta\bar{\tau} \ge 0$ due to (10). Hence there are two possible situations:

$$\text{i) } (c^T - x^{0^T}A)\Delta\bar{\tau} = 0,\ \Delta x^T A\Delta\bar{\tau} > 0,$$

$$\text{ii) } (c^T - x^{0^T}A)\Delta\bar{\tau} = 0,\ \Delta x^T A\Delta\bar{\tau} = 0. \tag{57}$$

In situation i), we have $\Delta\bar{\tau} \in \Delta K(x^0)$ and $\Delta x^T A\Delta\bar{\tau} > 0$. But this contradicts relations (43).

Consider situation ii). By construction, $t^k = \theta^k(\Delta\bar{\tau} + \eta^k)$, where $\theta^k = ||t^k||$, $||\eta^k|| \to 0$ as $k \to \infty$. Note that $B(\Delta\bar{\tau} + \eta^k) \le 0$. Hence, for any row b_m^T of the matrix B, we have an implication:

$$b_m^T\eta^k < 0 \Rightarrow b_m^T\Delta\bar{\tau} > 0.$$

Taking into account the last inequalities and that $||\eta^k|| \to 0$ as $k \to \infty$, we conclude that $B(\frac{1}{2}\Delta\bar{\tau} + \eta^k) \le 0$, for sufficiently large k. This implies that $\tilde{t}^k := \theta^k(\frac{1}{2}\Delta\bar{\tau} + \eta^k) \in K$ for sufficiently large k. Rewrite the vector $\tilde{t}^k$ as follows:

$$\tilde{t}^k = \theta^k(\Delta\bar{\tau} + \eta^k) - \frac{1}{2}\theta^k\Delta\bar{\tau} = t^k - \frac{1}{2}\theta^k\Delta\bar{\tau}.$$

Taking into account relations (57) and the following one:

$$t^T D\tau \geq 0 \text{ for all } t \in K,\ \tau \in \Delta K,$$

it is easy to show that

$$f(\tilde{t}^k, x^0) = f(t^k, x^0) - \theta^k {t^k}^T D\Delta\bar{\tau} \leq f(t^k, x^0),\ g(\tilde{t}^k) = g(t^k),\ \ ||\tilde{t}^k|| < ||t^k||.$$

But the last relations contradict condition (56). Hence, the situation ii) is impossible as well. Thus we can can conclude that $\varepsilon^* > 0$. The proposition is proved. □

Permissions

All chapters in this book were first published in SOIC, by International Academic Press; hereby published with permission under the Creative Commons Attribution License or equivalent. Every chapter published in this book has been scrutinized by our experts. Their significance has been extensively debated. The topics covered herein carry significant findings which will fuel the growth of the discipline. They may even be implemented as practical applications or may be referred to as a beginning point for another development.

The contributors of this book come from diverse backgrounds, making this book a truly international effort. This book will bring forth new frontiers with its revolutionizing research information and detailed analysis of the nascent developments around the world.

We would like to thank all the contributing authors for lending their expertise to make the book truly unique. They have played a crucial role in the development of this book. Without their invaluable contributions this book wouldn't have been possible. They have made vital efforts to compile up to date information on the varied aspects of this subject to make this book a valuable addition to the collection of many professionals and students.

This book was conceptualized with the vision of imparting up-to-date information and advanced data in this field. To ensure the same, a matchless editorial board was set up. Every individual on the board went through rigorous rounds of assessment to prove their worth. After which they invested a large part of their time researching and compiling the most relevant data for our readers.

The editorial board has been involved in producing this book since its inception. They have spent rigorous hours researching and exploring the diverse topics which have resulted in the successful publishing of this book. They have passed on their knowledge of decades through this book. To expedite this challenging task, the publisher supported the team at every step. A small team of assistant editors was also appointed to further simplify the editing procedure and attain best results for the readers.

Apart from the editorial board, the designing team has also invested a significant amount of their time in understanding the subject and creating the most relevant covers. They scrutinized every image to scout for the most suitable representation of the subject and create an appropriate cover for the book.

The publishing team has been an ardent support to the editorial, designing and production team. Their endless efforts to recruit the best for this project, has resulted in the accomplishment of this book. They are a veteran in the field of academics and their pool of knowledge is as vast as their experience in printing. Their expertise and guidance has proved useful at every step. Their uncompromising quality standards have made this book an exceptional effort. Their encouragement from time to time has been an inspiration for everyone.

The publisher and the editorial board hope that this book will prove to be a valuable piece of knowledge for researchers, students, practitioners and scholars across the globe.

List of Contributors

Iqbal Ahmad, Mijanur Rahaman and Rais Ahmad
Department of Mathematics, Aligarh Muslim University, India

Ram U. Verma
Department of Mathematics, University of North Texas, USA

G. J. Zalmai
Department of Mathematics and Computer Science, Northern Michigan University, USA

Qiuyu Wang
School of Software, Henan University, Kaifeng 475000, China

Wenjiao Cao
School of Mathematics and Statistics, Henan University, Kaifeng 475000, China

Zheng-Fen Jin
School of Mathematics and Statistics, Henan University of Science and Technology, Luoyang 471023, China

Ya Ju Fan and Chandrika Kamath
Lawrence Livermore National Laboratory, CA, USA

I.V. Rozora
Department of Applied Statistics, Faculty of Cybernetics, Taras Shevchenko National University of Kyiv, Ukraine

R.U. Khan, M.A. Khan and M.A.R. Khan
Department of Statistics and Operations Research, Aligarh Muslim University, India

N. K. Sahu
Dhirubhai Ambani Institute of Information and Communication Technology, Gandhinagar, India

N. K. Mahato
Pdpm Iiitdm, Jabalpur, India

Ram N. Mohapatra
University of Central Florida, Orlando, FL. 32816, USA

Plern Saipara and Parin Chaipunya
Kmutt Fixed Point Research Laboratory, Department of Mathematics, King Mongkut's University of Technology Thonburi, Thailand

Wiyada Kumam
Program in Applied Statistics, Department of Mathematics and Computer Science, Faculty of Science and Technology, Rajamangala University of Technology Thanyaburi, Thailand

Jalal Chachi
Department of Mathematics, Statistics and Computer Sciences, Semnan University, Iran

Ajit Chaturvedi and Taruna Kumari
Department of Statistics, University of Delhi, India

Yu. V. Kozachenko
Department of Probability Theory, Statistics and Actuarial Mathematics, Taras Shevchenko National University of Kyiv, Ukraine

M.Yu. Petranova
Department of Probability Theory and Mathematical Statistics, Vasyl' Stus Donetsk National University, Ukraine

Alexander Bulinski and Alexey Kozhevin
Faculty of Mathematics and Mechanics, Lomonosov Moscow State University, Moscow, Russia 119991

A.I.Kirjanen, O.A.Malafeyev and N.D.Redinskikh
Faculty of Applied Mathematics and Control Processes, Saint-Petersburg State University, Russia

Olga Kostyukova and Maryia Kurdina
Institute of Mathematics, National Academy of Sciences of Belarus, Minsk, Belarus

Tatiana Tchemisova
Center for Research and Development in Mathematics and Applications, University of Aveiro, Aveiro, Portugal

Index

S

T

V

www.ingramcontent.com/pod-product-compliance
Lightning Source LLC
Chambersburg PA
CBHW082016190326
41458CB00010B/3201
9781682856888